Stability in the Mechanics of Continua

International Union of Theoretical
and Applied Mechanics

Stability in the Mechanics of Continua

2nd Symposium, Nümbrecht, Germany,
August 31 – September 4, 1981

Editor
F. H. Schroeder

With 167 Figures

Springer-Verlag
Berlin Heidelberg New York 1982

Prof. Dr.-Ing. F. H. SCHROEDER
Fachbereich Bautechnik, Fachgebiet Baumechanik
Universität-GH-Wuppertal
Pauluskirchstraße 7, D-5600 Wuppertal 2

ISBN-13: 978-3-642-81813-4 e-ISBN-13: 978-3-642-81811-0
DOI: 10.1007/978-3-642-81811-0

Softcover reprint of the hardcover 1st edition 1982

Library of Congress Cataloging in Publication Data
Main entry under title:
Stability in the mechanics of continua.
(IUTAM-Symposium)
At head of title. International Union of Theoretical and Applied Mechanics.
Bibliography: p.
Includes index.
1. Continuum mechanics – Congresses.
2. Stability – Congresses.
I. Schroeder, F. H. (Frederick-Henry), 1925-.
II. International Union of Theoretical and Applied Mechanics
III. IUTAM Symposium on Stability in the Mechanics of the Continua (2nd: 1981 · Nümbrecht, Germany)
IV. Title.
V. Series: IUTAM-Symposien.
QY808.2.S7 531 82-5462 AACR2

2061/3020 5 4 3 2 1 0

Preface

In 1969 scientists came together for the first time to discuss problems of instability of continuous systems on a IUTAM - Symposium.

Since then twelve years of a fast and interesting development in the field of stability have passed.

Therefore it was necessary to arrange a meeting for scientists again in order to exchange experiences and opinions about this topic.

In 1978 the Bureau of IUTAM asked me to become chairman of the scientific committee of the Second IUTAM Symposium on Stability in the Mechanics of Continua and to organize this meeting.

The following scientists have been appointed as members of this scientific committee:

Prof. T. B. Benjamin
University of Oxford, Mathematical Institute
Oxford, GB

Prof. V. V. Bolotin
USSR Academy of Science, Mechanical Engineering Research Institute, Moscow, USSR

Prof. J. W. Hutchinson
Harvard University, Division of Applied Sciences
Cambridge, USA

Prof. K. Kirchgässner
Universität Stuttgart, Mathematisches Institut A
Stuttgart, FRG

Prof. A. Kulikovsky
Mechanics National Committee
Moscow, USSR

Prof. H. H. E. Leipholz
University of Waterloo, Department of Civil Engineering
Waterloo, Canada

Prof. R. C. Di Prima
Rensselaer Polytechnical Institute, Department of
Mathematical Science, Troy, USA

The symposium took place in Nümbrecht near Cologne from August the 31th to September the 4th, 1981 and it dealt with both solid and fluid continua.

The following volume contains the lectures in the succession as they were held during the symposium, devided into two parts:

1. Stability of Solid Continua and Mathematical Methods

2. Stability of Fluid Continua.

I regret that Professor Reissner was not able to give the opening lecture as it had been announced. Therefore he asked me to do this instead of him. His paper will be published as the second contribution in this volume.

On behalf of the scientific committee I want to thank the International Union of Theoretical and Applied Mechanics for suggesting the symposium and for the financial support as well as the other sponsors:

the University of Wuppertal

the Deutsche Forschungsgemeinschaft

the state of Nordrhein-Westfalen.

Moreover I want to thank the participants, especially the lecturers, and all the others who helped prepare and run the symposium.

Wuppertal, December 1981 F.H. Schroeder

Participants

Y. Aihara (Japan)
W. Andres (FRG)
G. Bertrand (FRG)
R. de Boer (FRG)
R. Bogacz (Poland)
O. Bruhns (FRG)
G. Capriz (Italia)
A. Cardon (Belgium)
J. Carr (UK)
P. Chossat (France)
G. Cognet (France)
A. Davey (UK)
K.S. Eckhoff (Norge)
S. Eilering (FRG)
I. Elishakoff (Israel)
A. Fix (FRG)
V. Garg (India)
P. Hagedorn (FRG)
A. v. Hagen (FRG)
P. Hall (UK)
A. Heinen (FRG)
D. Hoeborn (FRG)
K. Huseyin (Canada)
J.A. Johnson III (USA)
H.J. Jung (FRG)
R.C. Kar (FRG)
K. Kirchgässner (FRG)
S. Kisliakov (Bulgaria)
S.J. Knops (UK)
W.T. Koiter (Netherlands)
E.L. Koschmieder (USA)
A.N. Kounadis (Greece)
W.B. Krätzig (FRG)
K.H. Laermann (FRG)
A. Marinetti (Italia)
P. Meijers (Netherlands)
D. Meyer (FRG)
G. Mikhailov (USSR)
N. Nakagawa (Japan/FRG)
H. Nölke (FRG)
H. Obrecht (FRG)
G. Oliveto (Italia)
H. Petryk (Poland)
A. Pflüger (FRG)
R.H. Plaut (USA)
M. Potier-Ferry (France)
R. Pyrz (Poland)
R.S. Rivlin (USA)
G.J. Simitses (USA)
J. Sijbrand (Netherlands)
P. Szilard (FRG)
F.H. Schroeder (FRG)
B. Storakers (Sweden)
H. Stumpf (FRG)
S. Tento (USA)
C. Toklu (Turkey)
Z. Waszczyszyn (Poland)
U. Wittek (FRG)
W. Wunderlich (FRG)

Contents

Stability of Fluid Continua

The General Stability Equations in the Theory of Rods

F.H. SCHROEDER

Department of Civil Engineering
University of Wuppertal
Wuppertal, FRG

1. Introduction

In order to solve problems of stability it is necessary to consider large deformations of a solid. In the theory of rods it is mostly usual, to transfer the three-dimensional problem into an one-dimensional problem with the help of two methods of approximation.

first method: The displacement vector w in the rod space will be developed into an exponential series with respect to the cross section coordinates θ^2, θ^3

$$w = \overset{\circ}{v}(\theta^1,t) + \theta^2 \overset{2}{v}(\theta^1,t) + \theta^3 \overset{3}{v}(\theta^1,t) + \theta^2 \theta^3 \overset{23}{v}(\theta^1,t) + \dots$$

with the unknown vectors $\overset{\circ}{v}(\theta^1,t)$, $\overset{1}{v}(\theta^1,t)$, $\overset{2}{v}(\theta^2,t)$, $\overset{23}{v}(\theta^1,t)$, ... as functions of the rod-line parameter θ^1.

second method: The displacement vector w in the rod space will be expressed by the Cosserat vectors v and $\vec{\varphi}$. The vector v is the displacement vector of the rod-line and $\vec{\varphi}$ is the rotation vector of the rigid cross section. With respect to the large rotation $\vec{\varphi}$ only the time derivation $\dot{w}$ (not w) can be expressed by v and $\vec{\varphi}$:

$$\dot{w} = \dot{v}(\theta^1,t) + \dot{\vec{\varphi}}(\theta^1,t) \times b$$

with only two unknowns v and $\vec{\varphi}$. The vector $b = \theta^2 a_2 + \theta^3 a_3$ is the cross section vector.

With the use of the additional B e r n o u l l i - hypothesis the rotation vectors around the axes of the cross section can be expressed by the displacement vector v. In order that the problem is not overdetermined with respect to the equilibrium conditions it is not allowed to introduce constitutive equations for the shear forces.

The first method of development in series was formulated by E. V o l t e r r a and later on by several scientists. Since it is necessary to get the correct position of the shear centre, you must take into account the series elements up to order six in θ^2, θ^3 . This method is not usable, because of the very large mathematical effort. You get 82 differential equations for the unknown $\overset{ik}{v}(\theta^1)$.

The second method which reduces the displacement vector u in the rod space to the Cosserat vectors v and $\vec{\varphi}$ was used by diverse scientists. This method is only practicable in the theory of shells. The method is not practicable in the rod theory, because the needed shear centre can not be described with the help of Cosserat vectors. It is necessary to know this centre, in order to include the coupling between bending and torsion.

For these reasons the presented paper is based on a Cosserat theory with a warping vector. To get the position of the shear centre it is assumed that the cross section is warped by torsion

$$\breve{w} = \Omega(\theta^2, \theta^3) \cdot \varkappa_1(\theta^1)\, v_1$$

The function $\Omega(\theta^2, \theta^3)$ is the wellknown warping function according to de St. V e n a n t . $\varkappa_1$ is the relativ rotation-angle in direction v_1 of the rod axis. The equations of motion of the rod are calculated with the help of the principle of virtual work. The constitutive variables of the constitutive equations are derived from the deformation work.

The constitutive equations for the internal rod forces and moments are not found out by the Cosserat theory but by the three-dimensional constitutive equations of the theory of elasticity, because the elastic constants are known contrary to the Cosserat theory.

In the fundamental state, which is analysed with respect to its stability, finite rotations and displacement vectors are taken into consideration.

2. The geometry of the undeformed precurved rod at time t_o

Let us consider the following undeformed rod with the rod-line and the cross section

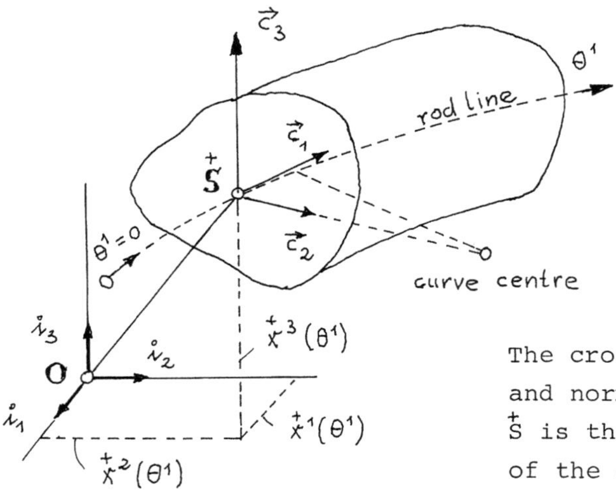

The geometrical form of the rod-line is given by the following vector equation:

$$\overset{+}{r} = \overset{+}{x}^m(\theta^1)\cdot \overset{\circ}{i}_m \quad \ldots (1)$$

θ^1 = length of the rod-line

The cross section is plain and normal to θ^1 -line. $\overset{+}{S}$ is the center of gravity of the cross section.

We erect the following wellknown base vectors in point $\overset{+}{S}$ of the rod-line:

tangent vector : $\vec{c}_1 = \dfrac{\partial \overset{+}{r}}{\partial \theta^1} = \overset{+}{r}_{,1}$ ... (2)

normal vector : $\vec{c}_2 = \overset{+}{r}_{,11} / |\overset{+}{r}_{,11}|$ ... normal to $\vec{c}_1$

binormal vector : $\vec{c}_3 = \vec{c}_1 \times \vec{c}_2$ ... normal to $\vec{c}_1; \vec{c}_2$ ... (3)

Setting (1) into (2) and (3) the upper vectors become

$$(i=1,2,3) \qquad \vec{c}_i = c_i^{\;m}\cdot \overset{\circ}{i}_m \qquad c_i^{\;m} \text{ are wellknown coefficients} \qquad (4)$$

Instead of $\vec{c}_1, \vec{c}_2$ we introduce in future the principle axis $\overset{+}{v}_2$, $\overset{+}{v}_3$ in the cross section, with the known angle ψ between $\vec{c}_3, \vec{c}_2$ and $\overset{+}{v}_3, \overset{+}{v}_2$.

We now consider a point $\overset{+}{\mathbf{P}}$ in the rod space

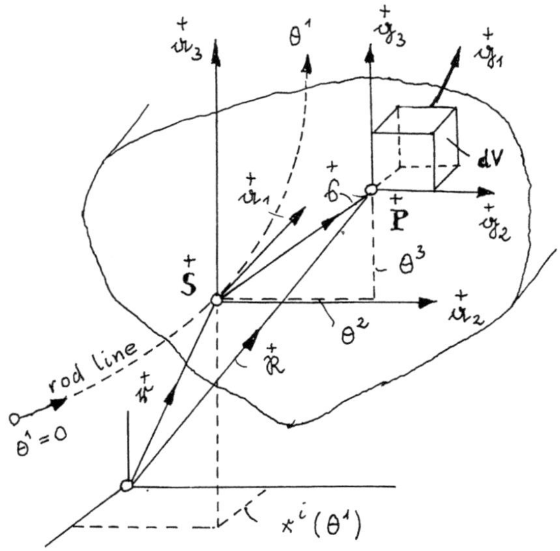

A point $\overset{+}{\mathbf{P}}$ of the rod space is given by the director

$$\overset{+}{b} = \theta^2 \overset{+}{v}_2 + \theta^3 \overset{+}{v}_3 \tag{5}$$

with the coordinates θ^2, θ^3

The rod space is measured out by the vector

$$\overset{+}{R} = \overset{+}{r}(\theta^1) + \overset{+}{b}(\theta^1, \theta^2, \theta^3) \equiv \overset{+}{R}(\theta^1, \theta^2, \theta^3) \tag{6}$$

By $\overset{+}{R}(\theta^1, \theta^2, \theta^3)$ the rod space is drawn through by a net of θ^i-lines.

The θ^i-lines form the infinitesimal element dV. The edge vectors, which are tangential vectors towards the θ^i-lines are given by

$$\overset{+}{g}_i = \frac{\partial \overset{+}{R}}{\partial \theta^i} = \overset{+}{R}_{,i} = \overset{+}{\mu}_i{}^m \cdot \overset{+}{v}_m \tag{7}$$

The base vector $\overset{+}{g}_1$ is not vertical to $\overset{+}{g}_2$ and $\overset{+}{g}_3$, if the rod-line is twisted by $\overset{+}{\varkappa}_1$.

Setting (6) into (7) we get the coefficients $\overset{+}{\mu}_i{}^m$

$$\overset{+}{\mu}_i{}^m = \begin{pmatrix} 1 - \theta^2 \overset{+}{\varkappa}_3 + \theta^3 \overset{+}{\varkappa}_2 & -\theta^3 \overset{+}{\varkappa}_1 & \theta^2 \overset{+}{\varkappa}_1 \\ 0 & 1 & 0 \\ 0 & 0 & 1 \end{pmatrix}$$

The curvature coefficients $\overset{+}{\varkappa}_1, \overset{+}{\varkappa}_2, \overset{+}{\varkappa}_3$ can be calculated by the derivative of $\overset{+}{v}_i$ with respect to θ^1. We get according to [1]:

$$\overset{+}{v}_{i,1} = e_i{}^{mn} \cdot \overset{+}{\varkappa}_n \cdot \overset{+}{v}_m \; ; \qquad e_i{}^{mn} = \text{permutation tensor}$$

$$\overset{+}{\varkappa}_1 = \tau + \psi_{,1} \, , \qquad \overset{+}{\varkappa}_2 = \varkappa \cdot \sin \psi \; ; \qquad \overset{+}{\varkappa}_3 = \varkappa \cdot \cos \psi \tag{8}$$

3. The geometry of the deformed rod at time t

Let us consider the undeformed and deformed rod in the following picture

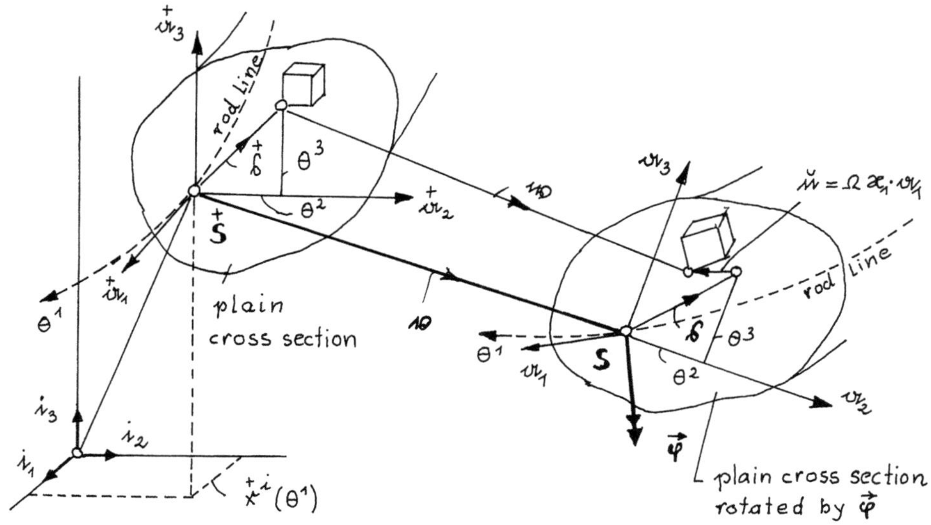

The deformation shall be described by kinematical vectors $\boldsymbol{v}$, $\vec{\varphi}$ of the rod-line

The displacement of the undeformed rod-line is given by:

$$\boldsymbol{v} = v^m(\theta^1, t)\, \overset{+}{\boldsymbol{v}}_m \qquad \text{... displacement vector} \tag{9}$$

The rigid rotation of the undeformed plain cross section may be given by:

$$\vec{\varphi} = \varphi^m(\theta^1, t)\, \overset{+}{\boldsymbol{v}}_m \qquad \text{... rotation vector} \tag{1o}$$

The cross section does not remain plain by $\breve{\boldsymbol{w}}$:

$$\breve{\boldsymbol{w}} = \Omega(\theta^2, \theta^3) \cdot \varkappa_1 \boldsymbol{v}_1 \qquad \text{... warping vector} \tag{11}$$

$\Omega(\theta^2, \theta^3)$ = known warping function ... de St. V e n a n t.

Ω satisfies the potential eq. $\Delta\Omega = \Omega_{,22} + \Omega_{,33} = o$

$\varkappa_1(\theta^1, t)$ = relativ torsion angle about the tangent

$\varkappa_1$ is defined by the derivative: $\vec{\varphi}_{,1} = \varkappa_1 \boldsymbol{v}_1 + \varkappa_2 \boldsymbol{v}_2 + \varkappa_3 \boldsymbol{v}_3$

The introduction of the warping vector is fundamental, because the coupling between bending and torsion of a dyssymmetrical cross section can be described by $\breve{\boldsymbol{w}}(\theta^\alpha, t)$ and <u>not</u> by $\boldsymbol{v}$ and $\vec{\varphi}$ alone.

It is important to know the relation between the base vectors v_i and $\overset{+}{v}_i$ of the deformed and undeformed rod. We try to find the following relations:

$$v_i = \Omega_i{}^m \overset{+}{v}_m \; ; \qquad \Omega_i{}^m = \text{unknown rotation tensor} \qquad (12)$$

In the following we show the calculation of $\Omega_i{}^m$ in a short manner. The detailed derivation is given in my paper [2]. The rate $\dot{v}_i$ with respect to time is caused by the rotation vector $\vec{\varphi}$. The following relation is valid:

$$\dot{v}_i = \dot{\vec{\varphi}} \times v_i \quad ; \quad \vec{\varphi} = \varphi^m \overset{+}{v}_m \qquad \text{can be large} \qquad (13)$$

On setting (12) into (13) one immediately finds the following tensor-differential-equation for the rotation tensor:

$$\dot{\Omega}_i{}^K = \left(\dot{\varphi}^s \cdot e_{sm}{}^K\right) \Omega_i{}^m = \dot{\varphi}_m{}^K \; \Omega_i{}^m \qquad (14)$$

The solution of this differential equation is known.

$$\boxed{\Omega_i{}^K = e^{\varphi_i{}^K} + c_i{}^K = a_1 \delta_i^K + a_2 \varphi_i{}^K + a_3 \varphi_i{}^m \cdot \varphi_m{}^K} \qquad (15)$$

The constants of integration $c_i{}^K$ are zero, which follows from the initial condition $v_i(0) = \overset{+}{v}_i$. The tensor function $e^{\varphi_i{}^K}$ is a polynom of order two in $\varphi_i{}^K$ according to Caley-Hamilton with the following coefficients a_i:

$$a_1 = 1 \; ; \quad a_2 = \frac{\sin \varphi}{\varphi} \; ; \quad a_3 = \frac{1 - \cos \varphi}{\varphi \cdot \varphi} \; \ldots \; \text{with } \varphi = \sqrt{(\varphi^1)^2 + (\varphi^2)^2 + (\varphi^3)^2}$$

The linearized rotation tensor $\Omega_i{}^K$ as function of φ^i has the form:

$$\Omega_i{}^K \cong \delta_i^K + \varphi_i{}^K = \begin{pmatrix} 1 & \varphi^3 & -\varphi^2 \\ -\varphi^3 & 1 & \varphi^1 \\ \varphi^2 & -\varphi^1 & 1 \end{pmatrix} \quad \ldots \qquad (16)$$

4. Deformation work and the constitutive variables

We integrate the wellknown deformation work of the infinitesimal dV over $d\Theta^1$, $d\Theta^2$, $d\Theta^3$ whereby we first integrate over the cross section $dF = d\Theta^2 \cdot d\Theta^3$ and afterwards over the rod-line $d\Theta^1$. We get

$$\dot{W} = \int_{\Theta^1} \Big[\iint_F \tau^{1m} \mathfrak{g}_m \cdot \dot{\mathfrak{w}}_{,1} \sqrt{g}\, dF + \iint_F \tau^{\alpha m} \mathfrak{g}_m \cdot \dot{\mathfrak{w}}_{,\alpha}\, dF \Big] d\Theta^1 \qquad \ldots(17)$$

In consequence of the assumptions made before, the displacement vector $\mathfrak{w}$ of the point **P** has a finite number of degrees of freedom. These are the kinematical vectors $\mathfrak{v}$, $\vec{\varphi}$ of the rod-line.

We see from the picture on page 3, that

$$\mathfrak{w} = \mathfrak{v} + \mathfrak{b} - \overset{+}{\mathfrak{b}} + \Omega \cdot (\varkappa_1 \mathfrak{r}_1)$$

The derivative with respect to time t produces the following equation:

$$\dot{\mathfrak{w}} = \dot{\mathfrak{v}} + \underbrace{\dot{\mathfrak{b}}}_{=\, \dot{\vec{\varphi}} \times \mathfrak{b}} - 0 + \Omega \cdot (\dot{\varkappa}_1 \mathfrak{r}_1 + \varkappa_1 \underbrace{\dot{\mathfrak{r}}_1}_{=\, \dot{\vec{\varphi}} \times \mathfrak{r}_1})$$

We set $\dot{\mathfrak{w}}$ into the deformation work (17) and integrate the work over the cross section dF. With respect to the equilibrium of moments we get the deformation work after a long consideration in the following final form:

$$\dot{W} = \int_{\Theta^1} \Big[\mathfrak{N} \cdot \dot{\vec{\varepsilon}} + \mathfrak{M} \cdot \dot{\vec{\varkappa}} + \mathfrak{B} \cdot \dot{\vec{\beta}} + \mathfrak{L} \cdot \dot{\vec{\gamma}} \Big] d\Theta^1 \qquad (18)$$

Here are $\mathfrak{N}$, $\mathfrak{M}$ and $\mathfrak{B}$, $\mathfrak{L}$ the internal rod forces and rod moments.

$$\left.\begin{aligned} \mathfrak{N} &= \iint_F \tau^{1m} \mathfrak{g}_m \sqrt{g}\, dF \\ \mathfrak{M} &= \iint_F \tau^{1m} \mathfrak{g}_m \times \mathfrak{b} \sqrt{g}\, dF \end{aligned}\right\} \quad (19)$$

$\mathfrak{N}$, $\mathfrak{M}$ are the classical rod forces and moments.

$$\left.\begin{aligned} \mathfrak{B} &= \iint_F \tau^{1m} \mathfrak{g}_m \sqrt{g} \cdot \Omega\, dF \\ \mathfrak{L} &= \iint_F \tau^{\alpha m} \mathfrak{g}_m \sqrt{g} \cdot \Omega_{,\alpha}\, dF \end{aligned}\right\} \quad (20)$$

$\mathfrak{B}$, $\mathfrak{L}$ are the Bimoments. They are more profound than the former ones.

In (18) the $\dot{\vec{\varepsilon}}$, $\dot{\vec{\kappa}}$ and $\dot{\vec{\beta}}$, $\dot{\vec{\gamma}}$ are the time derivated constitutive variables.

$$\left.\begin{array}{ll|l} \dot{\vec{\varepsilon}} & = \dot{\mathbf{r}}_{,1} + \mathbf{r}_{,1} \times \dot{\vec{\varphi}} & \dot{\vec{\beta}} = (\kappa_1 \mathbf{v}_1)^{\cdot}_{,1} + (\kappa_1 \mathbf{v}_1)_{,1} \times \dot{\vec{\varphi}} \\ \dot{\vec{\kappa}} & = \dot{\vec{\varphi}}_{,1} & \dot{\vec{\gamma}} = (\kappa_1 \mathbf{v}_1)^{\cdot} + (\kappa_1 \mathbf{v}_1) \times \dot{\vec{\varphi}} \end{array}\right\} \tag{21}$$

The time derivated constitutive variables follow without any geometrical consideration directly from the deformation work.

In the following we write the internal rod forces and moments in components:

$$\mathfrak{N} = N^m \mathbf{v}_m \quad , \quad \mathfrak{M} = M^m \mathbf{v}_m \; ; \quad \mathfrak{B} = B^m \mathbf{v}_m \; ; \quad \mathfrak{C} = C^m \mathbf{v}_m \tag{22}$$

and also the time derivated constitutive variables:

$$\dot{\vec{\varepsilon}} = \dot{\varepsilon}^m \mathbf{v}_m \; ; \quad \dot{\vec{\kappa}} = \dot{\kappa}_m \mathbf{v}^m ; \quad \dot{\vec{\beta}} = \dot{\beta}^m \mathbf{v}_m ; \quad \dot{\vec{\gamma}} = \dot{\gamma}_m \mathbf{v}^m \tag{23}$$

Therefore the work (18) can be written in components:

$$\dot{W} = \int_{\theta^1} \left(N^m \dot{\varepsilon}_m + M^m \dot{\kappa}_m + B^m \dot{\beta}_m + C^m \dot{\gamma}_m \right) d\theta^1 \tag{24}$$

The $\dot{\varepsilon}_m$, $\dot{\kappa}_m$ and $\dot{\beta}_m$, $\dot{\gamma}_m$ can be led back into the total time-differentials. With the help of (21) and (23) we get the remarkable result:

$$\begin{array}{ll|l} \dot{\varepsilon}_m & = (\mathbf{r}_{,1} \cdot \mathbf{v}_m)^{\cdot} & \dot{\beta}_m = \left((\kappa_1 \mathbf{v}_1)_{,1} \cdot \mathbf{v}_m \right)^{\cdot} \\ \dot{\kappa}_m & = \frac{1}{2} \left(e_m{}^{ik} \mathbf{v}_{i,1} \cdot \mathbf{v}_k \right)^{\cdot} & \dot{\gamma}_m = \left(\kappa_1 \mathbf{v}_1 \cdot \mathbf{v}_m \right)^{\cdot} \end{array} \quad \ldots \tag{25}$$

$e_m{}^{ik}$ is the permutation tensor with the terms +1,-1,o. After integration over the time t we immediately obtain the constitutive variables. The constitutive variables ε_m, κ_m and β_m, γ_m can be expressed with the help of

$$\mathbf{r} = \overset{+}{\mathbf{r}} + \mathbf{w} \quad ; \quad \mathbf{v}_i = \Omega_i{}^m \overset{+}{\mathbf{v}}_m \quad \text{with } \Omega_i{}^m (\varphi^k)$$

by the kinematical vectors $\mathbf{w}$ and $\vec{\varphi}$ of the rod-line. The constitutive variables contain the components of the kinematical vectors $\mathbf{w}$ and $\vec{\varphi}$ in nonlinear form.

We specify the constitutive variables, linearized in v^m and φ^m:

$$\varepsilon_1 = v^1|_1 \qquad \varkappa_1 = \varphi^1|_1 \qquad \beta_1 = (\varphi^1|_1)_{,1} \qquad \gamma_1 = \varphi^1|_1$$

$$\varepsilon_2 = v^2|_1 + \varphi^3 \;;\quad \varkappa_2 = \varphi^2|_1 \;;\quad \beta_2 = \overset{+}{\varkappa}_3 \cdot \varphi^1|_1 \;;\quad \gamma_2 = 0$$

$$\varepsilon_3 = v^3|_1 - \varphi^2 \qquad \varkappa_3 = \varphi^3|_1 \qquad \beta_3 = -\overset{+}{\varkappa}_2 \cdot \varphi^1|_1 \qquad \gamma_3 = 0$$

In the deformation work (24) the N^2, N^3 are the shear forces, so that the according constitutive variables ε_2, ε_3 are the shear deformations. If we neglect the shear deformations ε_2, ε_3 we get the wellknown linearized rotations φ^2, φ^3 of the cross section according to the Bernoulli-hypothesis as follows:

$$\varepsilon_2 = v^2|_1 + \varphi^3 = 0 \quad \rightarrow \quad \varphi^3 \cong - v^2|_1$$

$$\varepsilon_3 = v^3|_1 - \varphi^2 = 0 \quad \rightarrow \quad \varphi^3 \cong v^3|_1$$

In the Bernoulli-theory only the four kinematic variables v^1, v^2, v^3 and φ^1 are unknown. In order that the problem is not overdetermined with respect to the six equilibrium conditions, we have to introduce additional unknowns, here the two shear forces N^2, N^3. It is not allowed to formulate constitutive equations for the shear forces. This fact is the wellknown discrepancy of the Bernoulli-hypothesis, so that we prefer the Cosserat-theory with the six unknown kinematic variables v^m, φ^m of the rod-line.

In (24) the four rod forces and moments $\mathfrak{N}$, $\mathfrak{M}$ and $\mathfrak{L}$, $\mathcal{L}$ at once followed from the deformation work. We will show in the following chapter, that the same and <u>no other</u> vectors appear in the equations of motion. This is a remarkable result, because we can formulate hyper-elastic constitutive equations with the help of deformation work for the $\mathfrak{N}$, $\mathfrak{M}$ and $\mathfrak{L}$, $\mathcal{L}$, which are unknown forces and moments in the equations of motion.

5. The principle of virtual work and the equations of motion for a rod, deformed by the kinematical vectors $v, \vec{\varphi}$ of the rod-line

The equations of motion of the infinitesimal element dV, formed by the rod coordinates $\theta^1, \theta^2, \theta^3$ at time t is given by the following wellknown vector equation:

$$\left[\left(\tau^{im} g_m \sqrt{g} \right)_{,i} + Q^m g_m \varrho \sqrt{g} \right] d\theta^1 d\theta^2 d\theta^3 = \overset{+}{\varrho} \sqrt{\overset{+}{g}}\, \ddot{w}\, d\theta^1 d\theta^2 d\theta^3 \Big| \cdot \delta w \quad (26)$$

The displacement vector w of the element dV is given by

$$w = v + b - \overset{+}{b} + \Omega \cdot (x_1 v_1)$$

The differential with respect to time t contains the kinematical vectors v and $\vec{\varphi}$ of rod-line in the following form:

$$\dot{w}\, Dt = \dot{v}\, Dt + \dot{b}\, Dt + \Omega \cdot (x_1 v_1)^{\cdot}\, Dt$$
$$\dot{b} = \dot{\vec{\varphi}} \times b$$

Replacing the real time-increments $(\dots)^{\cdot}$ Dt by the virtual symbol " δ ", the virtual displacement vector is given by:

$$\delta w = \delta v + \delta \vec{\varphi} \times b + \Omega \cdot \delta (x_1 v_1) \quad (27)$$

A possible way to formulate the equations of motion for an infinitesimal rod element with a finite cross section can be shown with mathematical precision with the help of virtual work.

The virtual work:

We multiply (26) with δw and integrate over the cross section $dF = d\theta^2 \cdot d\theta^3$ and also over the length $d\theta^1$ of the rod-line. We immediately get the following virtual work at the rod element:

$$\int_{\theta^1} \iint_F \left(\tau^{1m} g_m \sqrt{g} \right)_{,1} \cdot \delta w \cdot dF \cdot d\theta^1 + \underbrace{\int_{\theta^1} \iint_F \left(\tau^{\alpha m} g_m \sqrt{g} \right)_{,\alpha} \cdot \delta w \cdot dF \cdot d\theta^1}_{\mathfrak{J}} \quad (28)$$
$$+ \int_{\theta^1} \iint_F \varrho\, Q^m \sqrt{g}\, g_m \cdot \delta w\, dF \cdot d\theta^1 = \int_{\theta^1} \iint_F \overset{+}{\varrho} \sqrt{\overset{+}{g}}\, \ddot{w} \cdot \delta w \cdot dF \cdot d\theta^1$$

The integral $\mathfrak{I}$ in (28) can be transformed with the use of product rule of differentiation $u' \cdot v = (u \cdot v)' - u \cdot v'$ into:

$$\mathfrak{I} = \iint_F \left(\tau^{\alpha m} g_m \sqrt{g} \cdot \delta w\right)_{,\alpha} d\theta^2 d\theta^3 - \iint_F \tau^{\alpha m} g_m \sqrt{g} \cdot \delta w_{,\alpha} \, d\theta^2 d\theta^3$$

After summation over $\alpha = (2,3)$, we integrate the first integral over $d\theta^2$, $d\theta^3$ and get:

$$\mathfrak{I} = \oint_S \underbrace{\left(\tilde{\tau}^{2m} \tilde{g}_m \sqrt{\tilde{g}} \, d\tilde{\theta}^3 + \tilde{\tau}^{3m} \tilde{g}_m \sqrt{\tilde{g}} \, d\tilde{\theta}^2\right)}_{L \,=\, P^m \tilde{g}_m \, ds \,=\, \text{surface forces}} \cdot \delta w - \iint_F \tau^{\alpha m} g_m \sqrt{g} \cdot \delta w_{,\alpha} \, d\theta^2 d\theta^3$$

The integration over $d\tilde{\theta}^3$ and $d\tilde{\theta}^2$ follows over the boundary curve $\tilde{\theta}^2(s)$ and $\tilde{\theta}^3(s)$ of the cross section, according the below picture:

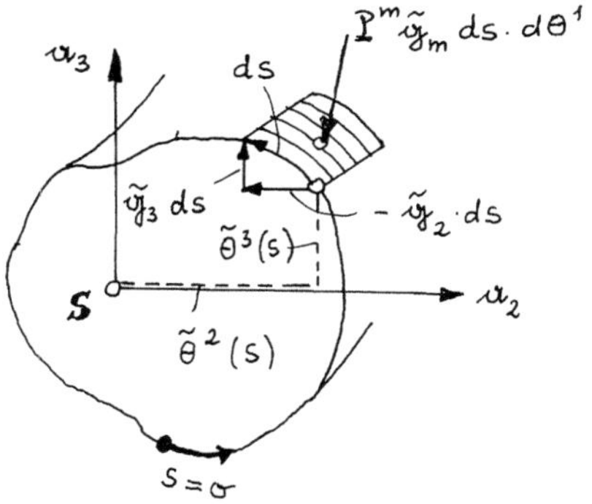

The above underlined surface integrand is equal to the surface forces as consequence of the boundary condition at the boundary element

The principle of virtual work (28) gets the following form, if we introduce the rod forces and moments $\mathfrak{N}, \mathfrak{M}, \mathfrak{B}, \mathfrak{C}$ and the above integral $\mathfrak{I}$. If we take also into account the equilibrium of the moments we get after a long consideration:

$$\int_{\theta^1} \left[\begin{array}{l} \delta w \cdot \left(\mathfrak{N}_{,1} + \overset{L}{\vec{Q}} \right) + \\ + \delta \vec{\varphi} \cdot \left(\mathfrak{M}_{,1} + \mathfrak{N} \times w_{,1} + \left(\mathfrak{B} \times x_1 v_1 \right)_{,1} + \overset{L}{\vec{M}} \right) + \\ + \delta \vec{\varphi}_{,1} \cdot \left(B^1|_1 \, v_1 - C^1 v_1 + \overset{\Omega}{M}{}^1 v_1 \right) \end{array} \right] d\theta^1 = 0 \qquad (30)$$

The $\overset{L}{\vec{Q}}$ and $\overset{L}{\vec{M}}, \overset{\Omega}{\vec{M}}$ are forces and moments, caused by the surface forces p^m and body forces Q^m.

$$\overset{L}{\vec{Q}} = \oint_S p^m \mathfrak{v}_m \, ds + \iint_F Q^m \mathfrak{g}_m \sqrt{g} \, dF = \overset{L}{Q}{}^m \mathfrak{v}_m$$

$$\overset{L}{\vec{M}} = \oint_S p^m \mathfrak{v}_m \times \left(\mathfrak{b} + \Omega \varkappa_1 \mathfrak{v}_1\right) ds + \iint_F Q^m \mathfrak{g}_m \sqrt{g} \times \left(\mathfrak{b} + \Omega \varkappa_1 \mathfrak{v}_1\right) dF = \overset{L}{M}{}^m \mathfrak{v}_m$$

$$\overset{\Omega}{\vec{M}} = \oint_S p^m \mathfrak{v}_m \cdot \Omega \, ds + \iint_F Q^m \mathfrak{g}_m \sqrt{g} \cdot \Omega \, dF = \overset{\Omega}{M}{}^m \mathfrak{v}_m$$

The body forces Q^m contain the inertial forces, caused by acceleration.

The equations of motion:

If we integrate in (3o) the lowest term by part, the principle of virtual work (3o) changes into:

$$\int_{\theta^1} \left[\begin{array}{l} \delta \mathfrak{w} \cdot \left(\mathfrak{N}_{,1} + \overset{L}{\vec{Q}} \right) + \\ + \delta \vec{\varphi} \cdot \left(\mathfrak{M}_{,1} + \mathfrak{N} \times \mathfrak{b}_{,1} + \left(\mathfrak{L} \times \varkappa_1 \mathfrak{v}_1 \right)_{,1} + \overset{L}{\vec{M}} - \right. \\ \left. - \delta \vec{\varphi} \cdot \left(B^1|_1 \mathfrak{v}_1 - C^1 \mathfrak{v}_1 + \overset{\Omega}{M}{}^1 \mathfrak{v}_1 \right)_{,1} \right. + \end{array} \right] \cdot d\theta^1 + \qquad (31)$$

$$+ \left\{ \delta \vec{\varphi} \cdot \left(B^1|_1 \mathfrak{v}_1 - C^1 \mathfrak{v}_1 + \overset{\Omega}{M}{}^1 \mathfrak{v}_1 \right) \right\}_{\theta^1 = 0}^{\ell} = 0$$

The above-mentioned relation is held by arbitrary $\delta \mathfrak{w}$ and $\delta \vec{\varphi}$ in the rod space and by arbitrary $\delta \vec{\varphi}$ at the boundary $\theta^1 = 0$ and $\theta^1 = \ell$. This at once leads to the equations of motion.

$$\text{Forces:} \quad \mathfrak{N}_{,1} + \overset{L}{\vec{Q}} = 0 \qquad (32)$$

$$\text{Moments:} \quad \left. \begin{array}{l} \mathfrak{M}_{,1} + \mathfrak{N} \times \mathfrak{b}_{,1} + \left(\mathfrak{L} \times \varkappa_1 \mathfrak{v}_1 \right)_{,1} + \overset{L}{\vec{M}} - \\ - \left(B^1|_1 \mathfrak{v}_1 - C^1 \mathfrak{v}_1 + \overset{\Omega}{M}{}^1 \mathfrak{v}_1 \right)_{,1} = 0 \end{array} \right\} \qquad (33)$$

From (31) we also obtain the dynamical boundary condition for the bimoments B^1 and C^1 at $\theta^1 = o$ and $\theta^1 = \ell$:

$$\mathfrak{v}_1 : \quad \left| B^1|_1 - C^1 + \overset{\Omega}{M}{}^1 \right|_{\theta^1=0;\ \theta^1=\ell} = \sigma \tag{34}$$

These equations are only valid, if $\varphi^1(o)$ and $\varphi^1(\ell)$ are not prescribed. If $\varphi^1(o)$ and $\varphi^1(\ell)$ are given, the $\delta\varphi^1(o)$ and $\delta\varphi^1(\ell)$ are zero, so that the boundary term $\{..\}_o^\ell$ in (31) is also zero.

<u>The equations of motion in components</u>:

If the rod forces and moments are given in components

$$\mathfrak{N} = N^m \mathfrak{v}_m ;\quad \mathfrak{M} = M^m \mathfrak{v}_m ;\quad \mathfrak{B} = B^m \mathfrak{v}_m ;\quad \overset{L}{\vec{Q}} = \overset{L}{Q}{}^m \mathfrak{v}_m \ldots \text{ etc.},$$

the foregoing vector equations of motion can be written in components as follows.

The equations of motion for the forces, according (32):

$$\mathfrak{v}_m : \quad \boxed{N^m|_1 + \overset{L}{Q}{}^m = \sigma} \tag{32a}$$

The equations of motion for the moments, according (33):

$$\mathfrak{v}_m : \quad \boxed{\begin{aligned} &M^m|_1 + N^s\left(\varepsilon_n + \delta_n^1\right)\cdot e_s{}^{mn} + \left(B^s \cdot \varkappa_1 \cdot e_{s1}{}^m\right)\Big|_1 \\ &- \left(\delta_1^m \left[B^1|_1 - C^1 + \overset{\Omega}{M}{}^1\right]\right)\Big|_1 + \overset{L}{M}{}^m = \sigma \end{aligned}} \tag{33a}$$

Here is $e_s{}^{mn}$ the permutation tensor with the terms +1,-1,o. The dynamical boundary conditions for the bimoments are:

$$\mathfrak{v}_1 : \quad \boxed{\left| B^1|_1 - C^1 + \overset{\Omega}{M}{}^1 \right|_{\theta^1=0;\ \theta^1=\ell} = \sigma} \tag{34a}$$

In these equations of motion, the behaviour of deformation is entirely given by the constitutive variables $\varepsilon_m, \varkappa_m$.

If we set the variables zero, we get the equations of motion for the undeformed rod. The covariant derivations can be carried out with $v_{s,1} = e_s^{mn}(\overset{+}{\varkappa}_n + \varkappa_n) v_m$. We get for example with $\mathfrak{N}_{,1} = (N^m v_m)_{,1} = N^m|_1 v_m$:

$$N^m|_1 = N^m_{,1} + N^s \cdot e_s^{mn} \cdot (\overset{+}{\varkappa}_n + \varkappa_n) \equiv N^m_{,1} + N^s \cdot \Gamma_{s1}^m \qquad \ldots(35)$$

It is e_s^{mn} the permutation tensor with the terms +1,-1,o.

6. The elastic constitutive equations for the rod forces and moments

Let us consider the deformation work of rods, according (24):

$$\dot{W} = \int_{\theta^1} (N^m \cdot \dot{\varepsilon}_m + M^m \cdot \dot{\varkappa}_m + B^m \cdot \dot{\beta}_m + C^m \cdot \dot{\gamma}_m)\, d\theta^1 \qquad (36)$$

We assume, that a strain energy function exists:

$$\Pi = \Pi^*(\varepsilon_m, \varkappa_m, \beta_m, \gamma_m)$$

For example Π may be given by the following anisotropic scalar function:

$$\Pi^* = \frac{1}{2} \overset{1}{A}{}^{ik} \varepsilon_i \cdot \varepsilon_k + \frac{1}{2} \overset{2}{A}{}^{ik} \varkappa_i \cdot \varkappa_k + \overset{3}{A}{}^{ik} \varepsilon_i \cdot \varkappa_k \qquad \ldots(37)$$

where $\overset{1}{A}{}^{ik}$, $\overset{2}{A}{}^{ik}$, $\overset{3}{A}{}^{ik}$, ... are the tensors of elasticity with open material constants.

The rod is named hyper-elastic, if the rod forces and moments satisfy the following constitutive equations:

$$N^m = \frac{\partial \Pi^*}{\partial \varepsilon_m}; \quad M^m = \frac{\partial \Pi^*}{\partial \varkappa_m}; \quad B^m = \frac{\partial \Pi^*}{\partial \beta_m}; \quad C^m = \frac{\partial \Pi^*}{\partial \gamma_m} \qquad (38)$$

The upper constitutive equations are very simple from the mathematical point of view. But they are of no use, because the material constants in (37) get always new values for differing cross sections. The elastic constants therefore have to be found again by complicate tests.

These difficulties do not exist, if we use the following well-known three-dimensional deformation work:

$$\dot{W} = \int_{\theta^1} \iint_F \tau^{ik} \cdot \dot{\gamma}_{ik} \cdot \sqrt{g}\, dF \cdot d\theta^1 \quad ; \qquad dF = d\theta^2\, d\theta^3 \tag{39}$$

The elastic three-dimensional strain energy function Π is given by

$$\Pi = \frac{1}{2} E^{ikrs} \cdot \gamma_{ik} \cdot \gamma_{rs} \quad ; \qquad E^{ikrs} = \text{tensor of elasticity} \tag{4o}$$

In this case the tensor of elasticity includes material constants, which are valid for every cross section.
The hyper-elastic constitutive equations for stresses has the following form:

$$\sqrt{g/\overset{+}{g}} \cdot \tau^{ik} = \frac{\partial \Pi}{\partial \gamma_{ik}} \tag{41}$$

According to that the deformation work becomes:

$$\dot{W} = \int_{\theta^1} \iint_F \tau^{ik}\, \dot{\gamma}_{ik} \sqrt{g}\, dF = \int_{\theta^1} \iint_F \frac{\partial \Pi}{\partial \gamma_{ik}} \cdot \dot{\gamma}_{ik} \sqrt{\overset{+}{g}}\, dF \cdot d\theta^1 \tag{42}$$

The Green-tensor γ_{ik} is

$$\gamma_{ik} = \frac{1}{2} \left(\mathbf{g}_i \cdot \mathbf{g}_k - \overset{+}{\mathbf{g}}_i \cdot \overset{+}{\mathbf{g}}_k \right)$$

This strain tensor can be expressed by the constitutive variables of the one-dimensional rod theory. With $\mathbf{g}_i = (\overset{+}{\mathbf{r}} + \mathbf{v} + \mathbf{b} - \overset{+}{\mathbf{b}} + \Omega\, \varkappa_1 \mathbf{a}_1)_{,i}$ we get after a short calculation, see paper [2]:

$$\gamma_{ik} = \hat{\gamma}_{ik} \left(\varepsilon_m, \varkappa_m, \beta_m, \gamma_m \right) \tag{43}$$

Or in detail:

$$\gamma_{11} = \varepsilon_1 - \theta^2 \left(\varkappa_3 + \varepsilon_1 \overset{+}{\varkappa}_3 - \varepsilon_3 \overset{+}{\varkappa}_1 \right) + \theta^3 \left(\varkappa_2 + \varepsilon_1 \overset{+}{\varkappa}_2 - \varepsilon_2 \overset{+}{\varkappa}_1 \right) + \Delta \gamma_{11}$$

$$2\, \gamma_{12} = \varepsilon_2 - \theta^3 \varkappa_1 + \Omega_{,2} \cdot \gamma_1 + \Omega \cdot \beta_2 + \Delta \gamma_{12}$$

$$2\, \gamma_{13} = \varepsilon_3 + \theta^2 \varkappa_1 + \Omega_{,3} \cdot \gamma_1 + \Omega \cdot \beta_3 + \Delta \gamma_{13}$$

$$\gamma_{22} = \Delta \gamma_{22} \quad ; \qquad \gamma_{23} = \Delta \gamma_{23} \quad ; \qquad \gamma_{33} = \Delta \gamma_{33}$$

The Δ-terms contain the quadratic variables ε_i, $\varkappa_i$, β_i, γ_i, which we do not specify here.

Setting (43) in (42) the deformation work $\dot{W}$ becomes:

$$\dot{W} = \int_{\theta^1} \iint_F \frac{\partial \Pi}{\partial \gamma_{ik}} \cdot \left(\frac{\partial \hat{\gamma}_{ik}}{\partial \varepsilon_m} \cdot \dot{\varepsilon}_m + \frac{\partial \hat{\gamma}_{ik}}{\partial \varkappa_m} \cdot \dot{\varkappa}_m + \frac{\partial \hat{\gamma}_{ik}}{\partial \beta_m} \cdot \dot{\beta}_m + \frac{\partial \hat{\gamma}_{ik}}{\partial \gamma_m} \dot{\gamma}_m \right) \sqrt{\overset{+}{g}}\, dF \cdot d\theta^1$$

Comparing this equation with the one-dimensional deformation work (36) the constitutive equations for the rod forces and moments get the following form:

$$\left.\begin{aligned} N^m &= \iint_F \frac{\partial \Pi}{\partial \gamma_{ik}} \cdot \frac{\partial \hat{\gamma}_{ik}}{\partial \varepsilon_m} \sqrt{\overset{+}{g}}\, dF \; ; & B^m &= \iint_F \frac{\partial \Pi}{\partial \gamma_{ik}} \cdot \frac{\partial \hat{\gamma}_{ik}}{\partial \beta_m} \sqrt{\overset{+}{g}}\, dF \\ M^m &= \iint_F \frac{\partial \Pi}{\partial \gamma_{ik}} \cdot \frac{\partial \hat{\gamma}_{ik}}{\partial \varkappa_m} \sqrt{\overset{+}{g}}\, dF & C^m &= \iint_F \frac{\partial \Pi}{\partial \gamma_{ik}} \cdot \frac{\partial \hat{\gamma}_{ik}}{\partial \gamma_m} \sqrt{\overset{+}{g}}\, dF \end{aligned}\right\} \quad (44)$$

We integrate these equations over the cross section $dF = d\theta^2 \cdot d\theta^3$. We use hereby the principle axes θ^2 , θ^3 of the cross section with the following properties:

$$F_3 = \iint_F \theta^3 \cdot dF = 0 \; ; \quad F_2 = \iint_F \theta^2 \cdot dF = 0 \; ; \quad F_{23} = \iint_F \theta^2 \cdot \theta^3 \cdot dF = 0$$

Setting Π (4o) and γ_{ik} (43) into (44) we get the following constitutive equations in ε_m, $\varkappa_m$, β_m, γ_m :

The constitutive equations for the rod forces are

$$N^1 = EF\,\varepsilon_1 + \Delta N^1 \; ; \quad N^2 = EF\,\varepsilon_2 + \Delta N^2 \; ; \quad N^3 = EF\,\varepsilon_3 + \Delta N^3 \quad (45)$$

The constitutive equations for the rod moments are given by

$$\begin{aligned} M^1 &= G \cdot J_T \cdot \varkappa_1 + G F_{3\Omega} \cdot \beta_2 + G F_{2\Omega} \cdot \beta_3 + \Delta M^1 \\ M^2 &= E F_{33} \cdot \left(\varkappa_2 - \varepsilon_1 \overset{+}{\varkappa}_2\right) + E F_{3\Omega} \cdot \beta_1 + \Delta M^2 \\ -M^3 &= E F_{22} \cdot \left(\varkappa_3 - \varepsilon_1 \overset{+}{\varkappa}_3\right) + E F_{2\Omega} \cdot \beta_1 + \Delta M^3 \end{aligned} \quad (46)$$

The ΔN^m and ΔM^m depend quadratic on ε_m, $\varkappa_m$, β_m, γ_m, and they are also proportional to $\overset{+}{\varkappa}_1$ of the pretwisted undeformed rod. The ΔN^m and ΔM^m are zero, if $\overset{+}{\varkappa}_1$ is zero, see [2]. They are not specified here.

The constitutive equations for the rod bimoments are

$$\boxed{\begin{aligned} B^1 &= EF_{2\Omega}\cdot\left(-\varkappa_3+\varepsilon_1\overset{+}{\varkappa}_3\right)+EF_{3\Omega}\cdot\left(\varkappa_2+\varepsilon_1\overset{+}{\varkappa}_2\right)+EF_{\Omega\Omega}\cdot\beta_1+\Delta B^1 \\ B^2 &= -GF_{3\Omega}\cdot\varkappa_1+GF_{\Omega\Omega,2}\cdot\varkappa_1+\Delta B^2 \\ B^3 &= +GF_{2\Omega}\cdot\varkappa_1+GF_{\Omega\Omega,3}\;\varkappa_1+\Delta B^3 \\ C^1 &= GF_{\Omega\Omega,2}\cdot\beta_2+GF_{\Omega\Omega,3}\cdot\beta_3+\Delta C^1\;;\qquad C^2=C^3=0 \end{aligned}} \tag{47}$$

The following cross section quantities are known:

$$F_{22}=\iint_F\theta^2\cdot\theta^2\cdot dF\;;\qquad F_{33}=\iint_F\theta^3\cdot\theta^3\cdot dF\;;\qquad F_{\Omega\Omega}=\iint_F\Omega\cdot\Omega\cdot dF$$

$$F_{2\Omega}=\iint_F\theta^2\cdot\Omega\cdot dF\;;\qquad F_{3\Omega}=\iint_F\theta^3\cdot\Omega\cdot dF\;;\qquad F_{\Omega\Omega,2}=\iint_F\Omega\cdot\Omega_{,2}\cdot dF$$

The cross section quantities $F_{2\Omega}$, $F_{3\Omega}$ are zero, if the shear centre coincides with the centre of gravity. That does not apply to unsymmetrical cross sections.

The hindered warping of the cross section is described by the quantities $F_{\Omega\Omega}$, $F_{\Omega\Omega,2}$, $F_{\Omega\Omega,3}$. This influence can be often neglected.
The bimoments $\Delta B^m, \Delta C^1$ are zero like ΔN^m, ΔM^m, if the undeformed rod is not pretwisted by $\overset{+}{\varkappa}_1$. They are not specified here.

The constitutive equations (45), (46), (47) of the precurved and pretwisted rod differ importantly from the constitutive equations of the straight lined rod by the $\overset{+}{\varkappa}_1$, $\overset{+}{\varkappa}_2$, $\overset{+}{\varkappa}_3$. It is sometimes usual, to transfer the wellknown constitutive equations of straight lined rods into equations of precurved and pretwisted rods, by exchanging the partial derivatives by covariant onses. This proceeding is unexact especially in the case of pretwisted rods by $\overset{+}{\varkappa}_1$.

7. The equations of linear stability of rods

To find out the behaviour of stability we consider a rod in three states: The starting state, the fundamental state, the disturbed state.

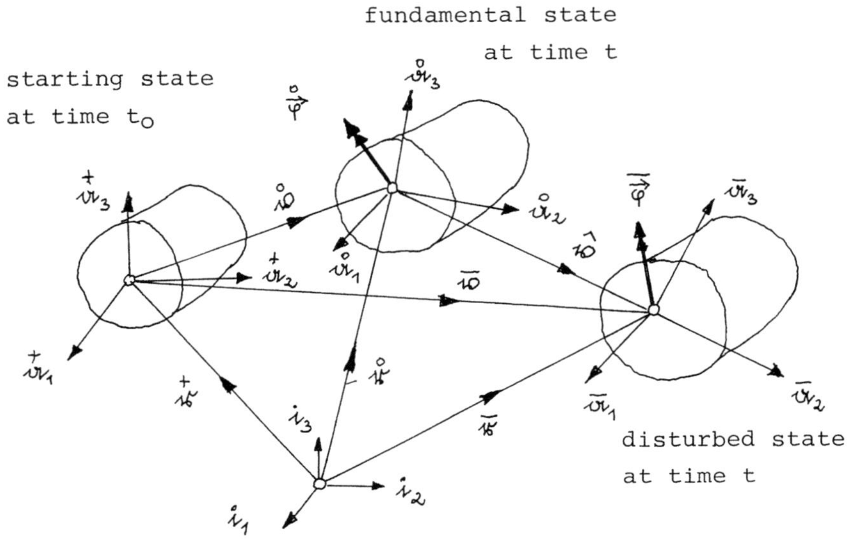

The starting state at time t_o is the state of the undeformed pretwisted and precurved rod by $\overset{+}{\varkappa}_1$ and $\overset{+}{\varkappa}_2$, $\overset{+}{\varkappa}_3$.

The fundamental state at time t is a known state of motion or equilibrium, which shall be investigated with respect to stability, whereby finite $\overset{\circ}{\mathbf{v}}$ and $\overset{\circ}{\vec{\varphi}}$ of the fundamental state are noted.

To understand the stability of the fundamental state, we disturb according to L j a p u n o f f the initial-conditions $\overset{\circ}{\mathbf{v}}(t_o)$, $\dot{\overset{\circ}{\mathbf{v}}}(t_o)$ and $\overset{\circ}{\vec{\varphi}}(t_o)$, $\dot{\overset{\circ}{\vec{\varphi}}}(t_o)$. As result of this disturbance the rod moves into the disturbed state

$$\begin{aligned} \bar{\mathbf{v}} &= \overset{\circ}{\mathbf{v}}(\theta^1,t) + \hat{\mathbf{v}}(\theta^1,t) \\ &= \overset{\circ}{\vartheta}{}^m \overset{+}{\mathbf{v}}_m + \hat{\vartheta}^m \overset{\circ}{\mathbf{v}}_m \end{aligned} \quad \Bigg| \quad \begin{aligned} \bar{\vec{\varphi}} &= \overset{\circ}{\vec{\varphi}}(\theta^1,t) + \hat{\vec{\varphi}}(\theta^1,t) \\ &= \overset{\circ}{\varphi}{}^m \overset{+}{\mathbf{v}}_m + \hat{\varphi}^m \overset{+}{\mathbf{v}}_m \end{aligned} \qquad (48)$$

without altering the time. The fundamental state is called linear stable, if the infinitesimal displacement vector $\hat{\mathbf{v}}$ and also the rotation vector $\hat{\vec{\varphi}}$ does not exceed a given magnitude.

For instance we consider a rod with time independent external forces, which have in different states of deformation always the same direction and magnitude at time t:

$$\overset{L}{\vec{Q}} = \overset{\circ}{\overset{L}{Q}}{}^{m} \overset{\circ}{v}_m = \overline{\overset{L}{Q}}{}^{m} \bar{v}_m \qquad ; \qquad \overset{\Omega}{\vec{M}} = \overset{\circ}{\overset{\Omega}{M}}{}^{m} \overset{\circ}{v}_m = \overline{\overset{\Omega}{M}}{}^{m} \bar{v}_m \qquad (49)$$
$$\overset{L}{\vec{M}} = \overset{\circ}{\overset{L}{M}}{}^{m} \overset{\circ}{v}_m = \overline{\overset{L}{M}}{}^{m} \bar{v}_m$$

In regard to these external forces, the fundamental state is in equilibrium. According to (32) and (33) the equations of the fundamental state are

$$\overset{\circ}{\mathfrak{N}}_{,1} + \overset{L}{\vec{Q}} = \sigma \qquad \text{and}$$

$$\overset{\circ}{\mathfrak{M}}_{,1} + \overset{\circ}{\mathfrak{N}} \times \overset{\circ}{\mathfrak{v}}_{,1} + \left(\overset{\circ}{\mathfrak{L}} \times \overset{\circ}{\mathfrak{x}}_{,1} \overset{\circ}{v}_1 \right)_{,1} + \overset{L}{\vec{M}} \qquad (50)$$
$$- \left(\overset{\circ}{v}_1 \overset{\circ}{B}{}^{1}|_1 - \overset{\circ}{v}_1 \overset{\circ}{C}{}^{1} + \overset{\circ}{v}_1 \overset{\circ}{\overset{\Omega}{M}}{}^{1} \right)_{,1} = \sigma$$

Since the external forces are time independent and they keep their direction, the disturbed state is also in equilibrium in this case, see [5]:

$$\bar{\mathfrak{N}}_{,1} + \overset{L}{\vec{Q}} = \sigma \qquad \text{and}$$

$$\bar{\mathfrak{M}}_{,1} + \bar{\mathfrak{N}} \times \bar{\mathfrak{v}}_{,1} + \left(\bar{\mathfrak{L}} \times \bar{\mathfrak{x}}_{,1} \bar{v}_1 \right)_{,1} + \overset{L}{\vec{M}} \qquad (51)$$
$$- \left(\bar{v}_1 \bar{B}{}^{1}|_1 - \bar{v}_1 \bar{C}{}^{1} + \bar{v}_1 \overline{\overset{\Omega}{M}}{}^{1} \right)_{,1} = \sigma$$

In order to derive the linear equations of stability for the unknown $\hat{\mathfrak{w}}$ and $\hat{\vec{\varphi}}$, we split the following vectors of the disturbed state into:

$$\bar{\mathfrak{v}} = \overset{\circ}{\mathfrak{v}} + \hat{\mathfrak{v}} \qquad \bar{v}_i = \overset{\circ}{v}_i + \hat{v}_i$$
$$\bar{\mathfrak{N}} = \overset{\circ}{\mathfrak{N}} + \hat{\mathfrak{N}} \quad ; \quad \bar{\mathfrak{M}} = \overset{\circ}{\mathfrak{M}} + \hat{\mathfrak{M}} \quad ; \quad \bar{\mathfrak{L}} = \overset{\circ}{\mathfrak{L}} + \hat{\mathfrak{L}} \qquad (52)$$

with the following infinitesimal components in direction $\overset{\circ}{v}_m$:

$$\hat{\mathfrak{v}} = \hat{\mathfrak{w}} = \hat{v}^{m} \overset{\circ}{v}_m \; ; \quad \hat{v}_i = \hat{\vec{\varphi}} \times \overset{\circ}{v}_i = \left(\hat{\varphi}^{s} \overset{\circ}{v}_s \right) \times \overset{\circ}{v}_i = \hat{\psi}_i{}^{m} \overset{\circ}{v}_m$$
$$\hat{\mathfrak{N}} = \hat{N} \overset{\circ}{v}_m + \overset{\circ}{N}{}^{m} \hat{v}_m \; ; \quad \hat{\mathfrak{M}} = \hat{M}{}^{m} \overset{\circ}{v}_m + \overset{\circ}{M}{}^{m} \hat{v}_m \; ; \quad \hat{\mathfrak{L}} = \cdots \qquad (53)$$
$$\hat{\Psi}_i{}^{m} = \hat{\varphi}^{s} e_{si}{}^{m}$$

We set (52) into (51). If we consider the equilibrium condition (5o) of the fundamental state, we get the following equations of stability for the unknown infinitesimal $\hat{\mathfrak{N}}, \hat{\mathfrak{M}}, \hat{\mathfrak{b}}, \hat{\mathfrak{L}}$:

$$\hat{\mathfrak{N}}_{,1} = \sigma \qquad \text{and} \qquad \cdot\cdot(54)$$

$$\hat{\mathfrak{M}}_{,1} + \hat{\mathfrak{N}} \times \mathring{\mathfrak{t}}_{,1} + \mathring{\mathfrak{N}} \times \hat{\mathfrak{t}}_{,1} + \left(\mathring{\mathfrak{b}} \times [\mathring{\varkappa}_1 \hat{\mathfrak{v}}_1 + \hat{\varkappa}_1 \mathring{\mathfrak{v}}_1] + \hat{\mathfrak{b}} \times [\mathring{\varkappa}_1 \mathring{\mathfrak{v}}_1] \right)_{,1}$$
$$- \left(\mathring{\mathfrak{v}}_1 \hat{B}^1|_1 + \hat{\mathfrak{v}}_1 \mathring{B}^1|_1 - \mathring{\mathfrak{v}}_1 \hat{C}^1 - \hat{\mathfrak{v}}_1 \mathring{C}^1 + \mathring{\mathfrak{v}}_1 \hat{\overset{\Omega}{M}}{}^1 + \hat{\mathfrak{v}}_1 \mathring{\overset{\Omega}{M}}{}^1 \right)_{,1} = \sigma \cdot\cdot(55)$$

In consideration of (53) we can transform these two vector equations into the equations of components in direction $\mathring{\mathfrak{v}}_m$, as follows

$$\mathring{\mathfrak{v}}_m: \quad \boxed{\hat{N}^m|_1 + \Delta Q^m = \sigma} \qquad (54a)$$

$$\mathring{\mathfrak{v}}_m: \quad \boxed{\begin{aligned} &\hat{M}^m|_1 + \hat{N}^s (\mathring{\varepsilon}_n + \delta_n^1)\, e_s{}^{nm} + \left(\hat{B}^s \cdot \mathring{\varkappa}_1 \cdot e_{s1}{}^m \right)\big|_1 \\ &- \left(\delta_1^m [\hat{B}^1|_1 - \hat{C}^1 + \hat{\overset{\Omega}{M}}{}^1] \right)\big|_1 + \Delta M^m = \sigma \end{aligned}} \qquad (55a)$$

These linear equations of stability have the same form as the equations of motion (32a) and (33a). In this case the ΔQ^m and ΔM^m are deflection forces and moments, caused by the forces $\mathring{N}^m$ and moments $\mathring{M}^m$, $\mathring{B}^m$, $\mathring{C}^m$ of the known fundamental state.
The deflection forces are given by:

$$\Delta Q^m = \left(\mathring{N}^s \hat{\Psi}_s{}^m \right)\big|_1 \qquad (56)$$

The deflection moments are:

$$\begin{aligned} \Delta M^m = &\left(\mathring{M}^s \hat{\Psi}_s{}^m \right)\big|_1 + \mathring{N}^\ell \left(\hat{\Psi}_\ell{}^s [\mathring{\varepsilon}_n + \delta_n^1] + \delta_\ell^s \cdot \hat{v}_n|_1 \right) \cdot e_s{}^{mn} + \\ &+ \left([\mathring{B}^\ell \hat{\Psi}_\ell{}^s \mathring{\varkappa}_1 + \mathring{B}^s (\mathring{\varkappa}_1 \hat{\Psi}_1{}^n + \hat{\varkappa}_1 \delta_1^n)] e_s{}^{mn} \right)\big|_1 - \\ &- \left(\hat{\Psi}_1{}^m [\mathring{B}^1|_1 - \mathring{C}^1 + \mathring{\overset{\Omega}{M}}{}^1] \right)\big|_1 \end{aligned} \qquad (57)$$

The covariant derivations, which are needed in the equations of stability have to be carried out in the same manner as in (35). For example we get:

$$\overset{\circ}{N}{}^{m}\big|_{1} = \overset{\circ}{N}{}^{m}{}_{,1} + \overset{\circ}{N}{}^{s} \cdot e_{s}{}^{mn} \left(\overset{+}{\varkappa}_{n} + \overset{\circ}{\varkappa}_{n} \right) \equiv \overset{\circ}{N}{}^{m}{}_{,1} + \overset{\circ}{N}{}^{s} \Gamma_{s1}{}^{m} \qquad (58)$$

The forces $\overset{\circ}{N}{}^{m}$ and moments $\overset{\circ}{M}{}^{m}$, $\overset{\circ}{B}{}^{m}$, $\overset{\circ}{C}{}^{1}$ of the fundamental state, which is examined for its stability, are known functions of θ^1. The finite deformation of the fundamental state is given by $\overset{\circ}{\varepsilon}_n, \overset{\circ}{\varkappa}_n$.

The constitutive equations for the unknown $\hat{N}^m$, $\hat{M}^m$, $\hat{C}^1$ follow from (45), (46), (47), if we replace here the ε_m, $\varkappa_m, \beta_m$ by the infinitesimal variables $\hat{\varepsilon}_m$, $\hat{\varkappa}_m, \hat{\beta}_m$. Setting these constitutive equations into the six equations of stability (54a), (55a), we get six linear homogeneous differential equations for the six unknown kinematical components $\hat{v}^m$ and $\hat{\varphi}^m$ of the rod-line.

In most cases the deformation of the fundamental state is neglected. This proceeding is not correct, because this influence is important, especially if bending and twisting moments exist in the fundamental state.

We demonstrate this influence for the following fundamental state:

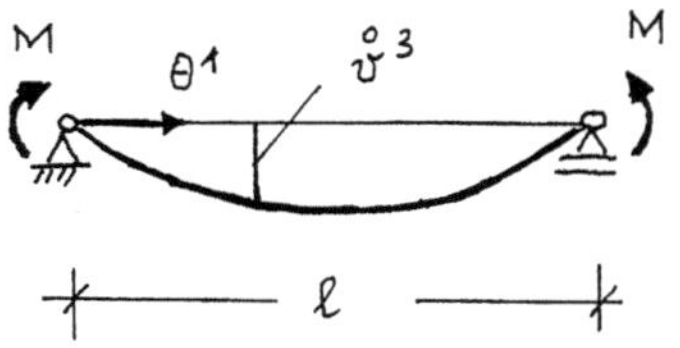

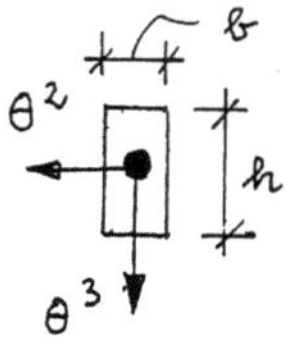

$\overset{\circ}{M}{}^2(\theta^1) = M = \text{const.}$

$\overset{\circ}{M}{}^3(\theta^1) = \overset{\circ}{M}{}^1(\theta^1) \equiv 0$

The solution for the critical moment $M = M^*$ is wellknown, if the deflection $\overset{\circ}{v}{}^3$ of the fundamental state is neglected. The classical solution according to P r a n d t l is:

$$M = M^* = \frac{\pi}{\ell} \cdot \sqrt{F_{22} \cdot J_T \cdot E \cdot G} \cong \frac{\pi}{\ell} \cdot \frac{b^2 \cdot h^2}{6} \cdot \sqrt{E \cdot G} \qquad (59)$$

This result is not correct, because for $h \to o$ the critical moment M^* tends to zero, so that a rod with a flat cross section (h→o) is unstable at once. This behaviour is clearly contrary to experience, because a rod with a flat cross section is never unstable by tipping.

If the deflection $\overset{\circ}{v}{}^3$ of the fundamental state is noted in the equations of stability, the critical moment $M=M^{**}$ differs importantly from the classical moment M^* (59). According to [3] and also to [4] the critical moment M^{**} is given in the following picture as a function of b/h.

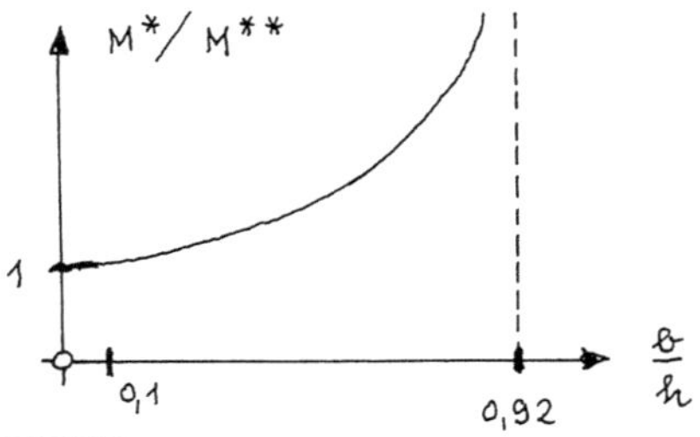

If b/h > o,92, the fundamental state cannot be unstable.

References

[1] Schroeder, F.-H., "Allgemeine Stabtheorie des dünnwandigen räumlich vorgekrümmten Stabes mit großen Verformungen", Ing.-Archiv 197o

[2] Schroeder, F.-H., "Cosserat-Theorie des vorgekrümmten Stabes mit endlichen Drehungen und Verschiebungen", lectures in the University Hannover, 1973-1975, manuscript will be published.

[3] Schroeder, F.-H., "Die linearen Stabilitätsgleichungen der Stabtheorie unter Berücksichtigung der Verformungen des Grundzustandes", Techn. Universitätsbibliothek Hannover, 1971

[4] Heinen, A./Eilering, S., "Einfluß der Verformungen des Grundzustandes auf die Stabilität gerader Stäbe", ZAMM 198o, Heft 12

[5] Schroeder, F.-H., "Die kinetische Stabilität elastisch deformierbarer Körper", Pflüger-Festzeitschrift Hannover, 1977.

On Lateral Beam Buckling and Finite-Deflection Plate Theory

E. REISSNER

Department of Applied Mechanics and Engineering Sciences
University of California, San Diego
La Jolla, CA 92093

Introduction

In this paper we supplement a recent derivation of finite-deflection theory of originally straight, narrow cross section beams, with and without consideration of differential bending in addition to the consideration of bending, twisting, transverse shearing and stretching [1], by an application of the more general of the two theories to the problem of lateral buckling of a cantilever beam, with a view towards obtaining corrections to the value of the Michell-Prandtl buckling load due to (1) end section warping restraint, (2) non-coincident centroid and shear center location and (3) differential sheet bending in the unbuckled state. While all three effects will, in general, be of no great practical significance for this class of beams the results reported are thought to be of interest, insofar as the effect of end section restraint is concerned because of the asymptotic procedure which is used, and insofar as the effect of differential sheet bending is concerned because the existence of this effect seems not to have been noticed previously.

Measures of Beam Stress and Strain in Terms of Corresponding Plate Measures

With x and y as cartesian coordinates in the midplane of an undeformed plate, which is taken to be symmetric about this midplane, we identify the x-axis with the spanwise direction and the y-axis with the chordwise direction of the one-dimensional beam the equations of which are obtained from the equations of the two-dimensional plate.

The results of a systematic analysis of the problem "plate into beam" which is described in [1], include a system of eleven defining relations for beam forces Q, beam moments M and differential beam moments R, in terms of plate stress resultants and couples,

$$\int (N_{xx}, N_{xy}, Q_x) dy = (Q_t, Q_s, Q_p) \ , \tag{1}$$

$$-\int (yN_{xx}, -yQ_x, M_{xx}, M_{xy}) = (M_s, M_{tQ}, M_p, M_{tM}) \ , \tag{2}$$

$$-\int (y^2 N_{xx}, -y^2 Q_x, yM_{xx}, yM_{xy}) = (R_s, R_{tQ}, R_p, R_{tM}) \ , \tag{3}$$

and an approximate system of expressions for plate strains, in terms of seven measures of one-dimensional beam strain

$$\varepsilon_{xy} = \gamma_s \ , \quad \kappa_{yx} = -\psi \ , \quad \kappa_{xy} = -\kappa_t - \psi \kappa_s y \ , \tag{4}$$

$$\kappa_{xx} = -\kappa_p - \psi' y \ , \quad \gamma_x = \gamma_p + (\kappa_t - \psi - \psi \kappa_t) y + \psi \kappa_s y^2 \ , \tag{5}$$

$$\varepsilon_{xx} = \gamma_t - (\kappa_s - \psi \gamma_p) y + \left(\psi \kappa_t - \frac{1}{2}\psi^2\right) y^2 \ , \tag{6}$$

which are a consequence of stipulated constraint conditions $\varepsilon_{yx} = \varepsilon_{yy} = \gamma_y = \kappa_{yy} = 0$. The integrals in (1) to (3) are definite integrals across the chord of the beam and the prime in (5) and in all that follows indicates differentiation with respect to the spanwise coordinate x.

A systematic analysis of the nonlinear two-dimensional equilibrium equations for plates leads to a system of seven one-dimensional beam equilibrium equations, which in the absence of distributed loads, can be written as follows:

$$(Q_t - \psi M_{tQ})' - \kappa_s Q_s - \kappa_p Q_p = 0 \ , \tag{7}$$

$$Q_s' + \kappa_s Q_t - \kappa_t Q_p = 0 \ , \tag{8}$$

$$(Q_p - \psi M_s)' + \kappa_p Q_t + \kappa_t Q_s = 0 \ , \tag{9}$$

$$(M_{tM} + M_{tQ} - \psi R_s)' - \kappa_p M_s + \kappa_s M_p - \gamma_p Q_s + \gamma_s Q_p = 0 \ , \tag{10}$$

$$M_p' - \kappa_s M_t + (\kappa_t - \psi) M_s + (1 + \gamma_t) Q_p - \gamma_p Q_t = 0 \ , \tag{11}$$

$$(M_s + \psi R_t)' + \kappa_p M_t - \kappa_t M_p + (1 + \gamma_t) Q_s - \gamma_s Q_t = 0 \ , \tag{12}$$

$$R_p' - \kappa_s R_t + (\kappa_t - \psi)R_s + (1+\gamma_t)M_{tQ} - M_{tM} - \gamma_p M_s = 0 , \tag{13}$$

where $R_t = R_{tM} + R_{tQ}$. The seven equilibrium equations (7) to (13) come out to be seven differential equations for the seven strain variables γ, κ and ψ - with the subscripts s and p indicating sheet and plate action, respectively, and with the subscript t used to indicate components tangential to the beam axis - upon evaluating plate constitutive equations which are in addition to the previously stated constraint relations

$$N_{xx} = C_{xx}\varepsilon_{xx} , \quad N_{xy} = C_{xy}\varepsilon_{xy} , \quad Q_x = C_Q\gamma_x , \tag{14}$$

$$M_{xx} = D_{xx}\kappa_{xx} , \quad M_{xy} = D_{xy}(\kappa_{xy} + \kappa_{xy}) . \tag{15}$$

To see this we note that the introduction of (14), (15) and (4) to (6) into equations (1) to (3) gives as beam constitutive relations

$$Q_t = C_0\gamma_t - C_1(\kappa_s - \psi\gamma_p) + C_2\left(\psi\kappa_t - \frac{1}{2}\psi^2\right) , \tag{16a}$$

$$-M_s = C_1\gamma_t - C_2(\kappa_s - \psi\gamma_p) + C_3\left(\psi\kappa_t - \frac{1}{2}\psi^2\right) , \tag{16b}$$

$$-R_s = C_2\gamma_t - C_3(\kappa_s - \psi\gamma_p) + C_4\left(\psi\kappa_t - \frac{1}{2}\psi^2\right) , \tag{16c}$$

where $C_n = \int y^n C_{xx}\,dy$,

$$Q_s = \gamma_s C_s , \tag{17}$$

$$Q_p = B_0\gamma_p + B_1(\kappa_t - \psi - \psi\gamma_t) + B_2\psi\kappa_s , \tag{18a}$$

$$M_{tQ} = B_1\gamma_p + B_2(\kappa_t - \psi - \psi\gamma_t) + B_3\psi\kappa_s , \tag{18b}$$

$$R_{tQ} = B_2\gamma_p + B_3(\kappa_t - \psi - \psi\gamma_t) + B_4\psi\kappa_s , \tag{18c}$$

where $C_s = \int C_{xy}$ and $B_n = \int y_n C_Q$, and

$$M_p = D_0\kappa_p + D_1\psi' , \quad R_p = D_1\kappa_p + D_2\psi' , \tag{19a, b}$$

$$M_{tM} = \frac{1}{4}D_{t0}(\kappa_t + \psi) + \frac{1}{4}D_{t1}\psi\kappa_s , \tag{20a}$$

$$R_{tM} = \frac{1}{4} D_{t1}(\kappa_t + \psi) + \frac{1}{4} D_{t2}\psi\kappa_s \ , \tag{20b}$$

where $D_n = \int y^n D_{xx}$ and $\frac{1}{4} D_{tn} = \int y^n D_{xy}$. In introducing the eleven constitutive relations in (16) to (20) into the seven equilibrium equations (7) to (13) it will be consistent to disregard all second degree terms in (16) to (20) in those portions of the equilibrium equations which are products of strain variables and stress variables so that, for example, we will have in equation (7) $\psi M_{tQ} = B_1\psi\gamma_p + B_2\psi(\kappa_t - \psi)$, effectively.

Lateral Cantilever Buckling

Indicating quantities pertaining to the unbuckled state, for which $\psi = \kappa_t = \kappa_p = \gamma_p = 0$, by overbars we have as equilibrium equations for the unbuckled state

$$\overline{Q}_t' - \overline{\kappa}_s\overline{Q}_s = 0 \ , \quad \overline{Q}_s' + \overline{\kappa}_s\overline{Q}_t = 0 \ , \quad \overline{M}_s' + (1+\overline{\gamma}_t)\overline{Q}_s - \overline{\gamma}_s\overline{Q}_t = 0 \ , \tag{21}$$

and as constitutive equations, upon identifying the x-axis with the line of centroids of the cross section, so that $C_1 = 0$ in equations (16),

$$\overline{Q}_t = C_0\overline{\gamma}_t \ , \quad \overline{M}_s = C_2\overline{\kappa}_s \ , \quad \overline{R}_s = -C_2\overline{\gamma}_t + C_3\overline{\kappa}_s \ , \quad \overline{Q}_s = C_s\overline{\gamma}_s \ . \tag{22}$$

As regards the boundary conditions for this state, we may express the conditions of an applied in-plane force P, of unchanging direction, at the end $x = 0$, with the end $x = L$ fully clamped, in the form

$$\overline{M}_s(0) = 0 \ , \quad \overline{Q}_t(L) = 0 \ , \quad \overline{Q}_s(L) = P \ . \tag{23}$$

The corresponding equations for the determination of the buckling load $P = P_B$ are obtained upon writing $Q_t = \overline{Q}_t + \Delta Q_t$, etc. and upon then linearizing the system (7) to (13) and (16) to (20) in terms of ΔQ_t, etc., ψ, κ_t, κ_p, γ_s, etc. In this way we obtain the four equilibrium equations

$$(Q_p - \psi M_s)' + \kappa_p\overline{Q}_t + \kappa_t\overline{Q}_s = 0 \ , \tag{24a}$$

$$(M_{tM} + M_{tQ} - \psi\overline{R}_s)' - \kappa_p\overline{M}_s + \overline{\kappa}_s M_p - \gamma_p\overline{Q}_s + \overline{\gamma}_s Q_p = 0 \ , \tag{24b}$$

$$M_p' - \overline{\kappa}_s M_t + (\kappa_t - \psi)\overline{M}_s + (1+\overline{\gamma}_t)Q_p - \gamma_p\overline{Q}_t = 0 \ , \tag{24c}$$

$$R_p' - \overline{\kappa}_s R_t + (\kappa_t - \psi)\overline{R}_s + (1+\overline{\gamma}_t) M_{tQ} - M_{tM} - \gamma_p \overline{M}_s = 0 \quad , \tag{24d}$$

together with a system of seven constitutive equations which are in addition to equations (19a, b) for M_p and R_p ,

$$Q_p = B_1\gamma_p + B_1(\kappa_t - \psi - \psi\overline{\gamma}_t) + B_2\psi\overline{\kappa}_s \quad , \tag{25a}$$

$$M_{tQ} = B_1\gamma_p + B_2(\kappa_t - \psi - \psi\overline{\gamma}_t) + B_3\psi\overline{\kappa}_s \quad , \tag{25b}$$

$$R_{tQ} = B_2\gamma_p + B_3(\kappa_t - \psi) \quad , \tag{25c}$$

$$M_{tM} = \frac{1}{4} D_{t0}(\kappa_t + \psi) + \frac{1}{4} D_{t1}\psi\overline{\kappa}_s , \quad R_{tM} = \frac{1}{4} D_{t1}(\kappa_t + \psi) \quad . \tag{26a, b}$$

The above system is complemented by the loading boundary conditions

$$M_p(0) = R_p(0) = M_{tM}(0) + M_{tQ}(0) = 0 \quad , \tag{27a, b, c}$$

and by the support boundary bonditions

$$Q_p(L) = 0 \; , \qquad \psi(L) = 0 \quad , \tag{27d, e}$$

where the condition for Q_p once again expresses the fact that the applied end load P retains its direction during the process of buckling.

The system (24), (25), (26) and (19) is of the fifth order in γ_p, κ_t, ψ and κ_p, consistent with the statement of the five boundary conditions in (27).

Before proceeding further we observe that the buckling problem as formulated takes account of the two classical secondary effects associated with the non-linearity of the base-state equations (21), and with the appearance of the factors $\overline{\kappa}_s$, $\overline{\gamma}_s$ and $\overline{\gamma}_t$ in the buckling equations (24), as discussed in [2]. It furthermore takes account of the secondary effect of lateral transverse shear deformations via the assumption of non-vanishing values of γ_p and $\kappa_t - \psi$ in equations (24) as also considered in [2]. In what follows we disregard these effects in order to focus on three other secondary effects which are additionally implied by the form of equations (24) to (27) and (19).

Simplified Buckling Equations

Neglect of the terms with $\overline{\kappa}_s$, $\overline{\gamma}_t$ and $\overline{\gamma}_s$ in (21), in conjunction with (23), gives as expressions for forces and moment of the unbuckled state

$$\overline{Q}_t = 0 , \quad \overline{Q}_s = P , \quad \overline{M}_s = -Px \quad . \tag{28a}$$

Additionally, the constitutive relations (22) give as expression for the differential sheet bending moment R_s

$$\overline{R}_s = r\overline{M}_s , \quad r = C_3/C_2 \ . \tag{28b}$$

The assumption of absent transverse shear deformability, in conjunction with the assumption of negligible pre-buckling deformations, is equivalent to setting

$$B_n = \infty ; \quad \gamma_p = 0 , \quad \kappa_t = \psi \ , \tag{29a, b}$$

in (25). Therewith Q_p and M_{tQ} are then *reactive* quantities, to be determined from what remains of the four equilibrium equations (24),

$$(Q_p + \psi x P)' + \psi P = 0 , \quad M_p' + Q_p = 0 \ , \tag{30a, b}$$

$$(M_{tM} + M_{tQ} + r\psi x P)' + xP\kappa_p = 0 \ , \tag{30c}$$

$$R_p' + M_{tQ} - M_{tM} = 0 \ . \tag{30d}$$

The four equations in (30) are complemented by the two constitutive relations (19a, b) for M_p and R_p, in conjunction with an expression for M_{tM} which follows from (26a) and (29b). For convenience sake we write these relations once more together,

$$M_p = D_0\kappa_p + D_1\psi' , \quad R_p = D_1\kappa_p + D_2\psi' , \quad M_{tM} = \frac{1}{2} D_{t0}\psi \ . \tag{31}$$

Derivation of a first integral. We multiply (30a) by a factor x and add the resulting relation to (30b) so as to have

$$xQ_p' + Q_p + (x^2\psi' + 2x\psi)P + M_p' = 0 \ . \tag{32}$$

From this, and the boundary condition (27a), follows the first integral relation

$$xQ_p + \psi x^2 P + M_p = 0 \ . \tag{33}$$

Reduction to one fourth order differential equation. Having (33) we introduce an auxiliary function S such that

$$Q_p + x\psi P = S \ , \quad M_p = -xS \ , \quad P\psi = -S' \ , \tag{34}$$

with the remaining differential equations being (30c, d) and (31), and where we will use (30d) and (31) in order to reduce (30c) to one fourth order equation for S. For the purpose of this reduction we first use (31) in conjunction with (34) to express κ_p and R_p in terms of S

$$\kappa_p = -\frac{xS}{D_0} + \frac{D_1}{D_0}\frac{S''}{P} \ , \quad R_p = -D_1\frac{xS}{D_0} - \frac{D_2 D_0 - D_1^2}{D_0}\frac{S''}{P} \ . \tag{35a, b}$$

We next use (30d) in order to write $M_{tQ} = M_{tM} - R_p'$ and we introduce this relation, together with M_{tM} from (31), and ψ from (34), into equation (30c) so as to obtain the fourth order differential equation

$$-\frac{D_2 D_0 - D_1^2}{D_0 D_{t0}} S^{IV} + \underline{S'' + \frac{P^2 x^2}{D_0 D_{t0}} S} - \frac{P}{D_t}\left(2\frac{D_1}{D_0} - r\right)(xS')' = \underline{0} \ , \tag{36}$$

with the four boundary conditions (27b-e) taking on the form

$$\underline{S(L) = 0} \ , \quad S'(L) = 0 \ , \quad S''(0) = 0 \ , \tag{37a, b, c}$$

$$-\frac{D_2 D_0 - D_1^2}{D_0 D_{t0}} S'''(0) + \underline{S'(0)} - \frac{PD_1}{D_0 D_{t0}} S(0) = \underline{0} \ . \tag{37d}$$

The known Michell-Prandtl part of the problem consists of the underlined portions in (36) and (37). It is apparent that modifications of the known theory are due to (1) the effect of the differential sheet bending moment R_s, by way of the coefficient r in (36), (2) the effect of shear center location, with $D_1/D_0 = a$ being the classical Taylor-Griffith expression for the distance between shear center and centroid, and (3) the effect of end section restraint against warping which is responsible for the appearance of the terms with factor $(D_2 D_0 - D_1^2)/D_0 D_{t0}$ in (36) and (37d).

Nondimensionalization. We introduce a nondimensional axial coordinate ξ and dimensionless parameters σ, ϵ, η_s, η_r by writing

$$\xi = \frac{x}{L}\,, \quad \sigma = \frac{PL^2}{\sqrt{D_0 D_t}}\,, \quad \epsilon = \sqrt{\frac{D_2 D_0 - D_1^2}{D_0\, D_{t0} L^2}}\,, \quad (\eta_a, \eta_r) = \frac{(a, r)}{L}\sqrt{\frac{D_0}{D_{t0}}}\,,$$

and we indicate differentiation with respect to ξ by dots. In this way, equations (36) and (37) become, with the notation $\eta_b = 2\eta_a - \eta_r$ for convenience sake,

$$-\epsilon^2 S^{\cdot\cdot\cdot\cdot} + S^{\cdot\cdot} + \sigma^2 \xi^2 S - \eta_b\, \sigma(\xi S^{\cdot})^{\cdot} = 0 \tag{38}$$

$$S(1) = 0\,, \qquad S^{\cdot}(1) = 0\,, \qquad S^{\cdot\cdot}(0) = 0\,, \tag{39a, b, c}$$

$$-\epsilon^2 S^{\cdot\cdot\cdot}(0) + S^{\cdot}(0) - \eta_a \sigma S(0) = 0\,. \tag{39d}$$

For the purpose of solving the boundary value problem (38) and (39) we will make use of the fact that the three parameters ϵ, η_a, η_b are all of the order chord over span and therewith small compared to unity, whereas the buckling parameter σ will be of order of magnitude unity in view of the fact that when $\epsilon = \eta_a = \eta_b = 0$ we have the Michell-Prandtl result $\sigma = \sigma_0 \approx 4.01$.

The asymptotic second order problem. Given that the small parameter ϵ^2 multiplies the highest derivative term only in (38) and (39) we will have an asymptotic solution of the problem in the form

$$S = S_e + S_i\,, \tag{40}$$

with an edge zone solution contribution S_e subject to

$$-\epsilon^2 S_e^{\cdot\cdot} + S_e = 0\,, \tag{41a}$$

and an interior solution contribution S_i subject to

$$S_i^{\cdot\cdot} + \sigma^2 \xi^2 S_i - \eta_b\, \sigma(\xi S_i^{\cdot})^{\cdot} = 0\,. \tag{41b}$$

Having (41a, b) we use the solution S_e for the purpose of deducing just two boundary conditions for S_i from the original four boundary conditions in (39), as follows. We write

$$S_e = \begin{cases} C_0\, e^{-\xi/\varepsilon} & , \quad 0 \le \xi \ll 1 \\ C_1\, e^{-(1-\xi)/\varepsilon} & , \quad 0 \ll 1-\xi \le 1 \end{cases} \quad . \tag{42}$$

With this the four relations in (39) become

$$C_1 + S_i(1) = 0\ , \quad \varepsilon^{-1} C_1 + S_i^{\bullet}(1) = 0\ , \quad \varepsilon^{-2} C_0 + S_i^{\bullet\bullet}(0) = 0\ , \tag{43a, b, c}$$

$$-\eta_a \sigma C_0 - \varepsilon^2 S_i^{\bullet\bullet\bullet}(0) + S_i^{\bullet}(0) - \eta_a \sigma S_i(0) = 0 \quad . \tag{43d}$$

Elimination of C_1 from (43a, b) gives as the first of two boundary conditions for S_i

$$S_i(1) - \varepsilon S_i^{\bullet}(1) = 0 \quad . \tag{44a}$$

Elimination of C_0 from (43c, d), and omission of terms which are small of higher order, gives as second boundary condition for S_i

$$S_i^{\bullet}(0) - \eta_a \sigma S_i(0) = 0 \quad , \tag{44b}$$

with the Michell-Prandtl case of the problem being included in equations (41b) and (44a, b), as the special case $\varepsilon = \eta_a = \eta_b = 0$.

Perturbation solution for the second order interior problems. Given that ε, η_a and η_b are small parameters we may determine their effect by using the leading terms of expansions

$$S_i = S_{i0} + \varepsilon S_{i\varepsilon} + \eta_a S_{ia} + \eta_b S_{ib} + \cdots \quad , \tag{45a}$$

$$\sigma = \sigma_0 (1 + c_\varepsilon \varepsilon + c_a \eta_a + c_b \eta_b + \cdots) \ , \tag{45b}$$

with the determination of S_{i0} and σ_0 being the classical problem

$$S_{i0}^{\bullet\bullet} + \sigma_0^2 \xi^2 S_{i0} = 0 \ , \quad S_{i0}^{\bullet}(0) = S_{i0}(1) = 0 \quad . \tag{46}$$

Effect of end section warping constraint. We determine the value of c_ε, in response to a question raised by Brock [3], by directly taking the solution of equation (41b), with $\eta_b = 0$, $\eta_a = 0$, which satisfies the boundary condition (44b) in the form $S_i = \xi^{1/2} J_{-1/4}\left(\frac{1}{2}\sigma\xi^2\right)$. Therewith we obtain from (44a) as equation for σ,

$$J_{-1/4}\left(\tfrac{1}{2}\sigma\right) - \varepsilon\left[\tfrac{1}{2} J_{-1/4}\left(\tfrac{1}{2}\sigma\right) + \sigma J'_{-1/4}\left(\tfrac{1}{2}\sigma\right)\right] = 0 \ , \tag{47a}$$

with $\sigma = \sigma_0$, where $J_{-1/4}\left(\frac{1}{2}\sigma_0\right) = 0$, when $\varepsilon = 0$. In order to obtain σ for the case of a nonvanishing $\varepsilon \ll 1$ we write for the first term in (47a)

$$J_{-1/4}\left(\tfrac{1}{2}\sigma\right) \approx \tfrac{1}{2}(\sigma - \sigma_0)\, J_{-1/4}\left(\tfrac{1}{2}\sigma_0\right) \ , \tag{47b}$$

and we write $\sigma = \sigma_0$ in the ε-terms in (47a), which then reduces to the remarkably simple form

$$\tfrac{1}{2}(\sigma - \sigma_0) - \varepsilon\sigma_0 = 0 \ , \tag{47c}$$

with this implying that the coefficient c_ε in (45b) is given by

$$c_\varepsilon = 2 \ . \tag{48}$$

We note that the introduction of (48) into (45b) gives for the special case of a cross section with coincident centroid and shear center location,

$$\sigma \approx \sigma_0\left(1 + \frac{2}{L}\sqrt{\frac{D_2}{D_{t0}}}\right) \ . \tag{49}$$

For the case of a homogeneous rectangular cross section of chord b and thickness h, for which $D_{t0} = (1/3)\,G b h^3$ and $D_2 = (1/144)\,E b^3 h^3/(1-\nu^2)$ this becomes

$$\sigma \approx \sigma_0\left(1 + \sqrt{\frac{E/G}{12(1-\nu^2)}}\ \frac{b}{L}\right) \ . \tag{50}$$

Effect of differential sheet bending. We now have the two homogeneous boundary conditions $S_{ir}(1) = S'_{ir}(0) = 0$ and the expansions in (45), with $\varepsilon = \eta_a = 0$, and $\eta_b = -\eta_r$ give as differential equation for S_{ir}

$$S_{ir}^{\cdot\cdot} + \sigma_0^2 \xi^2 S_{ir} = \sigma_0 (\xi S_{i0}^{\cdot})^{\cdot} + 2c_r \sigma_0^2 \xi^2 S_{i0} \tag{51}$$

where $S_{i0} = \xi^{1/2} J_{-1/4}\left(\frac{1}{2}\sigma_0 \xi^2\right)$. The solution of the non-homogeneous equation (51) subject to the two specified homogeneous boundary conditions leaves as equation for the determination of c_r,

$$\int_0^1 [\sigma_0 (\xi S_{i0}^{\cdot})^{\cdot} + 2c_r \sigma_0^2 \xi^2 S_{i0}] S_{i0} \, d\xi = 0 \quad . \tag{52}$$

With $\int_0^1 (\xi S_{i0}^{\cdot\cdot} + S_{i0}^{\cdot}) S_{i0} \, d\xi = -\sigma_0^2 \int_0^1 \xi^3 S_{i0}^2 \, d\xi + \frac{1}{2} [S_{i0}]_0^1$ and with $S_{i0}(1) = 0$ this becomes

$$c_r = \frac{\sigma_0^3 \int_0^1 \xi^3 S_{i0}^2 \, d\xi + \frac{1}{2}\sigma_0 [S_{i0}(0)]^2}{2\sigma_0^2 \int_0^1 \xi^2 S_{i0}^2 \, d\xi} \tag{53}$$

or, with $\frac{1}{2}\sigma_0 \xi^2 = x$ and $[x^{1/2} J_{-1/4}(x)]_{x=0} = 2^{1/4}/\Gamma(3/4)$

$$c_r = \frac{\int_0^{\sigma_0/2} x^{3/2} [J_{-1/4}(x)]^2 \, dx + 2^{-3/2} [\Gamma(3/4)]^{-2}}{\sqrt{2\sigma_0} \int_0^{\sigma_0/2} x [J_{-1/4}(x)]^2 \, dx} \quad . \tag{54}$$

The two integrals in (54) have the values

$$\int_0^{\sigma_0/2} (x^{3/2}, x) [J_{-1/4}(x)]^2 \, dx = (0.536, 0.650) \quad , \tag{55}$$ †

and therewith and with $\Gamma(3/4) \approx 1.22$ we obtain

$$c_r = 0.42 \quad , \tag{56}$$

For a numerical example, we may consider the case of a homogeneous triangular cross section with thickness varying from 0 to h_m across

† I am indebted to John Reissner for obtaining the value of the first integral, and for checking a previously known value of the second integral.

the chord. We now have $r = C_3/C_2 = 2b/15$ and $D_0/D_{t0} = 1/2(1-\nu)$ and therewith

$$\frac{\sigma}{\sigma_0} \approx 1 + 0.42\frac{r}{L}\sqrt{\frac{D_0}{D_{t0}}} = 1 + \frac{0.42}{15}\sqrt{\frac{2}{1-\nu}}\;\frac{b}{L} \quad , \tag{57}$$

The second term on the right of (57) represents a correction of about one percent when $\nu = 1/3$ and $b/L = 1/5$, confirming an expectation that this particular effect will in general be of negligible influence on the values of lateral buckling loads. We remark that a similar conclusion holds in regard to the effect of shear center location for cross sections of the kind considered here and we therefore do not, in this note, include the relevant calculations.

Acknowledgment. Preparation of this paper has been supported by the Office of Naval Research of the USA.

References

[1] E. Reissner, On the Derivation of One-Dimensional Differential Equations for Beams from Two-Dimensional Plate Theory, J. Eng. Math. (to appear).

[2] E. Reissner, On Lateral Buckling of End-loaded Cantilever Beams, J. Appl. Math. Phys. (ZAMP) 30, 31-40, 1979.

[3] J. E. Brock, A Remark on Lateral Buckling of a Uniform Beam, J. Appl. Mech. 46, 698-699, 1979.

On the Stability of Continua with Elastic-Inelastic Material Behaviour *

R. de Boer, D. Meyer

Fachgebiet Mechanik, FB 10
Universität - GH - Essen
Universitätsstr. 15
D-4300 Essen 1

1. Introduction

The treatment of stability problems in continuum mechanics is not practicable but by a theory of finite deformations (cf. [1], [2]). Thereby the description of the state of stress and deformation leads to great difficulties, because the material behavior is very complex. Besides elastic and elastic-plastic deformations there may occur any, which are depending on time explicitly. In this place we shall only deal with an important special case of viscoplasticity and that is the *rate sensitive materials*. *Rate sensitivity* of a material behavior means that the beginning and the quantity of inelastic strains are subject to the velocity of strain. Even if we restrict to these rather simple elastic-inelastic materials, the development of consistent constitutive equations gives problems. In the past this has been leading to the disregard of inelastic effects in the theory of simple materials. For that reason we stick to classical plasticity, describing the inelastic material behavior with a *rate type theory*. Such an approach seems to be suitable, particularly since, in the framework of a geometrically non--linear theory, inelastic effects can only be taken into account by such numerical methods, which are formulated in the increments.

Aside from the common principles of continuum mechanics, we

* Dedicated to Prof. Dr.-Ing. habil. Dr.-Ing. e.h. A. Pflüger on his seventieth birthday.

have to consider certain statements of stability, at which we distinguish between *material stability* and *stability of system*. These statements are partly known from the geometrically linear theory. In this context we draw attention to the postulates of *Drucker* and *Iljuschin*, which refer to *material stability*, and to the demand of uniqueness, which was used by *Melan* to develop the concept of the *plastic potential*.

2. The fundamental equations of continuum mechanics

We consider a continuum $\mathcal{B}$ with its material points X, of which the position in a reference placement is identified with the spatial vektor $\mathbf{X}$ (cf. fig. 2). At a deformation process the motion of the material points X is described by a smooth function $\boldsymbol{\chi}$

$$\mathbf{x} = \boldsymbol{\chi}(\mathbf{X}, t) \tag{2.1}$$

wherein the spatial vector $\mathbf{x}$ determines the location of the particles X of the body in an actual placement at any time t. The deformation gradient $\mathbf{F}$ is defined by

$$\mathbf{F} = \operatorname{Grad} \boldsymbol{\chi}(\mathbf{X}, t) = \frac{\partial \mathbf{x}}{\partial \mathbf{X}}\ , \tag{2.2}$$

i. e. figuring the gradient with respect to the local vector $\mathbf{X}$. With $\mathbf{F}$ we are able to establish *Green's deformation tensor*

$$\mathbf{C} = \mathbf{F}^{T}\mathbf{F}\ . \tag{2.3}$$

Now we suppose that the continuum $\mathcal{B}$ is subjected statically to surface stresses $\mathbf{p}$ and body forces $\rho\mathbf{b}$ and that it holds a static equilibrium state in an actual placement. In this the equilibrium condition reads

$$\int_{\partial\mathcal{B}} \{\mathbf{p}\}da + \int_{\mathcal{B}} \{\rho\mathbf{b}\}dv = \mathbf{0} \tag{2.4}$$

and consequently *Cauchy's first motion law*

$$\int_B \{\operatorname{div} \underset{\sim}{T} + \rho\underset{\sim}{b}\}dv = \underset{\sim}{O} \tag{2.5}$$

is derived by means of *Cauchy's fundamental theorem on the stress tensor* and *Green's transformation*. We have introduced the element *da* of the surface ∂B , the volume element *dv* of B, Cauchy's stress tensor $\underset{\sim}{T}$, all quantities with reference to the actual placement.

The derivation of a *rate type theory* asks for an incremental formulation of the fundamental equations. The material time--derivative of $\underset{\sim}{F}$ is denoted by a dot and reads

$$\dot{\underset{\sim}{F}} = \frac{\partial \dot{\underset{\sim}{x}}}{\partial \underset{\sim}{X}} = \frac{\partial \dot{\underset{\sim}{x}}}{\partial \underset{\sim}{x}} \frac{\partial \underset{\sim}{x}}{\partial \underset{\sim}{X}} \tag{2.6}$$

We introduce the *deformation velocity gradient* $\underset{\sim}{L}$

$$\underset{\sim}{L} = \frac{\partial \dot{\underset{\sim}{x}}}{\partial \underset{\sim}{x}} = \underset{\sim}{D} + \underset{\sim}{W} \tag{2.7}$$

which is decomposed additively in the symmetric *stretching tensor* $\underset{\sim}{D}$ and the skew *spin tensor* $\underset{\sim}{W}$. Then the material time--derivative of Green's deformation tensor $\underset{\sim}{C}$ is

$$\dot{\underset{\sim}{C}} = 2\ \underset{\sim}{F}^T \underset{\sim}{D}\underset{\sim}{F} \tag{2.8}$$

and that of Cauchy's first motion law

$$\int_B \{\operatorname{div} \dot{\underset{\sim}{P}} + \rho\dot{\underset{\sim}{b}}\}dv = \underset{\sim}{O} \tag{2.9}$$

with the abbreviation

$$\dot{\underset{\sim}{P}} = \dot{\underset{\sim}{T}} + \underset{\sim}{T}\ (\underset{\sim}{D} \cdot \underset{\sim}{I}) - \underset{\sim}{T}\underset{\sim}{L}^T \tag{2.10}$$

3. The incremental constitutive equations of elastic-inelastic materials

We derive the constitutive equations in an axiomatic way, proceeding on the following assumptions:

(a) The principles of the general material theory - - determinism, frame-indifference, local effect - shall prevail.

(b) The elastic region shall be separated from the inelastic region by a scalar-valued function.

(c) The stretching tensor $\underset{\sim}{D}$ may be split in an elastic and an inelastic part

$$\underset{\sim}{D} = \underset{\sim}{D}_E + \underset{\sim}{D}_P \; . \tag{3.1}$$

3.1 Constitutive equations of elastic materials

The usual descriptions of the constitutive equations of elastic materials are well-known from the literature (e. g. [1], [2]). Subject to the assumption (a) a practicable form of the constitutive equations of an elastic point can be denoted as follows

$$\underset{\sim}{S} = \underset{\sim}{G}(\underset{\sim}{C}) \; . \tag{3.2}$$

Therein $\underset{\sim}{S}$ is the *symmetric Piola-Kirchhoff stress tensor*

$$\underset{\sim}{S} = (\det \underset{\sim}{F})\underset{\sim}{F}^{-1}\underset{\sim}{T}\,\underset{\sim}{F}^{T-1} \; . \tag{3.3}$$

By means of eq. (3.3) and some transformations the material time-derivative of $\underset{\sim}{S}$ yields

$$\dot{\underset{\sim}{P}} = \overset{4}{\underset{\sim}{A}}\underset{\sim}{D}_E + \underset{\sim}{L}\underset{\sim}{T} \; . \tag{3.4}$$

Now we introduce *Jaumann's stress rate*

$$\overset{\circ}{\underset{\sim}{T}} = \dot{\underset{\sim}{T}} - \underset{\sim}{W}\underset{\sim}{T} + \underset{\sim}{T}\underset{\sim}{W} \tag{3.5}$$

and get from eq. (3.4) with eq. (2.10) and eq. (3.5)

$$\overset{\circ}{\underset{\sim}{T}} = \overset{4}{\underset{\sim}{B}}\underset{\sim}{D}_E \tag{3.6}$$

with $$\overset{4}{\underset{\sim}{B}} = (\underset{\sim}{T} \otimes \underset{\sim}{I})^{\overset{14}{T}} + (\underset{\sim}{I} \otimes \underset{\sim}{T})^{\overset{14}{T}} - (\underset{\sim}{T} \otimes \underset{\sim}{I}) + \overset{4}{\underset{\sim}{A}} \; . \tag{3.7}$$

The notation $\overset{14}{T}$ means a transposition of 4th order tensors produced by the interchange of the 1st and the 4th base vector.

3.2 Constitutive equations of inelastic materials

We proceed from the classical concept of plasticity (cf. [3]), in which the beginning of plastic deformations is depending on the fulfillment of a scalar-valued tensor function

$$F = \phi(\underset{\sim}{D}) \quad . \tag{3.8}$$

This eq. (3.8) coincides with the assumption (b) and represents an enlargement of the classical concept in so far as we admit a dependence of the actual yield limit on the stretching tensor $\underset{\sim}{D}$ in order to be able to describe rate sensitive properties of viscoplasticity. For the *loading function F* we put up

$$F = \hat{F}(\underset{\sim}{T},\ \underset{\sim}{N},\ k) \tag{3.9}$$

In this $\underset{\sim}{N}$ is a tensor, which is intended to describe anisotropic effects, meanwhile k is a parameter of the material, which can be determined by the one-dimensional tensile test.

If eq. (3.9) is fulfilled, i. e. the material point has passed over to an inelastic state, we have to investigate the altered loading function

$$F + \dot{F}\,Dt = \phi + \dot{\phi}Dt \tag{3.10}$$

belonging to the following incremental loading step. If eq. (3.10) holds, the inelastic behavior of the material persists.

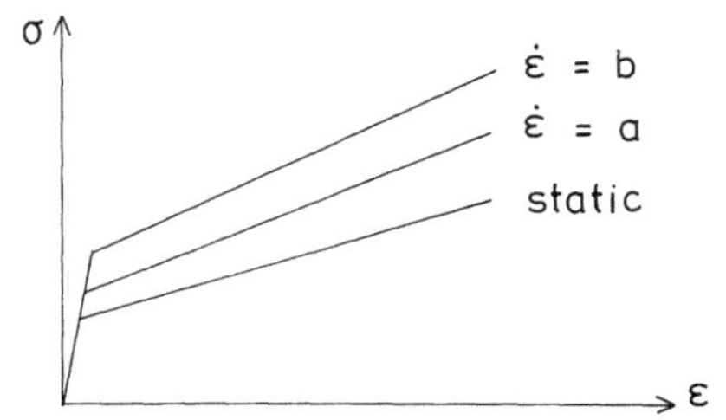

Fig. 1: Stress-strain diagramm with *rate sensitive* material behavior und linear hardening rule.

We suppose that ϕ does not depend on time explicitly, but merely fixes those significant lines of the material in the viscoplastic loading process, which belong to the particular deformation velocities (cf. fig. 1). This concept surely is sufficient referring to an incremental description and causes

$$\dot{\phi} = 0 . \tag{3.11}$$

Hence we gain from eq. (3.9), eq. (3.10) and eq. (3.11) the *condition of consistency*

$$\dot{F} = \frac{\partial F}{\partial \tilde{T}} \cdot \dot{\tilde{T}} + \frac{\partial F}{\partial \tilde{N}} \cdot \dot{\tilde{N}} + \frac{\partial F}{\partial k} \dot{k} = 0 . \tag{3.12}$$

Eq. (3.12) does not come up to the frame-indifference, because $\dot{\tilde{T}}$ and $\dot{\tilde{N}}$ are contradictory to this principle. If we confine ourselves to use flow rules with symmetric and frame-indifferent terms $\frac{\partial F}{\partial \tilde{T}}$ and $\frac{\partial F}{\partial \tilde{N}}$ and symmetric terms $\frac{\partial F}{\partial \tilde{T}} \tilde{T}$ and $\frac{\partial F}{\partial \tilde{N}} \tilde{N}$, we can replace eq. (3.12) by

$$\dot{F} = \frac{\partial F}{\partial \tilde{T}} \cdot \overset{\circ}{\tilde{T}} + \frac{\partial F}{\partial \tilde{N}} \cdot \overset{\circ}{\tilde{N}} + \frac{\partial F}{\partial k} \dot{k} = 0 \tag{3.13}$$

using the frame-indifferent Jaumann derivatives $\overset{\circ}{\tilde{T}}$ and $\overset{\circ}{\tilde{N}}$ now.

If the loading function eq. (3.8) is satisfied, we get in detail

$$F = \phi \begin{cases} \frac{\partial F}{\partial \tilde{T}} \cdot \overset{\circ}{\tilde{T}} \leq 0 : \tilde{D}_P = \tilde{0} ; \quad \overset{\circ}{\tilde{N}} = \tilde{0} , \quad \dot{k} = 0 \\ \qquad \text{(unloading and neutral loading process)} \\ \frac{\partial F}{\partial \tilde{T}} \cdot \overset{\circ}{\tilde{T}} > 0 : \tilde{D}_P \neq \tilde{0} ; \quad \overset{\circ}{\tilde{N}} \neq \tilde{0} ; \quad \dot{k} \neq 0 \\ \qquad \text{(active loading process)} \end{cases} \tag{3.14}$$

By means of the present terms we are able to formulate the constitutive equations, i. e. the correlation of the increments of stress and strain. We restrict to linear setups

$$\tilde{D}_P = \overset{4}{\tilde{P}}{}^{-1} \overset{\circ}{\tilde{T}} ; \quad \overset{\circ}{\tilde{N}} = \overset{4}{\tilde{\tilde{C}}} \overset{\circ}{\tilde{T}} ; \quad \dot{k} = \tilde{V} \cdot \overset{\circ}{\tilde{T}} \tag{3.15}$$

The tensors $\overset{4}{\underset{\sim}{P}}{}^{-1}$, $\overset{4}{\underset{\sim}{C}}$ and $\underset{\sim}{V}$ can be tied up with the *loading condition*, because $\underset{\sim}{D}_P$, $\overset{\circ}{\underset{\sim}{N}}$ und $\dot{k}$ disappear except in case of active loading processes. We set

$$\overset{4}{\underset{\sim}{P}}{}^{-1} = (\underset{\sim}{M} \otimes \frac{\partial F}{\partial \underset{\sim}{T}}) \; ; \quad \overset{4}{\underset{\sim}{C}} = \overset{4}{\underset{\sim}{L}}\overset{4}{\underset{\sim}{P}}{}^{-1} \; ; \quad \underset{\sim}{V} = (\overset{4}{\underset{\sim}{P}}{}^{-1})^T \underset{\sim}{H} \tag{3.16}$$

leaving $\overset{4}{\underset{\sim}{L}}$ and $\underset{\sim}{M}$ arbitrary frame-indifferent tensors.

3.3 Constitutive equations of elastic-inelastic materials

The constitutive equations of elastic and inelastic materials can be connected by means of the assumption (c). We have to suppose, however, that $\overset{4}{\underset{\sim}{B}}$ in eq. (3.6) is inversible. Then we receive

$$\underset{\sim}{D}_E = \overset{4}{\underset{\sim}{B}}{}^{-1} \overset{\circ}{\underset{\sim}{T}} \; . \tag{3.17}$$

With eq. (3.1), eq. (3.15) and eq. (3.17) we get

$$\underset{\sim}{D} = \underset{\sim}{D}_E + \underset{\sim}{D}_P = (\overset{4}{\underset{\sim}{B}}{}^{-1} + \overset{4}{\underset{\sim}{P}}{}^{-1})\overset{\circ}{\underset{\sim}{T}} \; . \tag{3.18}$$

We are able to invert eq. (3.18), because $\overset{4}{\underset{\sim}{B}}$ is supposed to be inversible and $\overset{4}{\underset{\sim}{P}}{}^{-1}$ is a simple tensor. The result is

$$\overset{\circ}{\underset{\sim}{T}} = \overset{4}{\underset{\sim}{B}}\underset{\sim}{D} + \overset{4}{\underset{\sim}{G}}\underset{\sim}{D} \tag{3.19}$$

with

$$\overset{4}{\underset{\sim}{G}} = \frac{-1}{\overset{4}{\underset{\sim}{B}}{}^{T} \cdot \overset{4}{\underset{\sim}{P}}{}^{-1} + 1} \overset{4}{\underset{\sim}{B}}\overset{4}{\underset{\sim}{P}}{}^{-1}\overset{4}{\underset{\sim}{B}} \; . \tag{3.20}$$

4. Stable behavior of inelastic material

Concerning the development of constitutive equations of inelastic materials, questions of material stability have always been playing an important part. *Material stability* stands for the uniqueness of the functional relation between stress- and deformation quantities or their increments respectively. In the framework of the *geometrically linear theory Drucker* has found a working hypothesis in the *stress space*. The consequence of his idea are two inequalities

$$(\underset{\sim}{T} - \underset{\sim}{T}_o) \cdot \dot{\underset{\sim}{E}}_p \geqq 0$$
$$\dot{\underset{\sim}{T}} \cdot \dot{\underset{\sim}{E}}_p \geqq 0 \tag{4.1}$$

with the linearised *distortion tensor* $\underset{\sim}{E}$ and $\underset{\sim}{T}_o$ marking the initial stress tensor of a loading cycle of stress. Similar results have been gained by *Iljuschin*, formulating a working hypothesis in the strain space (cf. [3]).

It is rather easy to produce an equivalent inequality in the framework of the *geometrically non-linear theory*. For that purpose we turn again to the consistency condition eq. (3.13). It is fulfilled identically in case of unloading and neutral loading process, as the loading condition eq. (3.14) shows. Having an active loading process, it produces the inequality

$$\frac{\partial F}{\partial \underset{\sim}{N}} \cdot \overset{\circ}{\underset{\sim}{N}} + \frac{\partial F}{\partial k} \dot{k} > 0 \, . \tag{4.2}$$

Replacing F by the ordinary loading functions of classical plasticity, for example Hill's hardening rule, this condition demands for a monotonously increasing slope of the stress-strain diagram in the one-dimensional tensile test. Hence it describes material stability.

Overmore eq. (4.2) enables us to determine the tensor $\underset{\sim}{M}$ of eq. (3.16) (cf. [3]).

5. Criterion of stability

We proceed from the assumption that the body is positioned in an actual placement, loaded by surface stresses $\underset{\sim}{p}$ and body forces $\rho\underset{\sim}{b}$. In this state the equilibrium condition eq. (2.5) holds. All boundary conditions are fulfilled. By an infinitesimal change of loading $\dot{\underset{\sim}{\bar{p}}}Dt$ and $(\rho\dot{\underset{\sim}{b}})Dt$ the body passes over to a state of equlibrium, called *ground state*, which corresponds to the boundary conditions as well. By this the equilibrium condition (cf. eq. 2.9) reads

$$\int_{\mathcal{B}} \{\mathrm{div}\,\dot{\bar{\underset{\sim}{P}}} + \rho\dot{\bar{\underset{\sim}{b}}}\}dv = \underset{\sim}{0} \; . \tag{5.1}$$

By a different infinitesimal raise of loading $\dot{\overset{*}{\underset{\sim}{p}}}Dt$ and $\rho\dot{\overset{*}{\underset{\sim}{b}}}Dt$ the body passes over to a *comparison state*, according to the boundary conditions equally.

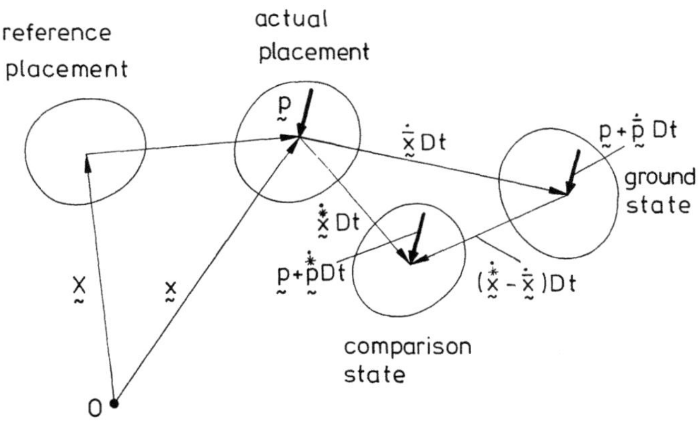

Fig. 2: Kinematics

Let us assume for the present that it is a state of equilibrium as well and write

$$\int_{\mathcal{B}} \{\mathrm{div}\,\dot{\overset{*}{\underset{\sim}{P}}} + \rho\dot{\overset{*}{\underset{\sim}{b}}}\}dv = \underset{\sim}{0} \; . \tag{5.2}$$

The loading process is supposed to be the same in both cases, i. e.

$$\begin{aligned} \dot{\overset{*}{\underset{\sim}{p}}}Dt &= \dot{\bar{\underset{\sim}{p}}}Dt \qquad \text{on } \partial\mathcal{B}_T{}^{*} \\ \rho\dot{\overset{*}{\underset{\sim}{b}}}Dt &= \rho\dot{\bar{\underset{\sim}{b}}}Dt \qquad \text{on } \mathcal{B} \; . \end{aligned} \tag{5.3}$$

---

* $\partial\mathcal{B}_T$: subsurface of $\partial\mathcal{B}$ with predetermined stresses

The geometrical boundary conditions are the same as well, i. e.

$$\dot{\overset{*}{\underset{\sim}{x}}}Dt = \dot{\bar{\underset{\sim}{X}}}Dt \qquad \text{on } \partial\mathcal{B}_u{}^{*} \,. \tag{5.4}$$

We subtract eq. (5.2) from eq. (5.1) and make use of the *principle of virtual displacement* with $(\dot{\overset{*}{\underset{\sim}{X}}} - \dot{\bar{\underset{\sim}{X}}})Dt$

$$\int_{\mathcal{B}} \{[\mathrm{div}\ (\dot{\overset{*}{\underset{\sim}{P}}} - \dot{\bar{\underset{\sim}{P}}}) + \rho(\dot{\overset{*}{\underset{\sim}{b}}} - \dot{\bar{\underset{\sim}{b}}})] \cdot (\dot{\overset{*}{\underset{\sim}{X}}} - \dot{\bar{\underset{\sim}{X}}})Dt\}dv = 0 \,. \tag{5.5}$$

After some transformations follows

$$\int_{\mathcal{B}} \{(\dot{\overset{*}{\underset{\sim}{P}}} - \dot{\bar{\underset{\sim}{P}}}) \cdot (\overset{*}{\underset{\sim}{L}} - \bar{\underset{\sim}{L}})\}dv = 0 \tag{5.6}$$

which is the necessary and sufficient condition of *indifference*.

A necessary condition of stability is the demand for uniqueness, i. e. the comparison state is no equibrium state, consequently we have to set an inequality sign in eq. (5.6). A necessary and sufficient *criterion of stability* finally is

$$\int_{\mathcal{B}} \{(\dot{\overset{*}{\underset{\sim}{P}}} - \dot{\bar{\underset{\sim}{P}}}) \cdot (\overset{*}{\underset{\sim}{L}} - \bar{\underset{\sim}{L}})\}dv > 0 \,. \tag{5.7}$$

Inserting the constitutive equation of elastic-inelastic materials eq. (3.19) we receive in the local form

$$\int_{\mathcal{B}} \{(\overset{4}{\underset{\sim}{A}}+\overset{4}{\underset{\sim}{G}})(\overset{*}{\underset{\sim}{D}}-\bar{\underset{\sim}{D}})\cdot(\overset{*}{\underset{\sim}{D}}-\bar{\underset{\sim}{D}})+(\underset{\sim}{T}\otimes\underset{\sim}{I})^{\overset{14}{T}}(\overset{*}{\underset{\sim}{L}}-\bar{\underset{\sim}{L}})\cdot(\overset{*}{\underset{\sim}{L}}-\bar{\underset{\sim}{L}})\}dv > 0 \,. \tag{5.8}$$

Setting $\overset{4}{\underset{\sim}{G}} = \overset{4}{\underset{\sim}{O}}$ we gain the criterion of purely elastic materials, but we cannot reduce to purely inelastic materials (cf. eq. (3.20). Let us now assume that the actual placement is without any stress, i. e. $\underset{\sim}{T} = \underset{\sim}{O}$. Hence we gain the criterion of *material stability*

$$\int_{\mathcal{B}} \{(\overset{4}{\underset{\sim}{A}} + \overset{4}{\underset{\sim}{G}})\ (\overset{*}{\underset{\sim}{D}} - \bar{\underset{\sim}{D}}) \cdot (\overset{*}{\underset{\sim}{D}} - \bar{\underset{\sim}{D}})\}dv > 0 \,, \tag{5.9}$$

---

* $\partial\mathcal{B}_u$: subsurface of $\partial\mathcal{B}$ with predetermined displacements

which is fulfilled if $(\overset{4}{\underset{\sim}{A}} + \overset{4}{\underset{\sim}{G}})$ is positively definite. If we suppose material stability, eq. (5.8) leads to the criterion of the *stability of system*

$$\int_B \{(\underset{\sim}{T} \otimes \underset{\sim}{I})^{\overset{14}{T}} (\overset{*}{\underset{\sim}{L}}-\bar{\underset{\sim}{L}}) \cdot (\overset{*}{\underset{\sim}{L}}-\bar{\underset{\sim}{L}})\}dv > - \int_B (\overset{4}{\underset{\sim}{A}}+\overset{4}{\underset{\sim}{G}})(\overset{*}{\underset{\sim}{D}}-\bar{\underset{\sim}{D}}) \cdot (\overset{*}{\underset{\sim}{D}}-\bar{\underset{\sim}{D}})\}dv \, . \qquad (5.10)$$

This condition is granted sufficiently by

$$\int_B \{(\underset{\sim}{T} \otimes \underset{\sim}{I})^{\overset{14}{T}} (\overset{*}{\underset{\sim}{L}} - \bar{\underset{\sim}{L}}) \cdot (\overset{*}{\underset{\sim}{L}} - \bar{\underset{\sim}{L}})\}dv > 0 \qquad (5.11)$$

which is valid in case of omnidirectional tensions. Let us finally remark that the fulfillment of both material and system stability is only a sufficient condition of total stability.

Acknowledgement: This work was carried out under a grant of "STIFTUNG VOLKSWAGENWERK", West-Germany

References:

[1] C. Truesdell and W. Noll: Non-Linear Field Theories of Mechanics, Handbuch der Physik, Vol. III/3, 1965.

[2] R. Hill: Uniqueness Criteria and Extremum Principles in Self-Adjoint Problems of Continuum Mechanics, J. Mech. Phys. Solids, 1962, Vol. 10.

[3] R. de Boer, W. Ehlers: Grundlagen der isothermen Plastizitäts- und Viskoplastizitätstheorie, Forschungsberichte aus dem Fachgebiet Bauwesen der Universität-GH-Essen, Heft 14, 1980.

Bounds to the Critical Stresses in Bifurcation of Cylindrical Specimens in the Case of Non-Associated Flow Laws

O. T. BRUHNS

Department of Mechanical Engineering
University of Kassel, Kassel, F. R. Germany

Summary

The bifurcation of cylindrical specimens under uniaxial stretching as well as lateral hydrostatic pressure is studied. The material properties are those of a frictional, plastically dilatant, non-associated elastic-plastic solid. Bounds to the critical stresses can be given by methods of bifurcation analysis which so far have been developed for elastic-plastic solids obeying normality.

Introduction

In the range of plastic deformations, some materials show a more or less significant deviation from "normality" in their constitutive relations. This is especially valid for a wide class of rock-like or soil-like materials. Moreover, there are also some doubts whether the normality rule can adequately describe the behaviour of metals close to regions of localized deformations or under adiabatic conditions. Non-associated plastic flow laws, as have been introduced by Mandel [1] and Mróz [2] are expected to give a better description of the behaviour of these materials.

In this paper the bifurcation problem governing the onset of axisymmetric necking/bulging of a circular cylindrical specimen in triaxial tension/compression is analysed. The material is assumed to be frictional, plastically dilatant according to a non-associated (i.e. "non-normal") flow-rule. As e.g. is indicated by Needleman [3] for such materials Hill's [4] rigorous theory of bifurcation, originally formulated for materials obeying normality, does not hold. Recently, in [5] , this theory therefore has been generalized to include a non-associated flow law. The present study is based on this work and gives an example of the applicability of respective bound theorems to determine upper and lower bounds to the primary bifurcation stresses of the specimen.

Constitutive Relations

The simplest generalization of the rate-constitutive equation of an elastic-plastic solid at yield point may be taken in the form

$$\overset{\Delta}{\underset{\sim}{\sigma}} = \underset{\sim}{L}\underset{\sim}{D} - \frac{\alpha}{H}\,(\underset{\sim}{\Lambda}^f \cdot \underset{\sim}{D})\underset{\sim}{\Lambda}^g \tag{1}$$

where $H > 0$, $\underset{\sim}{\sigma}$ is true (Cauchy) stress and $\overset{\Delta}{\underset{\sim}{\sigma}}$ is a (symmetric) objective stress-rate. $\underset{\sim}{D}$ herein is the strain-rate tensor and

$$\alpha = \begin{cases} 1 & \text{if } \underset{\sim}{\Lambda}^f \cdot \underset{\sim}{D} \geqq 0 \\ 0 & \text{if } \underset{\sim}{\Lambda}^f \cdot \underset{\sim}{D} < 0 \end{cases} . \tag{2}$$

If $h = H - \underset{\sim}{\Lambda}^f \cdot \underset{\sim}{\mu}^g > 0$ the unique inverse of (1) is

$$\underset{\sim}{D} = \underset{\sim}{M}\overset{\Delta}{\underset{\sim}{\sigma}} + \frac{\alpha_1}{h}\underset{\sim}{\mu}^g\,(\underset{\sim}{\mu}^f \cdot \overset{\Delta}{\underset{\sim}{\sigma}}) \quad , \quad \underset{\sim}{M} = \underset{\sim}{L}^{-1} \tag{3}$$

with

$$\alpha_1 = \begin{cases} 1 & \text{if } \underset{\sim}{\mu}^f \cdot \overset{\Delta}{\underset{\sim}{\sigma}} \geqq 0 \\ 0 & \text{if } \underset{\sim}{\mu}^f \cdot \overset{\Delta}{\underset{\sim}{\sigma}} < 0 \end{cases} , \quad \begin{array}{l} \underset{\sim}{\mu}^g = \underset{\sim}{M}\underset{\sim}{\Lambda}^g \\ \underset{\sim}{\mu}^f = \underset{\sim}{M}^T\underset{\sim}{\Lambda}^f \end{array} \tag{4}$$

Here, $\underset{\sim}{L}$ and $\underset{\sim}{M}$ are tensors of instantaneous elastic moduli and compliances; $\underset{\sim}{\Lambda}^f$ and $\underset{\sim}{\mu}^f$ represent the normals to the hyperplane surfaces separating the domains of elastic and plastic behaviour in the strain-rate and stress spaces, respectively, and $\underset{\sim}{\mu}^g$ and its dual $\underset{\sim}{\Lambda}^g$ specify the direction of the plastic part of the strain-rate tensor.

For an isotropic material, more specifically, Rudnicki and Rice [6] used a simple generalization of Prandtl-Reuss equation which, expressed in terms of $\underset{\sim}{\mu}^g$ and $\underset{\sim}{\mu}^f$ takes the form

$$\begin{array}{ll} \underset{\sim}{\mu}^f = \underset{\sim}{\sigma}'/2\tau_e + \beta\,\underset{\sim}{1}/3 , & \underset{\sim}{\sigma}' = \text{dev}\,\underset{\sim}{\sigma} \\ \underset{\sim}{\mu}^g = \underset{\sim}{\sigma}'/2\tau_e + \nu_o\,\underset{\sim}{1}/3 , & \tau_e^2 = \frac{1}{2}\,\underset{\sim}{\sigma}' \cdot \underset{\sim}{\sigma}' \end{array} , \tag{5}$$

where β is the plastic dilatancy factor and ν_o is the internal friction coefficient that measures the pressure sensitivity of yield. Furthermore, assuming elastic isotropy, (3) can be specified to give

$$\underset{\sim}{D} = \frac{1}{2G}\left[\underset{\sim}{\overset{\Delta}{\sigma}} - \frac{\nu}{1+\nu}\,\mathrm{tr}\,(\underset{\sim}{\overset{\Delta}{\sigma}})\,\underset{\sim}{1}\right] + \frac{\alpha}{h}\,\underset{\sim}{\mu}^{g}\,(\underset{\sim}{\mu}^{f}\cdot\underset{\sim}{\overset{\Delta}{\sigma}}) \quad . \tag{6}$$

Here, E, G, and ν can be regarded as usual elastic constants, where $E = 2G(1+\nu)$ and the compressibility of the material is expressed by

$$e = \frac{1+\nu}{3(1-2\nu)} \quad . \tag{7}$$

Rate Boundary-Value Problem and Uniqueness Condition

At a generic instant time during finite elastic-plastic deformation, at which all material constants are assumed to be known, a boundary-value problem is set up. The particular boundary conditions will consist of complementary components of velocity $\underset{\sim}{v}$ and nominal traction-rate $\dot{\underset{\sim}{T}}$, prescribed over the whole surface of the specimen. A velocity field will be a valid solution if it satisfies the velocity boundary conditions, is continuous and differentiable, and if - through (1) - it leads to a stress-rate field which is self-equilibrated and satisfies the boundary conditions of traction-rate.

In order to investigate the uniqueness of any fundamental solution of this boundary-value problem, a functional $R(\underset{\sim}{v},\phi_o)$ has been deduced [5] where $\phi_o = \underset{\sim}{\Lambda}^{f}\cdot\underset{\sim}{D}_o$ and $\underset{\sim}{D}_o$ is the strain-rate of the above mentioned fundamental solution. This functional, however, has a very complicated form and its direct application to estimate bifurcation stresses might lead to considerable difficulties. As is well known, in the situation when plastic flow laws admit a potential (i.e. when $\underset{\sim}{\Lambda}^{f} = \underset{\sim}{\Lambda}^{g}$), Hill [4] has proved that the elastic-plastic solid can be replaced by a slightly weaker, but more practicable "linear comparison solid", where $\alpha = 1$ for all possible strain-rates, and with the property that any solution of the rate boundary-value problem for the elastic-plastic solid is unique if the uniqueness of the analogous boundary value problem for the comparison solid is assured.

In this sense in [5] a one-parameter family of "linearized comparison solids" has been introduced

$$\overset{\Delta}{\underset{\sim}{\sigma}} = \underset{\sim\sim}{L}\underset{\sim}{D} - \frac{1}{4\psi H} \left[(\underset{\sim}{\Lambda}^{g} + \psi \underset{\sim}{\Lambda}^{f}) \cdot \underset{\sim}{D} \right] (\underset{\sim}{\Lambda}^{g} + \psi \underset{\sim}{\Lambda}^{f}) \tag{8}$$

that admits a potential and, what is more important from a practical point of view, has the property that if uniqueness is certain for the comparison solid then a bifurcation and instability is precluded for the underlying elastic-plastic solid. The respective functional $R_L(\underset{\sim}{v})$ has the properties of the well-known "exclusion" functional. The derived uniqueness criterion therefore may be used as a device to determine lower bounds to the magnitudes of primary bifurcation and instability stresses which ordinarily are unknown. To this end it is sufficient to solve the eigenvalue problem for linear field equations of the comparison solid. Since the comparison solid admits a potential, methods of bifurcation analysis that have been developed for elastic-plastic solids obeying normality can also be used to determine lower bounds to bifurcation and instability states in solids with non-associated flow laws. The still undetermined parameter $\psi > 0$ in (8) can be chosen such that these lower bounds give most optimal ones [7]

$$\psi = \sqrt{\frac{\underset{\sim}{\Lambda}^{g} \cdot \underset{\sim}{\mu}^{g}}{\underset{\sim}{\Lambda}^{f} \cdot \underset{\sim}{\mu}^{f}}} \tag{9}$$

Hill's bifurcation criterion can be deduced from the established one as a limiting case when normality holds (i.e. when $\underset{\sim}{\Lambda}^{f} = \underset{\sim}{\Lambda}^{g}$, $\psi = 1$).

However, due to the lack of variational principles, the first state for which the functional R_L becomes positive-semidefinite is ordinarily not the bifurcation state of the underlying elastic-plastic solid. Therefore a second linear solid is introduced whose constitutive relations have the same form as an elastic-plastic solid "in loading" (i.e. $\alpha = 1$)

$$\overset{\Delta}{\underset{\sim}{\sigma}} = \underset{\sim\sim}{L}\underset{\sim}{D} - \frac{1}{H} (\underset{\sim}{\Lambda}^{f} \cdot \underset{\sim}{D}) \underset{\sim}{\Lambda}^{g} \tag{10}$$

The first eigenstate of this solid may be identified with some bifurcation state of the underlying elastic-plastic solid. It is therefore an upper bound to the primary bifurcation state. Briefly spoken, the search for the genuine primary bifurcation state is replaced by a search for upper and lower bounds in the situation when normality fails to hold ($\underset{\sim}{\Lambda}^{f} \neq \underset{\sim}{\Lambda}^{g}$). Only in the case of normality (i.e. $\underset{\sim}{\Lambda}^{f} = \underset{\sim}{\Lambda}^{g}$, $\psi = 1$) (8) and (10) coincide and therefore the respective eigenstates give an exact solution to the underlying bifurcation problem.

Illustrative Example

Consider a homogeneous cylindrical specimen of circular cross section (with radius R and length L in the current configuration), uniformly stretched under axial loading σ_z, with the lateral surface subjected to uniform fluid pressure σ. The axial loading, either tensile or compressive, is taken to be described by imposing a uniform displacement on the circular faces of the specimen, with no shear traction applied on these faces. One state of deformation, consistent with the imposed boundary conditions, is the state of homogeneous deformations. At a generic instant the state of stress is given by

$$\sigma_{rr} = \sigma_{\theta\theta} = -\sigma, \qquad \sigma_{zz} = \sigma_z \tag{11}$$

in a cylindrical coordinate system. Bifurcation from this fundamental state is investigated here.

Similar problems have been solved by Miles [8] , Hutchinson and Miles [9] , Vardoulakis [10] and Bruhns and Mielniczuk [11]. The present investigation essentially differs from those in the constitutive relations of the considered material. According to our previous statements two different linear solids will be introduced. For the comparison solid (8) the rate-constitutive equations can be written as

$$\overset{\Delta}{\underset{\sim}{\sigma}} = 2G \Big\{ [\underset{\sim}{D} + (e - \tfrac{1}{3})\ \mathrm{tr}\ (\underset{\sim}{D})\ \underset{\sim}{1}]$$

$$- \frac{\delta}{2\psi} \Big[(1 + \psi) \frac{\underset{\sim}{\sigma}' \cdot D}{2\tau_e} + e\ \mathrm{tr}\ (\underset{\sim}{D})(\beta + \psi\nu_o)\Big]\ x \tag{12}$$

$$\Big[(1 + \psi) \frac{\underset{\sim}{\sigma}'}{2\tau_e} + e\ (\beta + \psi\nu_o)\ \underset{\sim}{1}\Big] \Big\} ,$$

where for abbreviation

$$\delta = \frac{1}{H/G} = \frac{1}{1 + h/G + 2\nu_o \beta e} \tag{13}$$

and

$$h = \frac{G_T}{1 - G_T/G} \tag{14}$$

with G_T the tangent shear modulus of the material measured in pure shear under hydrostatic pressure. The parameter ψ herein takes the form

$$\psi = \sqrt{\frac{1 + 2e\beta^2}{1 + 2e\nu_o{}^2}} \quad . \tag{15}$$

On the other hand the constitutive relations for the second solid read

$$\overset{\Delta}{\underset{\sim}{\sigma}} = 2G \Big\{ [\underset{\sim}{D} + (e - \tfrac{1}{3})\ \mathrm{tr}\ (\underset{\sim}{D})\ \underset{\sim}{1}]$$

$$- \frac{\delta}{2} \Big[\frac{\underset{\sim}{\sigma}' \cdot \underset{\sim}{D}}{2\tau_e} + \nu_o e\ \mathrm{tr}\ (\underset{\sim}{D})\Big] \Big[\frac{\underset{\sim}{\sigma}'}{2\tau_e} + e\beta \underset{\sim}{1}\Big] \Big\} \tag{16}$$

with δ defined as before.

The stress-rate fields of both solids as well as through (12) and (16) the respective velocity fields have to satisfy the e- quations of continuing equilibrium

$$\mathrm{div}\ \dot{\underset{\sim}{S}}^T = \mathrm{div}\ (\overset{\Delta}{\underset{\sim}{\sigma}} + \underset{\sim}{W}\underset{\sim}{\sigma} - \underset{\sim}{\sigma}\underset{\sim}{D}) = 0 \tag{17}$$

within the body (neglecting nominal body-force rates) and the boundary conditions

$$\text{on } z = 0,L: \qquad \dot{S}^{j3} = 0 \qquad (j = 1,2)$$
$$w = 0 \tag{18}$$
$$\text{on } r = R: \qquad \dot{S}^{k1} = \sigma \left[V^{1k} - \delta^{k1} \operatorname{tr}(\underset{\sim}{D})\right]$$

similar to those used in the investigations mentioned above. Herein $\underset{\sim}{\dot{S}}$ is the unsymmetrical nominal stress-rate, $\underset{\sim}{V}$ is the velocity gradient with $\underset{\sim}{W} = \frac{1}{2}(\underset{\sim}{V} - \underset{\sim}{V}^T)$ the antisymmetric spin tensor, and (u,v,w) are the components of the velocity of the additional motion in the r, θ, z directions, respectively.

Constitutive relations (12) and (16), equilibrium equations, and boundary conditions set up two linear eigenvalue problems for the critical stresses σ and σ_z. For axisymmetric modes of deformation the field equations as well as the boundary conditions can be expressed in terms of velocity components u and w. Introducing now a velocity field

$$\begin{aligned} u(r,z) &= u(r)\cos(\alpha z) \\ w(r,z) &= w(r)\sin(\alpha z) \end{aligned} \tag{19}$$

with

$$\alpha = n\,\frac{\pi}{L} \quad , \; n = 1,2,3... \tag{20}$$

such that boundary conditions on the faces z = 0,L are satisfied, and ordinary systems of equations are obtained. Thus, as has been shown in [9] for uniaxial tension of an incompressible elastic-plastic material obeying normality, both eigenvalue problems can be reduced to a single equation for u

$$L^2(u) + 2b\gamma^2 L(u) + c\gamma^4 u = 0 \quad , \quad \gamma = \alpha R \tag{21}$$

where

$$\begin{aligned} 2b &= (a_1 a_4 + a_2 a_6 - a_3 a_5)/a_2 a_4 \\ c &= a_1 a_6 / a_2 a_4 \end{aligned} \tag{22}$$

The operator in (21) is defined by

$$L(u) = \frac{\partial}{\partial \zeta}\left[\frac{1}{\zeta}\,\frac{\partial(\zeta u)}{\partial \zeta}\right] \tag{23}$$

where $\zeta = r/R$. Fluid pressure boundary conditions on the lateral surface ($\zeta = 1$) can be expressed as

$$(a_1+a_3)\gamma^2 u+a_2 L(u) = 0$$

$$a_2 a_4(b_1-b_2) \frac{\partial}{\partial\zeta} [\zeta L(u)] + \gamma^2 \left[(b_1-b_2)(a_1 a_4-a_3 a_5) + \right. \tag{24}$$

$$\left. + (1+b_1)a_3 a_6\right]\frac{\partial}{\partial\zeta} (\zeta u) - a_3 a_6\gamma^2 u = 0$$

where

$$a_1 = -(1+\theta+\theta_z) \quad , \qquad a_4 = a_1+2$$

$$a_2 = 2(1+\theta+k_1+1/2k_3) \qquad \theta = \sigma/2G \quad , \quad \theta_z = \sigma_z/2G$$

$$a_3 = 1+\theta-\theta_z+2k_1+k_3-2k_2 \; , \quad a_5 = -a_3-2k_2+3k_3 \tag{25}$$

$$a_6 = -2(1-\theta_z+k_1-k_3-k_4)$$

$$b_1 = k_1+1/2k_3+\theta \quad , \qquad b_2 = k_2$$

and the different material properties of the two solids are expressed by different values of the material coefficients k_i.

As indicated in [9] the general solution to (21) which satisfies boundedness condition at $\zeta = 0$ is

$$u(\zeta) = c_1 J_1(\gamma\rho\zeta) + c_2 J_1(\gamma\bar{\rho}\zeta) \tag{26}$$

where $\rho^2 = b+i\sqrt{c-b^2}$ and the bar implies complex conjugation. Substitution of (26), finally, into the boundary conditions (24) provides the eigenvalue equation for the critical values of θ_z and θ.

Using power series and asymptotic expansions of the above Bessel functions of the first kind, two limiting cases of the solution of this eigenvalue equation can be investigated. In one case, the term $\gamma \to 0$ (long wavelength limit); while in the other case $\gamma \to \infty$ (short wavelength limit) is taken.

Long wavelength limit:

$$\theta_z + \theta = \frac{3}{2} + k_2 - k_4 - \frac{(k_2+1/2)(1+3k_3)}{1+2k_1+k_3+2\theta} \tag{27}$$

Short wavelength limit:

$$\theta_z + \theta = \left\{ 2+k_2-k_4 - \frac{(k_2+1)(1+3/2k_3)}{1+k_1+1/2k_3+\theta} \right\} \Big/ \left\{ 1- \sqrt{\frac{(1-\theta_z-\theta)(1-\theta_z+k_1-k_3-k_4)}{(1+\theta_z+\theta)(1+\theta+k_1+1/2k_3)}} \right\} \tag{28}$$

Introducing here the specific material properties of the two solids, comparison solid (CS) and second linear solid (SS), respectively, upper and lower bounds to the bifurcation stresses of the underlying solid will be given.

Two typical results of (27) and (28) based on

$$\nu = 0.3, \; \nu_o = 1/2\sqrt{3}, \quad \beta = 0 \quad \text{and} \; \theta = 0$$

are shown with Fig. 1, where Fig. 1(a) is valid for slender specimens with $\gamma \to 0$ and Fig. 1(b) is valid for stubby ones for which $\gamma \to \infty$. For both cases the solutions for the comparison solid (CS) - upper curves - and for the second linear solid (SS) - lower curves - give the respective bounds to the critical region in the parameter space ($\sigma_z/2G$, G_T/G).

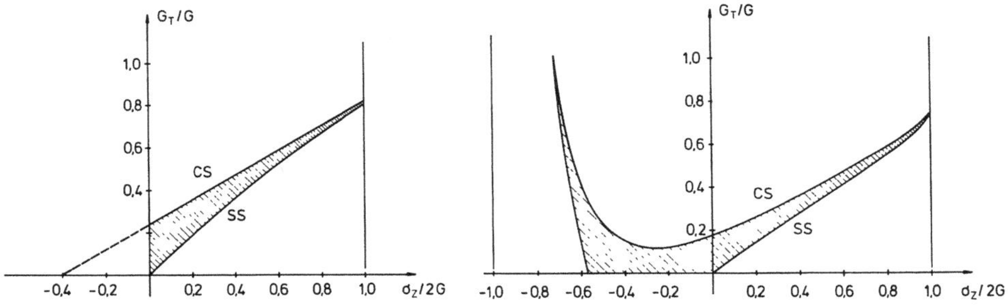

a) b)

Fig. 1. Bounds to the critical stresses in bifurcation of slender (a) and stubby specimens (b) with material properties $\nu = 0.3$, $\nu_o = 1/2\sqrt{3}$, $\beta = 0$, and $\theta = 0$

References

1. Mandel, J.: Conditions de stabilité et postulat de Drucker. Rheology and Soil Mechanics, Proc. IUTAM Symp. Grenoble, 1964 (Ed. by Kravtchenko J. and Sirieys, P.M.), Springer, Berlin (1966).

2. Mróz, Z.: On forms of constitutive laws for elastic-plastic solids. Arch. Mechanics 18 (1966) 3-35.

3. Needleman, A.: Non-normality and bifurcation in plane strain tension and compression. J. Mech. Phys. Solids 27 (1979) 231-254.

4. Hill, R.: A general theory of uniqueness and stability in elastic-plastic solids. J. Mech. Phys. Solids 6 (1958) 236-249.

5. Raniecki, B. and Bruhns, O.T.: Bounds to bifurcation stresses in solids with non-associated plastic flow law at finite strain. J. Mech. Phys. Solids 29 (1981) 153-172.

6. Rudnicki, J.W. and Rice, J.R.: Conditions for the localization of deformation in pressure-sensitive dilatant materials. J. Mech. Phys. Solids 23 (1975) 371-394.

7. Raniecki, B.: Uniqueness criteria in solids with non-associated plastic flow laws at finite deformations. Bull. Acad. Polon. Sci., ser. sci. techn. 27 (1979) 391-399.

8. Miles, J.P.: Fluid-pressure eigenstates and bifurcation in tension specimens under lateral pressure. J. Mech. Phys. Solids 21 (1973) 145-162.

9. Hutchinson, J.W. and Miles J.P.: Bifurcation analysis of the onset of necking in an elastic/plastic cylinder under uniaxial tension. J. Mech. Phys. Solids 22 (1974) 61-71.

10. Vardoulakis, I.: Bifurcation analysis of the triaxial test on sand samples. Acta Mech. 32 (1979) 35-54.

11. Bruhns, O. and Mielniczuk, J.: Zur Theorie der Verzweigungen nicht-isothermer elastoplastischer Deformationen. Ing.-Archiv 46 (1977) 64-74.

On Stability of Interaction between Continuous and Lumped Systems in Stationary Relative Motion

R. BOGACZ

Polish Academy of Sciences
Institute of Fundamental Technological Research, Warsaw

Summary

In the present paper stability of relative motion with constant velocity of a continuous and lumped systems is concerned. The configuration of instability regions was determined on the basis of Generalized Mikhaylov stability criteria. It is shown that in the case of at least two points of mutual interaction with a lumped system there exists a set of instability regions with qualitatively new lay-out. Also the influence of viscosity on critical velocity of relative motion of the two continuous systems is investigated.

1. Introduction

The stability of a relative motion of continuous and lumped systems constitutes an important engineering problem and has extensively been investigated. In the present paper, an extension is made of earlier results obtained for interaction with concentrated masses [1], densily distributed oscillators [2] and two interacting continuous systems in relative motion [3]. Application of such systems in electronophononics was demonstrated in [4].

In the second part of the paper, we shall formulate the problem and the fundamental equations as well as the boundary condition. The third part contains a description of the solution method. In the fourth part we shall determine the regions of instability of motion for various cases of interaction and the paper ends with a set of conclusions.

2. Statement of the Problem

We shall consider a model of the system consisting of a taut beam interacting with an elastic semi-space, with viscous surrounding and at a few points with a lumped system.

In general all subsystems are in the relative motion with a constant velocity. A sketch of the model is shown in Fig. 1.

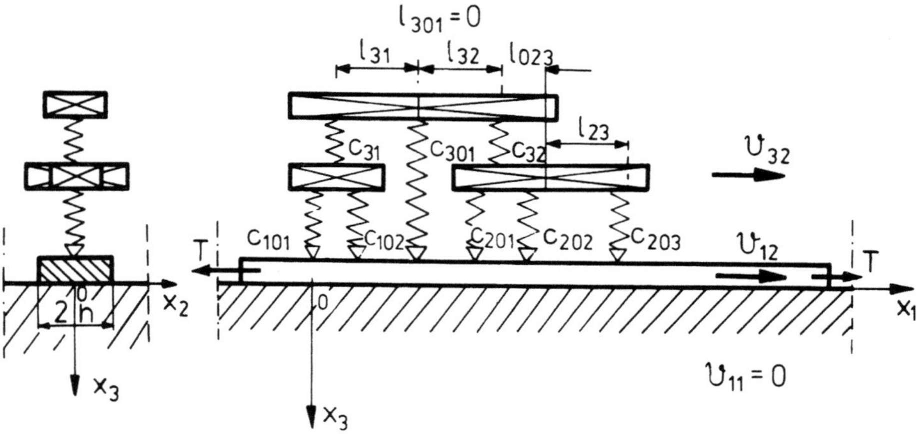

Fig. 1

v_{12} is a velocity of relative motion between the beam and the semi-space of elastic body in x_2 direction, v_{23} is a velocity of motion of the lumped system along the beam in x_2 direction, m_{oi} and M_i denote respectively the mass and the moment of inertia of the i-th lump, c_{oij} is the elasticity of the ij-th spring, T is the tension in the beam and EI and ρ_o denote respectively the rigidity and the linear mass density of beam. The elastic semi-space is characterized by velocities of elastic longitudinal and transversal waves; $a_1^2 = (\lambda + 2\mu)/\varrho$, $a_2^2 = \mu/\varrho$.

The motion of the lumped system will be described in a moving coordinate system (ξ, ζ) related to the system of coordinates of the semi-space (x_1, x_2, x_3) as follows

$$\xi - x_1 - v_{13} = 0, \quad \zeta - x_3 = 0, \quad v_{13} = v_{12} + v_{23} \tag{1}$$

The set of equations of motion of the lumped subsystem can be written in the form:

$$m_j \frac{dy_j}{dt^2} + \sum_{s=1}^{n} c_{js}(y_j - y_s) = P_j(t) \quad , \; j = 1,2,\ldots,n \tag{2}$$

$$P_j(t) = c_{ojo}(y_{jo}-y_j) \tag{2a}$$

where y_j is a displacement of j-th coordinates, y_{jo} denote the prescribed displacements which is a function of coordinates of the interaction points between the lumped and the continuous system z_{js}, l_{js}, $j=1,2,\dots,n$; $s=1,2,\dots,m$, s denotes the number of point of interaction of j-th lump.

$$\begin{aligned} J_{jo}(t) &= G_j(z_{11},\dots,z_{1m},\dots,z_{n1},\dots,z_{nm}) \\ P_j(t) &= F_j(P_1,P_2,\dots,P_n,l_{11},\dots,l_{1m},\dots,l_{n1},\dots,l_{nm}) \end{aligned} \tag{3}$$

The equation of motion of the beam, which is valid excluding points of interactions with the lumped system, can be written in ξ , ζ system in the following form

$$\frac{EI}{\rho_o}\frac{\partial^4 w}{\partial \xi^4} - \left(v_{23}^2\frac{\partial^2 w}{\partial \xi^2} + 2v_{23}\frac{\partial^2 w}{\partial \xi \partial t} - \frac{\partial^2 w}{\partial t^2}\right) = -\frac{1}{\rho_o}\left[P_1(\xi,t)+P_2(\xi,t)\right] \tag{4}$$

where $P_2(\xi,t)=\sum_{j=1}^{n} P_j(t)\big|_{\xi=\xi_j}$ and the $\frac{1}{\rho_o}P_1(\xi,t)$ is assumed to act on the beam in the plane of the contact with elastic semi-space. The equation of motion of elastic body have a form

$$a_2^2\,\nabla^2\vec{u} + \left(a_1^2 - a_2^2\right)\text{grad div}\,\vec{u} - \ddot{\vec{u}} = 0 \tag{5}$$

In addition, we shall assume that friction at interaction surfaces is negligibly small and the normal stress σ_{33} is constant for $x_2 \in (-h,+h)$, then, the surface conditions are

$$\begin{aligned} \sigma_{33}(x_1,x_2,o,t) &= \frac{1}{\rho_o}p_o(x_1,t)H(h-|x_2|) \\ \sigma_{12}(x_1,x_2,o,t) &= 0 \\ \sigma_{13}(x_1,x_2,o,t) &= 0 \end{aligned} \tag{6}$$

similarly as (4) the equation (5) and the boundary conditions (6) can be also written in a moving system of coordinates connected with the lumped system.

The additional boundary conditions will express the conformity of displacements and forces at the contact points

$$z_{js}(t) = W(\xi,t)\big|_{\xi=\xi_{js}}, \quad W(\xi,t) = u_3(x_1,o,o,t)\big|_{x_1=\xi-v_{13}t} \tag{7}$$

$$\vec{u}(x_1,x_2,x_3,t)\big|_{\overrightarrow{x_3\to\infty}} 0, \qquad W(\xi,t)\big|_{\overrightarrow{\xi\to\pm\infty}} 0 \tag{8}$$

In the case of the linear continuous systems, when the superposition principle can be used to solve the problem it is sufficient to define the following relation for a given value of concentrated force P

$$W(\xi,t) = R(\xi,\lambda)P_o e^{i\lambda t} \tag{9}$$

We shall proceed now to the solution of this system and the discussion of its stability.

3. Solution of the Problem, Characteristic Equations

Similarly as in [5] we shall seek a stationary solution of the set of equations satisfying the given boundary conditions in the form of the following travelling waves

$$y_j(t) = \sum_{n=1}^{m} B_{nj} e^{\lambda t+\theta}, \quad \vec{u} = \sum_{n=1}^{s} e^{\lambda t} \iint_{-\infty}^{\infty} e^{ir_1(x_1-V_1t)} A_n(r_1,r_2) e^{ir_2x_2-\gamma_n x_3} dr_1\, dr_2 \tag{10}$$

Substituting the solution into (5) in form (10) and satisfying the condition of radiation (8) and the boundary conditions (6) we obtain for $x_3=0$

$$\sigma_{33}(x_1,x_2,o,t) = e^{\lambda t}\iint_{-\infty}^{\infty} f(r_1,r_2)\, e^{ir_1(x_1-V_1t)+ir_2x_2}\, dr_1 dr_2 \tag{11}$$

$$u_3(x_1,x_2,o,t) = e^{\lambda t}\iint_{-\infty}^{\infty} f(r_1,r_2)\,\varkappa(r_1,r_2)\, e^{ir_1(x_1-V_1t)+ir_2x_2}\, dr_1 dr_2 \tag{12}$$

where $\gamma_1^2 = r_1^2+r_2^2-\dfrac{V_1^2}{a_1^2}$; $\gamma_2^2 = r_1^2+r_2^2-\dfrac{V_1^2}{a_2^2}$

$$\mathcal{H}(r_1, r_2) = \frac{1}{4}\,\frac{(1-r_1)^2 v_1^2 a_2^{-2}\sqrt{r_1 + r_2 - v_1^2 a_2^{-2}(1-r_1)^2}}{\left[r_1^2 + r_2^2 - v_1^2\, 2^{-1} a_2^{-2}(1-r_1)^2 - (r_1^2 + r_2^2)\right]\sqrt{r_2^2 + r_1^2 - v_1^2 a_1^{-2}(1-r_1)^2}} \tag{13}$$

The solution of the equation of motion of the beam is sought in the form

$$W(x_1, t) = e^{\lambda t}\int_{-\infty}^{\infty} W_o(r_1)\, e^{ir_1(x_1 - V_1 t)}\, dr_1 = u_3(x_1, o, o, t) \tag{14}$$

Writing the following functions in the form

$$H(h - |x_2|) = \frac{1}{\pi}\int_{-\infty}^{\infty} \frac{\sin \frac{1}{2} r_2 h}{r_2}\, e^{ir_2 x_2}\, dr_2, \quad \delta(x_1) = \frac{1}{2}\int_{-\infty}^{\infty} e^{ir_2 x_1}\, dr_1 \tag{15}$$

the displacement of the beam subjected to the moving force $P = P_o e^{\lambda t}$ takes the form

$$z(x_1, t) = P_o R(x_1, \lambda) = \frac{P}{2\pi}\int_{-\infty}^{\infty} \frac{2\frac{V_1^2}{EJ\lambda}\, e^{ir_1 x_1}\, dr_1}{r_1^4 + (1-r_1)^2 \frac{\varrho_o V_1^2}{EJ\lambda^2} + \frac{\varepsilon_o V_1^4}{\lambda^3}(1-r_1) - \frac{b V_1^4}{\pi EJ\lambda^4}\int_{-\infty}^{\infty} \mathcal{H}(r_1, r_2)\frac{\sin r_2 \frac{b}{4}}{r_2}\, dr_2} \tag{16}$$

The approximate method of calculations of such a kind of integrals can be found for instance in [5]. Substituting (10) in Eqs. (2) and (3), and utilizing relation (16) we obtain the characteristic equations $\Phi(\lambda, v_{12}, v_{23}) = 0$ which provides a basis for the stability analysis of the considered system. To simplify the problem let us consider the limiting case when the mass density of the elastic semi-space tends to zero.

3.1. Case of the beam on an elastic foundation

If the mass density of the beam is much greater than the mass density of the elastic semi-space the pressure acting on the beam can be expressed as follows

$$P_1(\xi, t) = \varepsilon_o\left(\frac{\partial W}{\partial t} - v_{23}\frac{\partial W}{\partial \xi}\right) + c_o W \tag{17}$$

Let $W_i(\xi, t)$, $i=1,2$ denote $W_1(\xi, t) = W(\xi, t)|_{\xi>0}$ and $W_2(\xi, t) = W(\xi, t)|_{\xi<0}$. Let $\xi = 0$ is the point of interaction with the lumped system. In the particular case relation (16) significantly simplifies to give

$$EI\left(\frac{\partial^3 W_1}{\partial \xi^3} - \frac{\partial^3 W_2}{\partial \xi^3}\right) + P_o e^{\lambda t} = 0 \tag{18}$$

with additional conditions for $\xi=0$

$$W_1=W_2, \quad \frac{\partial W_1}{\partial \xi} = \frac{\partial W_2}{\partial \xi}, \quad \frac{\partial^2 W_1}{\partial \xi^2} = \frac{\partial^2 W_2}{\partial \xi^2} \tag{19}$$

and $W_1=W_2=0$ for $\xi \to \pm\infty$.

In this case the solution takes the form

$$W_1=\left(c_i e^{r_1\xi} + c_2 e^{r_2\xi}\right)e^{\lambda t}, \quad W_2=\left(c_3 e^{r_3\xi} + c_4 e^{r_4\xi}\right)e^{\lambda t}$$

$$W=W_1 H(-\xi)+W_2 H(\xi)$$

where r_j, $j=1,2,3,4$ satisfying the following equation

$$EIr^4+\rho_o v_{23} r^2+\left(2\rho_o v_{23}\lambda-\varepsilon_o v_{23}\right)r+c_o+\rho_o\lambda^2 = 0 \tag{20}$$

Further simplification of the problem, which has also significant applications, are obtained assuming EI=0.

3.2. Case of the string

The equation of motion of the string, which is valid excluding points of interaction form

$$\frac{\partial^2 W}{\partial t^2} - 2v_{23}\frac{\partial^2 W}{\partial \xi \partial t} + \left(v_{23}^2-a^2\right)\frac{\partial^2 W}{\partial \xi^2} + \left(\varepsilon_1+\varepsilon_2\right)\left(a\frac{\partial W}{\partial t} - v_{23}\frac{\partial W}{\partial \xi}\right) - \varepsilon_2 v_{12}\frac{\partial W}{\partial \xi} + cW = 0 \tag{21}$$

where $a^2=T\rho_o^{-1}$ is the velocity of elastic waves in the string not interacting with the surrounding medium $c=c_o\rho^{-1}$ and ε_1 and ε_2 denote respectively coefficients of internal and external damping. The force $P_j(t)$ acting in the point $\xi=0$ expressed by means of the parameters of the string take form

$$P_j(o,t)=T\left(\frac{\partial W}{\partial \xi} - \frac{\partial W}{\partial \xi}\right)\Big|_{\xi=0} \tag{22}$$

and the solution assumes the form

$$W(\xi,t)=\sum_{n=1}^{2} c_{ns} e^{r_s\xi+\lambda t} \tag{23}$$

where

$$r_{1,2}=\frac{v_{23}(2\lambda+a\varepsilon_1+a\varepsilon_2)+v_{12}\varepsilon_2\pm\sqrt{\left[v_{23}(2\lambda+a\varepsilon_1+a\varepsilon_2)+v_{12}\varepsilon_2\right]^2+4\left(a^2-v_{23}^2\right)\left[\lambda^2+a(\varepsilon_1+\varepsilon_2)\lambda+c\right]}}{2\left(a^2-v_{23}^2\right)}$$

Substituting the above form of solution into the equation of motion (21) we obtain the following additional characteristic equation of interaction between the string and the viscous surrounding

$$\Phi\,(\lambda, v_{12})=\lambda^2+(\varepsilon_1+\varepsilon_2-2v_{12}r)\lambda+c+\left(v_{12}^2-a^2\right)r=0 \qquad (24)$$

The above case of interaction between the two continuous systems is relatively simple and can be studied as the limiting case only. More important for practical application is the case of the interaction of two or more unlimited, viscoelastic beams sliding on each other with constant velocity. The limiting case, when the rigidity EI of one of the beams is sufficiently small is a similar model to that of a belt conveyer system. The further simplification when the tension force $T\to 0$ is analogous to the case of the tubes conveying fluid.

The motion of the beams is described by equation (4) putting $v_{23}=0$ for the first beam and $v_{23}=v_{12}$ for the second beam. Requiring conformity of pressures and displacements given by the relationships

$$P_1^{(1)}(\xi,t)+P_1^{(2)}(\xi,t)=0,\quad W^{(1)}(\xi,t)+W^{(2)}(\xi,t)=0 \qquad (25)$$

and satisfying the equations of motion we obtain the following form of characteristic equation

$$\gamma\,\left\{\left[1-i\varepsilon_1(V_2+v_{12})\right]R_1^2-\left(V_2+v_{12}\right)^2\right\}+\left(1-i\varepsilon_2V_2\right)R_2^2-V_2^2=0 \qquad (26)$$

In the case of the beam sliding on the surface of elastic semi-space the right-hand part of the characteristic equation (26) is much more complex, but it is easy to see from equation (13) that this part is equal to zero for V_2 equal velocity of Rayleigh waves.

4. Instability of Solution, Determination of Instability Regions

In order to determine the ranges of velocities in which the system is unstable in the sense of Lapunov with measure

$$\rho(t) = \sup_{-\infty<x<\infty}|W| + \sup_{-\infty<x<\infty}\left|\frac{\partial W}{\partial t}\right| + \sup_{-\infty<x<\infty}\left|\frac{\partial W}{\partial x_j}\right| + \sqrt{\sum_{j=1}^{n} y_j^2}$$

we are seeking for regions in which there exist λ satisfying the characteristic equation with a positive real part. Negative real part of all roots λ guarantee stability of this kind of solution.

Assuming in the characteristic equations $\lambda = i\omega = ikV$ and $r = r_0 + ik$, we shall adopt a generalized Mikhaylov criterion [2] . The criterion is based on the analysis of the configuration of curves described by the following equations

$$\operatorname{Re}\Phi(ikV, v_{12}, v_{23}) = 0, \quad \operatorname{Im}\Phi(ikV, v_{12}, v_{23}) = 0 \tag{27}$$

We proceed now to illustrate the above procedure by means of examples.

4.1. Instability of interaction between two continuous systems in relative motion with a constant velocity.

Let us consider the case of a linear taut string interacting with viscoelastic surrounding and moving with a constant velocity v_{12}. Characteristic equation for this case is given by Eq. (24). Configuration of the curves representing $\operatorname{Re}\Phi = 0$ and $\operatorname{Im}\Phi = 0$ is drawn in Fig. 2.

According to the Generalized Mikhaylov criterion [2] for $|v_{12}| < v_{cr}$ the system is stable. The region of instability is given by the following formula

$$|v_{12}| > a\frac{\varepsilon_1 + \varepsilon_2}{\varepsilon_2} \tag{28}$$

and it is easy to see the critical value of v_{12} belongs to the set $v_{12cr} \in [a, \infty]$.

In the case of the two viscoelastic beams configuration of the characteristic curves is shown in Fig. 3. Thus, there exists a possibility of change of the critical velocity of the relative

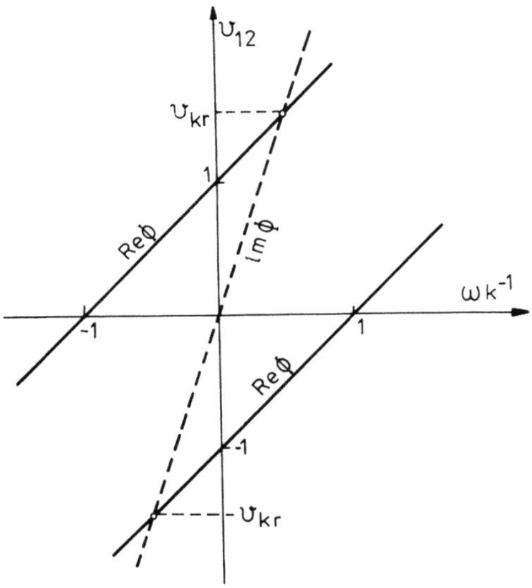

Fig. 2

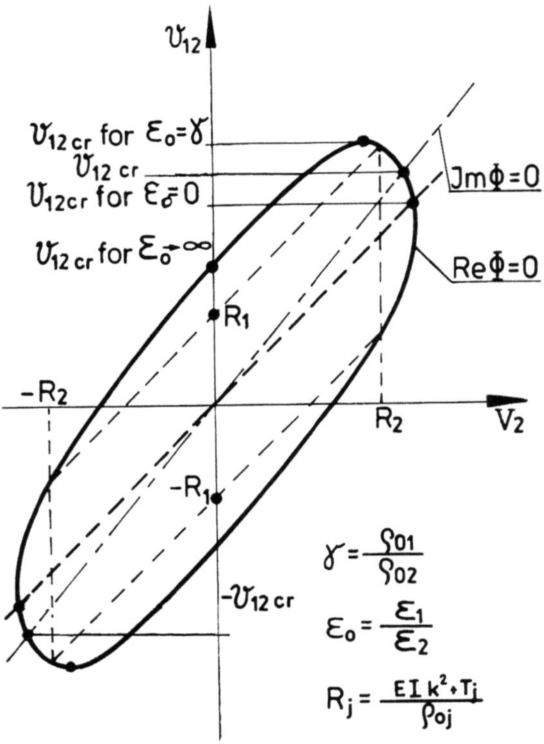

Fig. 3

motion of the beams within the range

$$\sqrt{R_1^2+\frac{1}{\gamma}R_2^2} \leqslant v_{12cr} \leqslant \sqrt{(\gamma+1)\left(R_1^2+\frac{1}{\gamma}R_2^2\right)}$$

for $R_1 \leqslant R_2$.

The critical value of relative motion is given by formula

$$v_{12cr} = \pm\left(\varepsilon_0^{-1}+1\right)\sqrt{\frac{\gamma R_1^2+R_2^2}{\gamma+\varepsilon_0^{-2}}} \qquad (29)$$

In the particular case, when e.g. $R_1=0$ the characteristic equation describes the case of the tubes conveying fluid. The comparison of the results with [1] shows that reduction of the stability problem to the determination of resonance regions for linear ordinary differential equation with periodic coefficients could lead to significant errors.

We now shall determine the instability regions for case of interaction with the lumped system.

4.2. Instability regions for moving lumped system interacting in two points with a continuous system

In order to assess the influence of each parameter of the system on the distribution lay-out of instability regions we confine the consideration to two degrees of freedom system with mass m and moment of inertia M, two points of interaction and to the symmetric case.

The equation of motion (2) takes the form

$$\begin{aligned} m\ddot{y}_1 + 2c_1 y_1 - c_1(z_1+z_2) &= 0 \\ M\ddot{y}_2 + 2c_1 l(y_1 - z_1) + 2c_1 l^2 y_2 &= 0 \end{aligned} \qquad (30)$$

and the relations (3) as follows

$$\begin{aligned} z_1+z_2+Q_1(\lambda)(P_1+P_2) &= 0 \\ z_1-z_2+Q_2(\lambda)(P_1-P_2) &= 0 \end{aligned} \qquad (31)$$

where $Q_1 = \frac{1}{m}\frac{2\alpha_1+\lambda^2}{\alpha_1\lambda^2}$, $Q_2 = \frac{\alpha_2+\lambda^2}{\alpha_1\lambda^2}$, $\alpha_1 = \frac{c_1}{m}$, $\alpha_2 = \frac{c_1 l^2}{2M}$

The characteristic equation for v_{23} greater than the velocity of elastic waves takes the form

$$Q_1(\lambda) - Q_2(\lambda)\,R(\lambda,1) + 2Q_1(\lambda)\,Q_2(\lambda) = 0$$

or

$$\Lambda^3 + \varkappa_1 f(\Lambda, \bar{v}_{23})\Lambda^2 + \varkappa_1\Lambda = 0 \qquad (32)$$

where $f(\Lambda, \bar{v}_{23})$ in case of the simplest continuous system - i.e. the string - is given by formula

$$f(\Lambda,\bar{v}_{23}) = \frac{\exp\frac{-\Lambda}{1-\bar{v}_{23}} + \left[2+\exp\left(\frac{-\Lambda}{1-|\bar{v}_{23}|}\right)\right] - H(\bar{v}_{23}^2-1)\,\exp\left(\frac{-\Lambda}{1-|\bar{v}_{23}^2|}\right)}{2\gamma_o(\bar{v}_{23}^2-1)^{-1}}$$

where $\Lambda = 2\lambda l$, $\varkappa_1 = 4\alpha_1 l^2$ $\bar{v}_{23} = v_{23}a^{-1}$, $\gamma_o = \frac{8\rho l}{m}$, $\Lambda = \sigma + i\,\theta$.

The curves representing Im =0 and possible instability regions according to the condition

$$\sigma \geqslant 0 \Rightarrow \frac{\bar{v}_{23}^2-1}{2\gamma_o}\, e^{\frac{-\sigma}{\bar{v}_{23}+1}}\left(\cos\frac{\theta}{\bar{v}_{23}+1} - e^{-\frac{2\sigma}{\bar{v}_{23}^2-1}}\cos\frac{\theta}{\bar{v}_{23}-1}\right) \leqslant 0 \qquad (33)$$

are shown in Fig. 4.

The influence of the value of parameters m and M on the shape of the instability regions containing the curves ReΦ=0 is shown in Fig. 5. The regions are determined using the following estimation

$$\left| \exp -\left(\frac{\Lambda}{|v_{23}-1|}\right) - \exp\left(-\frac{\Lambda}{|v_{23}+1|}\right)\right| \leqslant 2 \qquad (34)$$

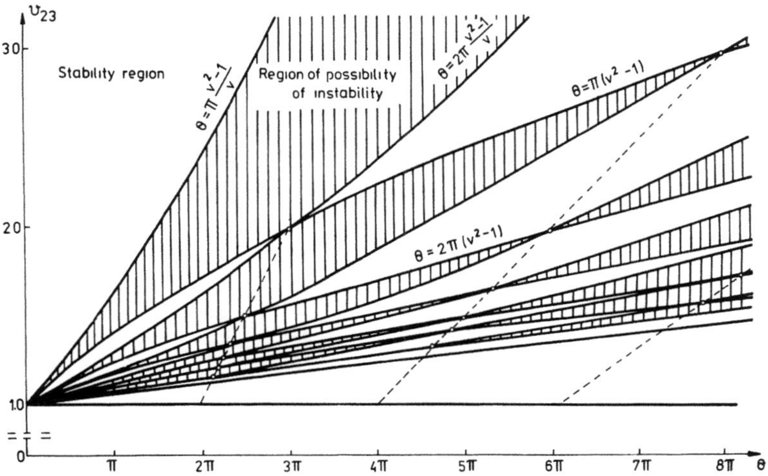

Fig. 4

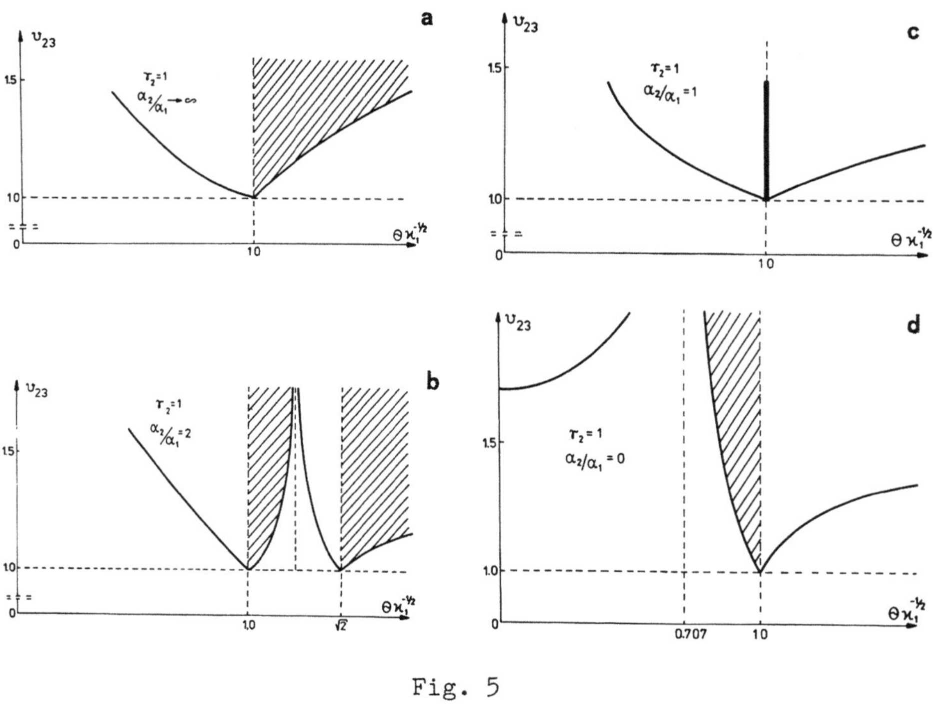

Fig. 5

Four cases of estimation of instability regions are shown in Fig. 5: a/ $M \to 0\ (\alpha_2 \to \infty)$, b/ $\alpha_2 = 2\alpha_1$, c/ $\alpha_1 = \alpha_2$, d/ $M \to \infty \left(\alpha_2 \to 0,\ Q_2 \to \frac{1}{c_1}\right)$ The superposition of regions shown in Fig. 4 and Fig. 5 yields estimates on the ranges of velocity of motion in which the system is unstable. The estimation is shown in Fig. 6.

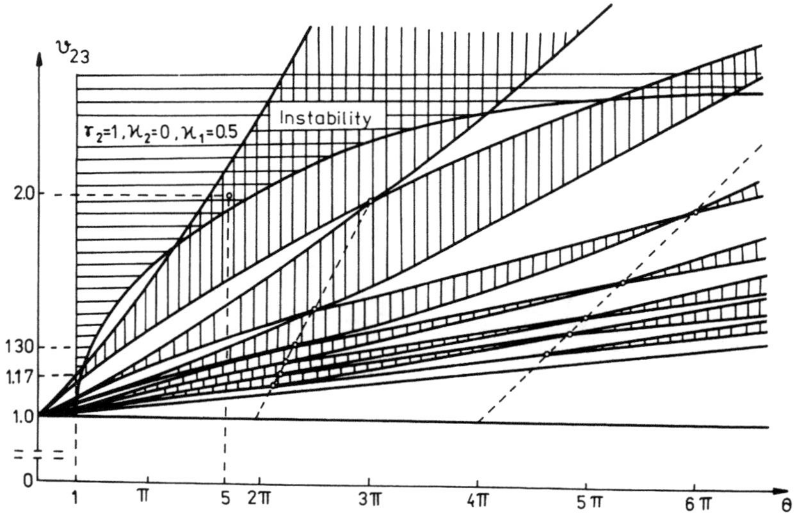

Fig. 6

5. Concluding Remarks

It follows from our considerations that viscosity exerts an essential influence on the critical value of velocity of relative motion of the continuous systems. In the case of interaction of the continuous systems with the lumped system the obtained lay-out of instability regions is qualitatively new. As follows from Fig. 4 and Fig. 6 for two at least points of interaction in general case a set of instability regions is always infinite but countable. The results obtained for the comparatively simple model are supposed to remain valid for more complex systems, and the range of importance of effects of a few points of mutual interaction, greater than for application to wave dynamics.

References

1. Nelson,H.D., Conver, R.A., Dynamic instability of beam carrying moving masses, Journal of Applied Mechanics, 38, 4, (1971) 1003-1006.

2. Bogacz,R., Interaction between a moving set of nonlinear oscillators and a travelling wave, Proceedings of Vibration Problems t. 9, 1, 1968.

3. Bogacz,R., Influence of viscosity on the stability of a relative motion of two media, Archives of Mechanics, 24, 4, 1971.

4. Kaliski,S., The perfect self-excited piesoquartz resonator with an extremal electron stream, Proceedings of Vibration Problems t. 9, 4, 1968.

5. Sneddon,I., The use of integral transforms, New York, McGraw Hill 1972.

Duality and Stability Questions for the Linearized Traction Problem with Live Loads in Elasticity

G. CAPRIZ & P. PODIO-GUIDUGLI

Istituto Matematico "L. Tonelli" and
Istituto di Scienza delle Costruzioni
Università di Pisa, Italy

Summary

The study of bifurcation of solutions of a functional equation by perturbation methods requires the determination of the kernels of the approximating linear operator and its adjoint. For the traction problem under live loads in elasticity, even the formal specification of the adjoint linear problem is not straightforward. We deal with this question here; we determine the type of dependence of the loads on the displacement gradient which allows us to construct an adjoint linear problem. Moreover, we find conditions sufficient for self-adjointness which coincide with the conditions of "weak conservativeness" for the loads first proposed by Sewell [1][2] and discussed at length by Knops & Wilkes [3], Chapts Q and R.

The linearized traction problem

We begin with a rough and succinct account of relevant background developments, which we will expand and make precise elsewhere.

Consider a non-linear traction problem for a three-dimensional elastic body under live loads. The elastic response and the functional dependence of the loads on the displacement must be chosen so as to satisfy requirements of physical plausibility and to match, besides, with conditions of "well-posedness" of the ensuing analytical problem. To formulate such conditions is of course the main open problem in finite elasticity. To obtain some hints on the correct direction to take, especially for uniqueness and stability questions, one may think of setting up a formal perturbation scheme, where a one-parameter family of loading operators is considered, modeling the gradual changes in the action of the environment on the body, and where a one-parameter family of equilibrium displacement fields is sought. In this way, a hierarchy of non-homogeneous linear problems is obtained, all ruled by the same operator L ; for the k-th of these, the datum depends in general on the solutions of the preceding k-1 problems. Now, one expects that well-posedness conditions on L could be arrived at by adapting standard techniques; and that consequent conditions on the original non-linear problem would ensue: more precisely, that L^T , the adjoint operator of L , could be constructed, the kernels of L and L^T inspected, and the theorem of alternative applied, as customary in perturbative methods, to derive conditions of solvability relevant to the description of branching of solutions.

With the above program in mind, which has proven to be viable in the case of dead loads (cf. e.g. [4] and the literature quoted therein), we state hereafter the differential version of the homogeneous linearized traction problem in statics.

Let a bounded regular region $\mathcal{B}$ of three-dimensional euclidean space, whose boundary $\partial\mathcal{B}$ has outward unit normal n , be occupied by a continuous body at equilibrium. Let the vector u(x) represent the displacement of a point $x\varepsilon\mathcal{B}$ in a small incremental deformation of $\mathcal{B}$. In cartesian components, the field u(·) is governed by the equation

$$(Eu)_i := (\mathbb{S}_{ijhk}u_{h,k})_{,j} + \mathbb{b}_{ihk}u_{h,k} + B_{ih}u_h = 0 \text{ in } \mathring{\mathcal{B}}\,, \qquad (1)$$

and subject to the boundary condition

$$(Fu)_i := (\mathbb{S}_{ijhk}n_j - \mathbb{C}_{ihk})u_{h,k} - C_{ih}u_h = 0 \text{ in } \partial\mathcal{B}. \qquad (2)$$

Briefly, the operator L := (E,F) is the pair of the field and boundary operators E and F , defined by (1) and (2) (where a possible dependence of the coefficients on x is understood).

$\mathbb{S}_{ijhk}$ are the components of the fourth-order elasticity tensor $\mathbb{S}$, and involve both the ground state of stress $S^{(0)}$ and the linear approximation of the elastic response function $\mathbb{D}$ in the following way:

$$\mathbb{S}_{ijhk} = \delta_{ih} S^{(0)}_{jk} + \mathbb{D}_{ijhk} \ . \tag{3}$$

In view of (3), the tensor $\mathbb{S}$ is symmetric if and only if the material is hyperelastic and, therefore, $\mathbb{D}$ is symmetric; however it is safer to presume that

$$\mathbb{S}_{ijhk} \neq \mathbb{S}_{hkij} \ , \tag{4}$$

in general.
The vector

$$b_i(u) := \mathbb{b}_{ihk} u_{h,k} + B_{ih} u_h \tag{5}$$

specifies the body force exerted by the environment at $x \varepsilon \mathring{\mathcal{B}}$ when $\mathcal{B}$ is subject to the displacement $u(\cdot)$. Likewise, the vector

$$t_i(u) := \mathbb{c}_{ihk} u_{h,k} + C_{ih} u_h \tag{6}$$

is the surface traction exerted by the environment at $x \varepsilon \partial \mathcal{B}$. Although problem (1), (2) is thought of here as the linear approximation, within a suitable perturbation scheme, of a non-linear traction problem with live loads, it coincides formally with the corresponding problem in the theory of small deformations superimposed on large (cf. [5], §§68-70, and [3]).

Variational formulation

If we want problem $Lu = 0$ to be "well posed", the boundary operator F cannot be arbitrarily given, as well known examples show; rather, one expects well-posedness to prevail only if F obeys certain compatibility conditions, perhaps such as to involve E as well (vid. [6], Chap. 2, and especially Sects. 1.4 and 8.3). As we shall see, appropriate compatibility conditions emerge spontaneously when, as a first step toward the definition of an adjoint operator, we attempt to give the problem a variational form. Naively, one could try to arrive at a weak formulation by multiplying equation $(1)_2$ by a test function v and integrating by parts; this procedure, however, would not yield $(2)_2$ as a natural condition. Rather, guidance and inspiration in the search must be sought from the so-called problems of regular oblique derivative (vid. [7], Lect. 11; [8], Chap. II, §§4 and 5; [6], Chap. 2, §9.5).

Let u be any function in the domain of E. Define the modified differential operator

$$(\tilde{E}u)_i := (\tilde{\mathbb{S}}_{ijhk}u_{h,k})_{,j} + \mathbb{b}_{ihk}u_{h,k} + B_{ih}u_h , \tag{7}$$

with

$$\tilde{\mathbb{S}} = \mathbb{S} - \mathbb{T} \quad , \qquad \tilde{b} = \mathbb{b} + \mathbb{a} \quad , \tag{8}$$

and introduce the following bilinear form

$$\tilde{a}(u,v) := \int_{\mathcal{B}}(-v_{i,j}\tilde{\mathbb{S}}_{ijhk}u_{h,k} + v_i\tilde{\mathbb{b}}_{ihk}u_{h,k} + v_iB_{ih}u_h) + \int_{\partial\mathcal{B}} v_iC_{ih}u_h \; . \tag{9}$$

It is not difficult to show that, if one selects $\mathbb{T}$ and $\mathbb{a}$ so that

$$\mathbb{T}_{ijhk} = - \mathbb{T}_{ikhj} \quad \text{and} \quad \mathbb{a}_{ihk} = \mathbb{T}_{ijhk,j} \quad , \tag{10}$$

then

$$\tilde{E}u \equiv Eu \tag{11}$$

for any u in the domain of E. Morover, if

$$\mathbb{c}_{ihk}n_k = 0 \qquad \text{over} \quad \partial\mathcal{B}_o \; , \tag{12}$$

then Green's formula holds

$$\tilde{a}(u,v) = \int_{\mathcal{B}} v\cdot Eu - \int_{\partial\mathcal{B}} v\cdot Fu \; . \tag{13}$$

The main tool for the proof is a representation formula for fields $\mathbb{T}$ with the symmetry $(10)_1$ which satisfy also the following condition:

$$\mathbb{T}_{ijhk}n_j = \mathbb{c}_{ihk} \tag{14}$$

over $\partial\mathcal{B}$. The formula reads

$$\mathbb{T}_{ijhk} + \mathbb{c}_{ihk}n_j - \mathbb{c}_{ihj}n_k + A_{ih}\,\mathbb{e}_{jkr}n_r \quad , \tag{15}$$

where A is an arbitrary second-order tensor and $\mathfrak{e}$ is the Ricci alternator. The definition of $\mathbf{T}$ is then smoothly extended throughout $\mathcal{B}$.

Under hypothesis (12), the differential problem $(1)_2$, $(2)_2$ is formally equivalent to its variational formulation:

Find u such that $\tilde{a}(u,v) = 0$ for any test function v.

The compatibility condition (12), which plays a central role in our present developments, has a simple interpretation: the surface traction t , defined by (6), cannot involve the normal derivative of the displacement.

Indeed, if $m^{(\alpha)}$, $\alpha = 1,2$ are mutually orthogonal unit tangent vectors at a point $x \varepsilon \partial\mathcal{B}$, one can split the gradient of u into its normal and surfacial parts:

$$u_{h,k} = \frac{\partial u_h}{\partial n} n_k + \frac{\partial u_h}{\partial m^{(\alpha)}} m_k^{(\alpha)} \quad , \tag{16}$$

(with $\frac{\partial u}{\partial n}$ and $\frac{\partial u}{\partial m^{(\alpha)}}$ the normal and tangential derivatives of u , respectively), and notice that

$$\mathfrak{C}_{ihk} u_{h,k} \equiv \mathfrak{C}_{ihk} \frac{\partial u_h}{\partial m^{(\alpha)}} m_k^{(\alpha)} \tag{17}$$

when (12) prevails. Accordingly, the boundary condition $(2)_2$ can be written in a form which puts in evidence the occurrence of both normal and tangential derivatives:

$$(Fu)_i = (\mathfrak{S}_{ijhk} n_j n_k) \frac{\partial u_h}{\partial n} + (\mathfrak{S}_{ijhk} n_j - \mathfrak{C}_{ihk}) m_k^{(\alpha)} \frac{\partial u_h}{\partial m^{(\alpha)}} - C_{ih} u_h = 0 \, . \tag{18}$$

The necessity of giving (18) a non-formal meaning sets the minimal objective for the regularization at the boundary of weak solutions. We remark that the matrix

$$\mathfrak{S}_{ih} := \mathfrak{S}_{ijhk} n_j n_k$$

is invertible, if $\mathbb{S}$ is strongly elliptic, i.e., if

$$\mathbb{S}_{ijhk} a_i a_h d_j d_k > 0$$

for any non-null vectors a, d . On the other hand, to guarantee that the surface traction depends on the displacement gradient it is sufficient to assume that there exist indices i,h and a tangential direction $m^{(\alpha)}$ such that

$$\mathbb{C}_{ihk} m_k^{(\alpha)} \neq 0 \quad . \tag{19}$$

The adjoint operator

Our goal now is to arrive at a duality relation of the type

$$\int_{\mathcal{B}} v \cdot Eu - \int_{\mathcal{B}} u \cdot E^T v = \int_{\partial\mathcal{B}} v \cdot Fu - \int_{\partial\mathcal{B}} u \cdot F^T v \quad , \tag{20}$$

that is to say, a Green's formula such as to suggest directly definitions of field and boundary operators E^T and F^T to be taken as formally adjoint to operators E and F , respectively. Repeated integration by parts and use of (10) allow us to give (9) the form

$$\begin{aligned} \tilde{a}(u,v) &= \int_{\mathcal{B}} u_h \left((\mathbb{S}^T_{hkij} v_{i,j})_{,k} - {}^t\mathbb{b}_{hik} v_{i,k} + (B^T_{hi} - {}^t\mathbb{b}_{hik,k}) v_i \right) + \\ &- \int_{\partial\mathcal{B}_o} u_h \left(\mathbb{S}^T_{hkij} v_{i,j} n_k - (C^T_{hi} + {}^t\mathbb{b}_{hik} n_k) v_i + (\mathbb{T}_{ikhj} v_i)_{,j} n_k \right) . \end{aligned} \tag{21}$$

(here,

$$\mathbb{S}^T_{hkij} = \mathbb{S}_{ijhk} \; , \quad {}^t\mathbb{b}_{hik} = \mathbb{b}_{ihk} \; , \quad B^T_{hi} = B_{ih} \;) \; .$$

Manipulation of the last boundary integrand in (21) gives, with the help of (12), (15) and by exploiting the symmetry of the surface gradient of the normal,

$$(\mathbb{T}_{ikjh} v_i)_{,j} n_k = \mathbb{C}_{ihj} v_{i,j} + \mathbb{C}_{ihk,j} (\delta_{kj} - n_k n_j) v_i \quad . \tag{22}$$

Thus, if we insert (22) into (21) and recall (13), we have that the duality relation (20) holds with

$$(E^T v)_h := (\mathbb{S}^T_{hkij} v_{i,j})_{,k} - {}^t\mathbb{b}_{hik} v_{i,k} + (B^T_{hi} - {}^t\mathbb{b}_{hik,k}) v_i \quad , \tag{23}$$

$$(F^T v)_h := (\mathbb{S}^T_{hkij} n_k + {}^t\mathbb{c}_{hij}) v_{i,j} \tag{24}$$

$$- (C^T_{hi} + {}^t\mathbb{b}_{hik} n_k - {}^t\mathbb{c}_{hik,j} (\delta_{kj} - n_k n_j)\ ) v_i$$

$$= (\mathbb{S}^T_{hkij} n_k n_j) \frac{\partial v_i}{\partial n} + (\mathbb{S}^T_{hkij} n_k - {}^t\mathbb{c}_{hij}) m_j^{(\alpha)} \frac{\partial v_i}{\partial m^{(\alpha)}} +$$

$$- (C^T_{hi} + {}^t\mathbb{b}_{hik} n_k - (\frac{\partial}{\partial m^{(\alpha)}}\, {}^t\mathbb{c}_{hij}) m_j^{(\alpha)}) \quad .$$

It is important to notice that we are now in a position to formulate a differential problem

$$L^T v = 0 \quad , \quad \text{with} \quad L^T := (E^T, F^T) \quad ,$$

formally adjoint to problem

$$Lu = 0$$

in the sense of Lions & Magenes [6], §2.5, and that L^T, under the hypotheses, does not depend on the auxiliary tensor T.

Self-adjointness and stability

Conditions sufficient for self-adjointness can be read off at once by juxtaposition of (1) and (23), and of (18) and $(24)_2$. These are:

$$\mathbb{S}_{ijhk} = \mathbb{S}_{hkij} \quad ; \tag{25}$$

$$\mathbb{b}_{ihk} = -\mathbb{b}_{hik} \quad , \quad B_{ih} = B_{hi} - \mathbb{b}_{hik,k} \quad ; \tag{26}$$

$$\mathbb{c}_{ihk} = -\mathbb{c}_{hik} \quad , \quad C_{ih} = C_{hi} + \mathbb{b}_{hik} n_k - \frac{\partial \mathbb{c}_{hij}}{\partial m^{(\alpha)}} m_j^{(\alpha)} \tag{27}$$

The algebraic consistency of conditions $(26)_1$, $(26)_2$, and $(27)_1$, $(27)_2$, respectively, is easy to check.

Condition (25) is guaranteed when the ground stress is spherical and the material is hyperelastic (cf. (3)); condition (26) when the body force is of the dead type. In this instance, the self-adjointness conditions are $(27)_1$ and the reduced form of $(27)_2$:

$$C_{ih} = C_{hi} - \left(\frac{\partial}{\partial m^{(\alpha)}} \mathbb{C}_{hij}\right) m_j^{(\alpha)} \tag{28}$$

It is worth noticing that conditions $(27)_1$ and (28) are precisely the conditions determined by Sewell [1],[2] for the loads to be, in the terminology of Knops & Wilkes, weakly conservative (cf. [3], §§66 and 67; in particular, footnote 4 on page 260).

An important example where conditions $(27)_1$ and (28) can be fulfilled is that of hydrostatic pressure loading, when

$$t_i(u) = -\Pi(n_i\delta_{hk} - \delta_{ik}n_h)u_{h,k} - n_i\frac{\partial\Pi}{\partial u_h}u_h \quad , \tag{29}$$

with $\Pi = \Pi(x,u) > 0$ the pressure.
By (29),

$$\mathbb{C}_{ihk}a_k = -\Pi(n_i a_h - a_i n_h) \tag{30}$$

so that, taking first $a = n$ and then $a = m^{(\alpha)}$ in (30), one can verify directly that both (12) and (19) are satisfied. Verification of $(27)_1$ is equally straightforward, whereas (28) sets restrictions on the choice of the pressure function which have been analyzed by Beatty [9] (cf. also [3], §67).

As Knops & Wilkes remark in the beginning of their §68, for weakly conservative loads the stability analyses of an equilibrium placement in a class of small incremental displacements become as a rule identical to those for dead loading, at the sole cost of modifying the energy functional by addition of a suitable potential for the loads. However, special care is needed for the traction problem; likewise, the continuous dependence issue is much harder to deal with. Here we content ourselves with the remark that our characterization (12) of surface tractions which allow for a variational formulation substantiates and puts into perspective the analogous choice of Spector [10], who confines his attention to surface tractions depending at most on the surfacial gradient of displacement.

References

1. Sewell, M. J. : On the calculation of potential functions defined on curved boundaries. Proc. Roy. Soc. London., Ser. A286 (1966) 402-411.
2. Sewell, M. J. : On configuration-dependent loading. Arch. Rational Mech. Anal. 23 (1967) 327-351.
3. Knops, R. J. ; Wilkes, E. W. : Theory of elastic stability. In PH VIa/3, C. Truesdell ed. · Berlin · Heidelberg · New York : Springer 1973.
4. Capriz, G.; Podio-Guidugli, P. : The role of Fredholm conditions in Signorini's perturbation method. Arch. Rational Mech. Anal. 70 (1979) 261-288.
5. Truesdell, C.; Noll, W. : The non-linear field theories of mechanics. In PH III/3, S. Flügge ed. · Berlin · New York : Springer 1965.
6. Lions, J. L.; Magenes, E. : Problèmes aux limites non homogènes et applications, Vol. 1. Paris : Dunod 1968.
7. Fichera, G. : Linear elliptic differential systems and eigenvalue problems. Lecture Notes in Mathematics, 8 · Berlin · Heidelberg · New York : Springer 1965.
8. Lions, J. L. : Equations differentielles operationnelles et problèmes aux limites · Berlin · Göttingen · Heidelberg : Springer 1961.
9. Beatty, N. F. : Stability of hyperelastic bodies subject to hydrostatic loads. Int. J. Non-Linear Mech. 5 (1970) 367-383.
10. Spector, S. J. : On uniqueness in finite elasticity with general loading. J. Elasticity 10 (1980) 145-161.

Crossover Effect in Reflected Waves at the Interface between an Elastic Bar and a Viscoelastic Half-Space

N. Nakagawa *

Alexander von Humboldt Researchfellow
Institute of Mechanics, University of Hannover, Hannover, FRG

R. Kawai

Dept. of Mechanical Engineering, Kobe University, Kobe, Japan

O. Mahrenholtz

Institute of Mechanics, University of Hannover, Hannover, FRG

Summary

The stress waves reflected at the interface between an elastic bar and a viscoelastic half-space are studied. First the case of a small circular area of the surface of radius r_0, over which a uniform stress is acting, is considered and the response for the displacement is obtained. Next, by using this response function, the reflection characteristics for stress waves at the interface are considered. As a numerical example, a 3-parameter model is chosen for the viscoelastic material and the amplitudes and phases of the reflected waves are calculated. It is noted that a "crossover effect" is obtained.

1. Introduction

The dispersion and reflection of waves present very important problems. Kolsky and Lee [1] have treated the reflections of stress pulses at the interface between viscoelastic and elastic rods. Boucher and Kolsky [2] considered the reflection of pulses at the interface between an elastic rod and an elastic half-space, and investigated the crossover effect. Moffett [3] studied experimentally the crossover effect by using a viscoelastic rod. When incident waves are reflected at the interface of two viscoelastic rods and when the rod in which the incident wave travels has a lower viscosity than that in which the transmitted wave travels, then incident waves of higher frequency than some critical value are reflected in-phase. On the other hand incident waves with frequency lower than this critical value are reflected 180^0 out-of-phase. This abrupt change is called the "crossover effect".

---

* On leave from Department of Mechanical Engineering, Kobe University, Kobe, Japan.

In the present paper the behavior of stress waves at the interface between an elastic bar and a viscoelastic half-space is investigated. As a numerical example, a 3-parameter model is considered and the crossover effect in the viscoelastic media is studied. Reflected waves in an elastic-viscoelastic bar system are also studied.

2. Analysis

2.1 Equation

First the axisymmetric problem of a viscoelastic half-space, on which a uniform stress is loaded, is considered (shown in Fig.1). The equations of motion are

$$\left.\begin{aligned} \rho\ddot{u}_r &= \sigma_{rr,r} + \frac{1}{r}\sigma_{r\theta,\theta} + \sigma_{rz,z} + \frac{1}{r}(\sigma_{rr} - \sigma_{\theta\theta}), \\ \rho\ddot{u}_z &= \sigma_{zr,r} + \frac{1}{r}\sigma_{z\theta,\theta} + \sigma_{zz,z} + \frac{1}{r}\sigma_{zr}. \end{aligned}\right\} \quad (1)$$

A comma preceding a suffix denotes differentiation with respect to the corresponding coordinate and a dot denotes differentiation with respect to time. The constitutive equation of a linear isotropic viscoelastic medium is given by [6]

$$\sigma_{ij} = 2\mu \otimes \dot{\varepsilon}_{ij} + \lambda \otimes \dot{\varepsilon}_{kk}\delta_{ij}, \quad (2)$$

where ε_{ij}, σ_{ij} and δ_{ij} are strain tensor, stress tensor and Kronecker δ, respectively. For convenience, we use the notation

$$f(t) \otimes g(t) = \int_{-\infty}^{t} f(t-\tau)\, g(\tau)\, d\tau. \quad (3)$$

A repeated suffix denotes summation over the range of the suffix. When Lame constants λ and μ have the same relaxation function $Y_o(t)$, Poisson's ratio $\nu=\lambda/2(\lambda+\mu)$ becomes constant and we obtain

$$\sigma_{ij} = Y_o(t) \otimes (\, 2\mu_o\dot{\varepsilon}_{ij} + \lambda_o\dot{\varepsilon}_{kk}\delta_{ij}), \quad (4)$$

where

$$\lambda(t) = \lambda_o Y_o(t), \quad \mu(t) = \mu_o Y_o(t).$$

In the cylindrical coordinate system (r,θ,z), Eq.(4) may be written as

$$\left.\begin{aligned} \sigma_{rr} &= Y_o(t) \otimes [\, 2\mu_o\dot{\varepsilon}_{rr} + \lambda_o(\dot{\varepsilon}_{rr} + \dot{\varepsilon}_{\theta\theta} + \dot{\varepsilon}_{zz}\,)\,], \text{ etc.} \\ \sigma_{r\theta} &= Y_o(t) \otimes 2\mu_o\dot{\varepsilon}_{r\theta}, \text{ etc.} \end{aligned}\right\} \quad (5)$$

Introducing the expressions for the strain components in the cylindrical coordinate system and assuming rotational symmetry and zero displacement in the θ-direction, we obtain

$$\left.\begin{aligned}
\sigma_{rr} &= Y_o(t) \otimes [\ 2\mu_o \dot{u}_{r,r} + \lambda_o \dot{\Delta}\ ],\\
\sigma_{\theta\theta} &= Y_o(t) \otimes [\ 2\mu_o \frac{1}{r} \dot{u}_r + \lambda_o \dot{\Delta}\ ],\\
\sigma_{zz} &= Y_o(t) \otimes [\ 2\mu_o \dot{u}_{z,z} + \lambda_o \dot{\Delta}\ ],\\
\sigma_{rz} &= \mu_o Y(t) \otimes (\ \dot{u}_{z,r} + \dot{u}_{r,z}\ ),\\
\sigma_{r\theta} &= \sigma_{\theta z} = 0,
\end{aligned}\right\} \quad (6)$$

where

$$\Delta = \frac{1}{r}(\ ru_r)_{,r} + u_{z,z}. \quad (7)$$

Substitution from Eq.(6) in the equations of motion (1) yields

$$\left.\begin{aligned}
\rho \ddot{u}_r &= Y_o(t) \otimes [\ (\lambda_o + 2\mu_o)\dot{\Delta}_{,r} + 2\mu_o \dot{W}_{,z}\ ],\\
\rho \ddot{u}_z &= Y_o(t) \otimes [\ (\lambda_o + 2\mu_o)\dot{\Delta}_{,z} - \frac{2\mu o}{r}(r\dot{W})_{,r}\ ],
\end{aligned}\right\} \quad (8)$$

where

$$W = \frac{1}{2}(\ u_{r,z} - u_{z,r}\ ). \quad (9)$$

Δ and W are the dilatation and rotation, respectively. From Eqs. (7), (8) and (9) we obtain

$$\rho \ddot{\Delta} = (\ \lambda + 2\mu\ ) \otimes [\ \frac{1}{r}(r\dot{\Delta}_{,r})_{,r} + \dot{\Delta}_{,zz}\ ], \quad (10)$$

$$\rho \ddot{W} = \mu \otimes [\ \{ \frac{1}{r}(r\dot{W})_{,r} \}_{,r} + \dot{W}_{,zz}\ ]. \quad (11)$$

Equations (10) and (11) are the governing equations for the dilatation and rotation waves, respectively.
We introduce the nondimensional quantity

$$\left.\begin{aligned}
&R = r/r_o, \qquad Z = z/r_o, \qquad U_i = u_i/r_o,\\
&\Sigma_{ij} = \sigma_{ij}/\mu_o, \qquad T_1 = c_1 t/r_o, \qquad T_2 = c_2 t/r_o,
\end{aligned}\right\} \quad (12)$$

where

$$c_1 = \sqrt{(\lambda_o + 2\mu_o)/\rho}, \qquad c_2 = \sqrt{\mu_o/\rho}. \quad (13)$$

The Fourier transformation of f(T) is given by

$$f^{f}(\Omega)= \int_{-\infty}^{\infty} f(T)\ \exp(-j\Omega T)\ dT. \tag{14}$$

The Hankel transform of f(R) is defined by

$$f_{(n)}(s)= \int_{-\infty}^{\infty} f(R)\ R\ J_n(sR)\ dR, \tag{15}$$

where J_n is the Bessel funktion of the first kind, of oder n. Using the nondimensional quantity and applying the Fourier and the Hankel transform to Eqs.(4), (8), (10) and (11), we obtain

$$\left.\begin{aligned} k^2\Omega_1^2G^2\Sigma^f_{zz(o)} &= 4s\,W^f_{(1),z} - k^2(2s^2-k^2\Omega_1^2G)\Delta^f_{(o)} \quad , \\ k^2\Omega_1^2G^2\Sigma^f_{Rz(1)} &= 2k^2s\Delta^f_{(o),z} - 2(2s^2-k^2\Omega_1^2G)\ W^f_{(1)} \quad , \end{aligned}\right\} \tag{16}$$

$$\left.\begin{aligned} k^2\Omega_1^2G\ U^f_{R(1)} &= k^2s\Delta^f_{(o)} - 2\ W^f_{(1),z} \ , \\ k^2\Omega_1^2G\ U^f_{R(1)} &= -k^2\Delta^f_{(o),z} + 2s\ W^f_{(1)} \ , \end{aligned}\right\} \tag{17}$$

and

$$\left.\begin{aligned} \Delta^f_{(o),zz} &= (s^2-\Omega_1^2\ G)\ \Delta^f_{(o)} \quad , \\ W^f_{(1),zz} &= (s^2-\ k^2\Omega_1^2\ G)\ W^f_{(1)} \quad , \end{aligned}\right\} \tag{18}$$

where G is the nondimensional complex compliance and

$$Y= \frac{1}{G} = j\Omega_1\ Y^f_o(\Omega_1) \ , \tag{19}$$

$$\Omega_i= \omega\ r_o/\ c_i \qquad (\ i= 1,\ 2\ ), \tag{20}$$

$$k= c_1/c_2 = \sqrt{(\lambda_o+ 2\mu_o)/\mu_o} = \sqrt{2(1-\nu)/(1-2\nu)}.$$

The boundary conditions at Z=∞ are

$$\Delta = W = 0 \qquad (\ Z= \infty\ ). \tag{21}$$

We consider first the case of a small circular area of the surface of radius 1, over which a uniform normal stress p(T) is acting. The boundary conditions at Z=0 are then

$$\left.\begin{aligned} \Sigma_{RZ} &= 0, \\ \Sigma_{ZZ} &= \begin{cases} p(T_1) & (\ 0 \leq R \leq 1\ ) \\ 0 & (\ 1 < R\ ). \end{cases} \end{aligned}\right\} \tag{22}$$

The solutions of Eq.(18) are then

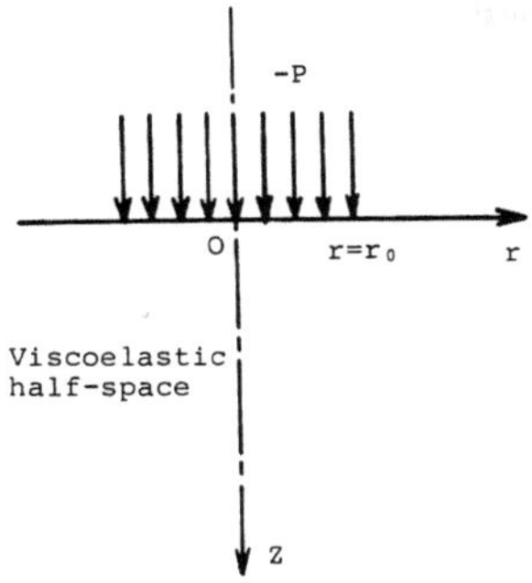

Fig.1. Viscoelastic half-space subjected to uniform stress

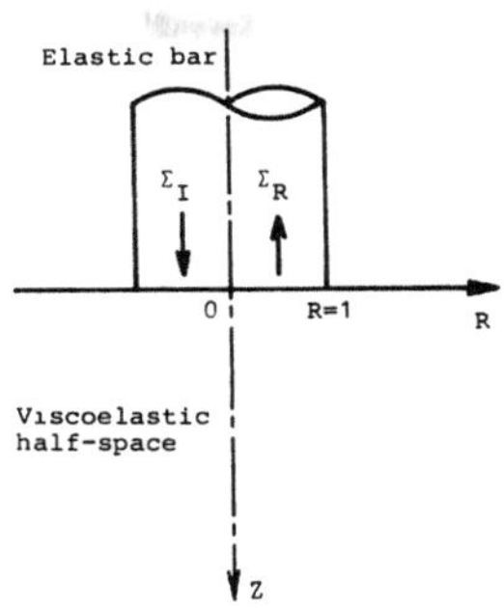

Fig.2. Reflection of stress wave at the interface

$$\left.\begin{aligned} \Delta^f_{(o)} &= B_1 \exp(-\sqrt{s^2 - \Omega_1^2 G}\ \ Z), \\ W^f_{(1)} &= B_2 \exp(-\sqrt{s^2 - k^2\Omega_1^2 G}\ \ Z). \end{aligned}\right\} \quad (23)$$

Introducing Eq.(23) into Eq.(16), we obtain the constants B_1 and B_2. From Eq.(17) we obtain the displacement at Z=O as

$$U_Z^f(O) = p^f k^2 G \int_0^\infty \frac{\sqrt{s^2-G}}{F(s)} J_1(\Omega_1 s)\ J_o(\Omega_1 sR)\ ds, \quad (24)$$

where

$$F(s) = (2s^2-k^2G)^2 - 4s^2\sqrt{s^2-G}\ \sqrt{s^2-k^2G}\ .$$

Introducing the mean displacement D in the area over which the stress is applied, Eq.(24) may be rewritten as

$$\frac{D^f}{p^f} = (\ 2k^2G^2/\Omega_1\ ) \int_0^\infty \frac{\sqrt{s^2 - G}}{sF(s)} [\ J_1(\Omega_1 s)\ ]^2 ds, \quad (25)$$

where

$$D = \frac{1}{\pi} \int_0^1 U_Z(O)\ 2\pi R\ dR.$$

2.2 Reflection Characteristics of Stress Waves at the Surface of a viscoelastic Half-Space

The reflection characteristics for stress waves at the surface between an elastic rod and a viscoelastic half-space, shown in Fig.2, are considered. It is assumed that the incident and reflected waves, which are denoted by Σ_I and Σ_R, in an elastic rod behave as one-dimensional wave motions.

The boundary conditins are

$$D = U_I + U_R \quad \text{(continuity of displacement)}, \tag{26}$$

$$p = \Sigma_I + \Sigma_R = -\frac{Ee\ c_1}{\mu_0\ c_e}(\dot{U}_I - \dot{U}_R) \quad \text{(equilibrium of the force)}, \tag{27}$$

where

$$c_e = \sqrt{E_e/\rho_e}.$$

The suffixes I and R denote the incident and reflected waves, respectively. Fourier transformation of Eqs.(26) and (27) yields

$$D^f = U_I^f + U_R^f, \tag{28}$$

$$p^f = -j\Omega_1\sqrt{k^2\ (3k^2-4)/(k^2-1)}\ \eta\ (U_I^f - U_R^f). \tag{29}$$

The coefficient of reflection is defined by

$$R_c = \frac{\Sigma_R^f}{\Sigma_I^f} = -\frac{\dot{U}_R^f}{\dot{U}_I^f} = \frac{\eta - \alpha}{\eta + \alpha}, \tag{30}$$

where

$$\eta = \frac{\rho\ c_0}{\rho_e c_e}, \tag{31}$$

$$\alpha = -j\Omega_1\sqrt{k^2(3k^2-4)/(k^2-1)}\ \frac{D^f}{p^f}. \tag{32}$$

η and α are called the first impedance ratio and the second impedance ratio, respectively.
When Ω_1 becomes infinite, Eq.(30) becomes

$$R_c|_{\Omega_1=\infty} = \frac{\eta - \alpha_\infty}{\eta + \alpha_\infty}, \tag{33}$$

$$\alpha_\infty = \sqrt{(3k^2-4)/(k^2-1)k^2}\ .$$

Eq.(33) corresponds to the coefficient of reflection to the elastic bar which has the second impedance ratio α_∞.† When Ω_1 becomes zero and the incident wave is a step wave, we obtain the reflected wave Σ_R in the elastic bar.

$$\Sigma_R = \frac{1}{2\pi}\int_{-\infty}^{\infty} \frac{1}{j\Omega_1}\,|R_c|\;\exp(j\theta)\,\exp(j\Omega_1 T)\,d\Omega_1 \qquad (34)$$

where

$$\theta = \arg R_c.$$

3. Numerical Results and Discussions

In order to investigate the influence of viscosity, a 3-parameter model of the type shown in Fig. 3 is considered. It is assumed that the incident wave is given by a Heaviside step function.

The stress distributions in the viscoelastic bar for the elastic-viscoelastic bar system with equal cross-sectional areas are given in Figs. 4 - 6. These were calculated by both a finite difference method and the Laplace transform method [7]. The results for $\eta = 0.3$ and $m = 5$ are shown in Fig. 4. The influences of the first impedance ratio η on the stress wave in the viscoelastic bar are indicated in Figs. 5 and 6 for $\eta = 0.3$ and 0.5. When the first impedance ratio η decreases, the stress wave becomes small.

The characteristics of stress waves in the elastic rod-viscoelastic half-space system are shown in Figs. 7-10. The influences of the first impedance ratio and viscosity on the reflected

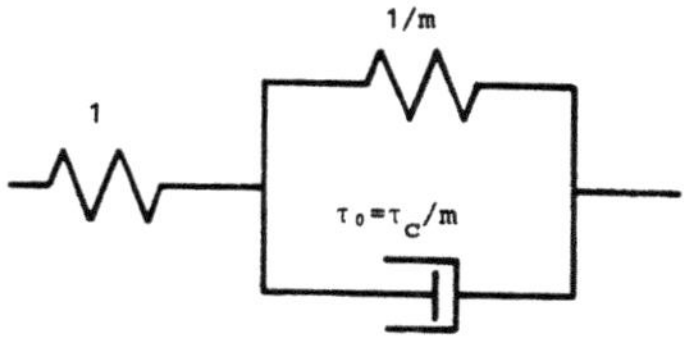

Fig.3. 3-parameter model

† In the case of the elastic bars which have cross-sectional areas A_1, A_2 and mechanical impedances $\rho_1 c_1$, $\rho_2 c_2$,

$$\eta = \rho_2 c_2/\rho_1 c_1, \qquad \alpha = A_1/A_2.$$

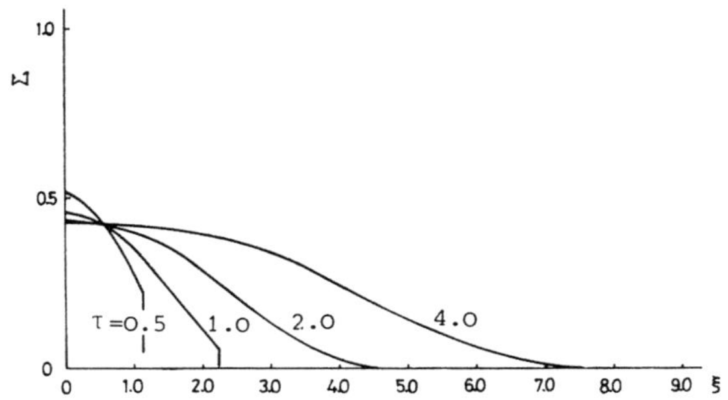

Fig.4. Stress distribution in the viscoelastic bar(η=0.3,m=5), $\tau=t/\tau_0$, $\xi=\sqrt{\rho m}x/\tau_0$

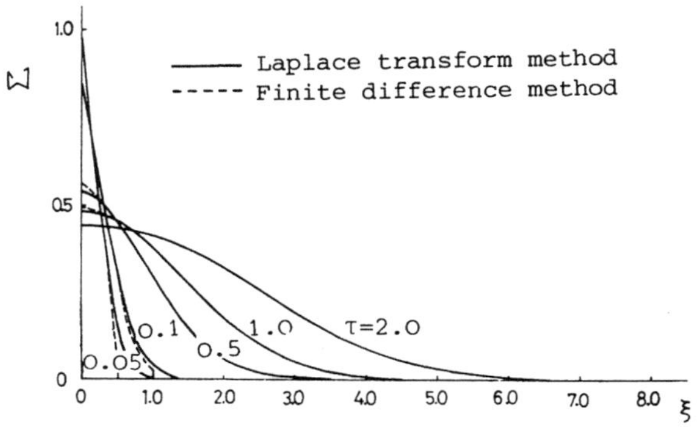

Fig.5. Stress distribution in the viscoelastic bar(η=0.3,m=100), $\tau=t/\tau_0$, $\xi=\sqrt{\rho m}x/\tau_0$

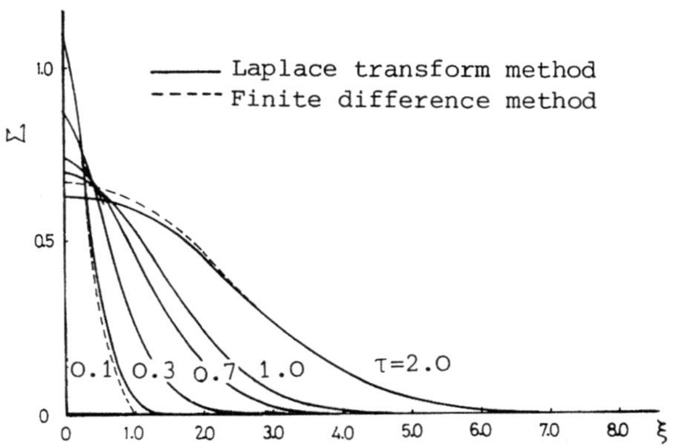

Fig.6. Stress distribution in the viscoelastic

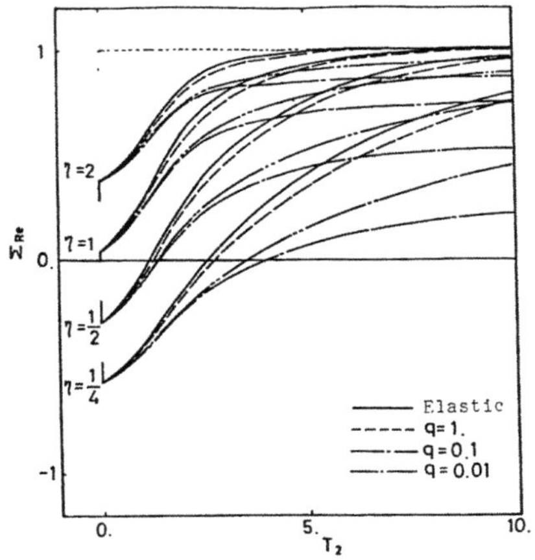

Fig.7. Variation of reflected stress waves (ν=0.25, $mr_0/c_1\tau_C$ = 0.1)

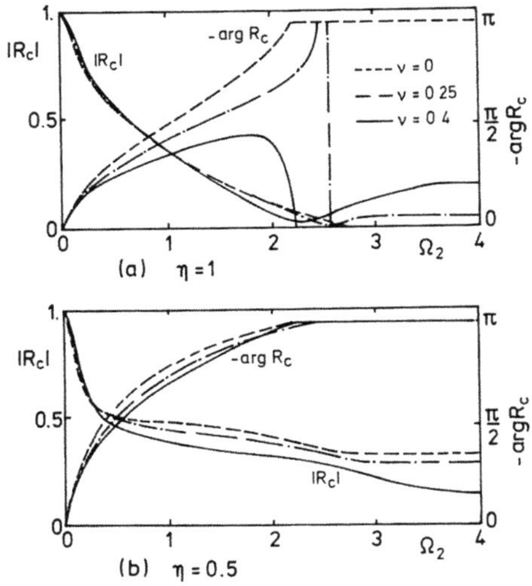

Fig.8. Frequency characteristics of reflection ($r_0/c_1\tau_C$=0.1, m=1)

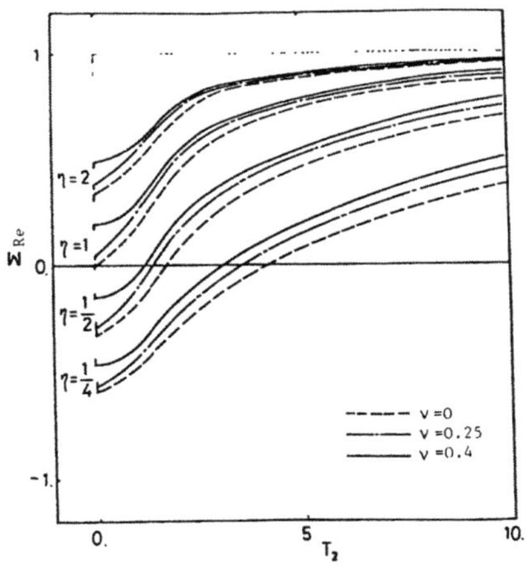

Fig.9.Variation of stress waves ($r_0/c_1\tau_C$=0.1, m=1)

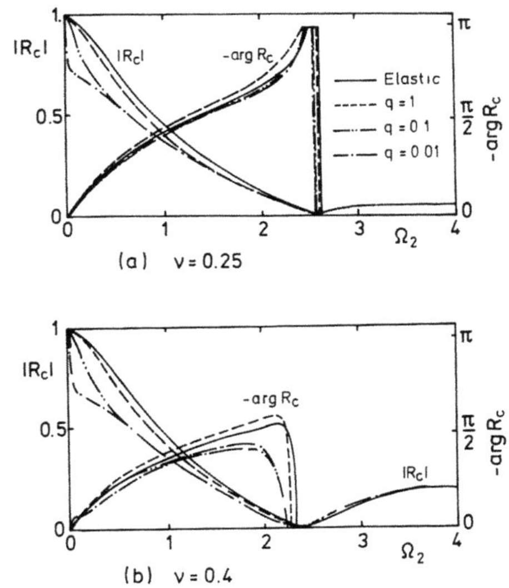

Fig.10. Frequency characteristics of the coefficient of reflection $R_C(\eta=1,\ mr_0/c_1\tau_C=0.1)$

stress wave in the elastic bar are shown in Fig. 7. It is seen that q decreases as the viscosity increases.

The coefficients of reflection R_C and the phase ($-\arg R_C$) are shown in Fig. 8 as function of Ω_2 for different values of

Poisson's ratio ν and the first impedance ratio η . The cross-over effect appears in Fig. 8(a) for $\eta = 1$ and $\nu = 0.25$.
A critical frequency Ω_c exists, such that there is an abrupt change in the pahse of amount π , for sinusoidal wave trains reflected at the interface, as Ω_2 passes through Ω_c . When two elastic bars of equal cross-sections are in contact, the coefficients of reflection become zero for $\eta = 1$, in the one-dimensional analysis.
Stress waves in the incident elastic bar are shown in Fig. 9. The reflected stress waves are initially negative when the mechanical impedance ratio η is small and positive for larger values of η . The reflection coefficients of the stress wave for $\eta = 1$ are illustrated in Fig. 10 for various values of Poisson's ratio ν and viscosity q ($= r_o / c_1 \tau_c$). It is seen that the critical frequency Ω_c depends on the viscosity.

References

1. Kolsky, H.; Lee, S.S.: The propagation and reflection of stress pulses in linear viscoelastic media. Brown Univ. Tech. Rept. No. 5 (1962).

2. Boucher, S.; Kolsky, H.: Reflection of pulses at the interface between an elastic rod and an elastic half-space. J. Acoust. Soc. Amer. 52 (1972) 884 - 898.

3. Moffett, B.: Experimental demonstration of a viscoelastic "crossover" effect. J. Acoust. Soc. Amer. 53 (1973) 1749 - 1751.

4. Nakagawa, N.; Kawai, R.; Akao, M.: Behavior of impact waves at the interface of an elastic-viscoelastic bar. Bull. Japan Soc. Mech. Eng. 20 (1977) 1402 - 1408.

5. Nakagawa, N.; Kawai, R.; Sasaki, T.: Stress Analysis of viscoelastic bars subjected to impulsive loading. VDI-Berichte 313 (1978) 525 - 530.

6. Valanis, K.C.: Exact and variational solutions to a general viscoelastic-kinetic problem. J. Appl. Mech. 33 (1966) 888 - 892.

7. Nakagawa, N.; Kawai, R.: Transient wave propagation in viscoelastic bar. Memoirs of Faculty of Eng., Kobe University 26 (1980) 1 - 17.

The Stability Equations of the Consistent Nonlinear Elastic Shell Theory with Moderate Rotations

H. STUMPF

Institut für Mechanik II
Ruhr-Universität Bochum, W. Germany

Summary

Using energy considerations the governing equations in Lagrangean description are derived for the consistent nonlinear elastic shell theory with moderate rotations. To determine the critical bifurcation or snap-through buckling loads the associated full set of stability equations including boundary conditions is given, which is a set of homogeneous linear partial differential equations with nonlinear coefficients. These coefficients depend on the membrane forces and rotations of the nonlinear prebuckling deformations. The symmetry of the stability equations is proved by using some methods of functional analysis. If the prebuckling deformations are rotation-free, the stability equations can be simplified considerably. To obtain approximate solutions the deformations in the neighbourhood of a critical state are linearized leading to a quadratic eigenvalue problem.

Nomenclature

$\mathbf{a}_\alpha (\alpha = 1,2)$	=	surface base vectors
$\mathbf{n}$	=	unit normal to the surface
$a_{\alpha\beta}$	=	surface metric tensor
$b_{\alpha\beta}$	=	surface curvature tensor
$\epsilon_{\alpha\beta}$	=	alternating tensor
u_α and u^α, resp.,	=	covariant and contravariant, resp., displacement components tangent to the middle surface
u_3	=	displacement component normal to the middle surface
$H^{\alpha\beta\lambda\mu}$	=	modified tensor of elastic moduli
dS	=	element of surface area
ds	=	element of arc length
$\mathbf{t} = t^\alpha \mathbf{a}_\alpha$	=	unit vector tangent to boundary
$\nu = \nu^\alpha \mathbf{a}_\alpha$	=	unit vector normal to boundary
$M^{\nu\nu} = M^{\alpha\beta}\nu_\alpha \nu_\beta$, $M^{t\nu} = M^{\alpha\beta} t_\alpha \nu_\beta$	=	components of stress couple vector at boundary
$F_j = M^{t\nu}(s_j + 0) - M^{t\nu}(s_j - 0)$	=	concentrated force at corner points $s_j (j = 1,\dots)$ of the boundary

J_p = total potential energy

$D^{(n)}J_p$ = n-th Gâteaux differential of the total potential energy

Introduction

In the theory of quasi-shallow shells the stability equations form a set of homogeneous linear partial differential equations with nonlinear coefficients, which depend on the prebuckling deformations. These coefficients are defined by the solution of the nonlinear governing equations for the prebuckling fundamental equilibrium state. The determination of the critical buckling load can be simplified considerably, if the prebuckling deformations are rotation-free or if the rotations are negligibly small. In this case the nonlinear stability problem is reduced to a linear eigenvalue problem. In the literature the assumption of rotation-free prebuckling deformations is commonly used also for shell deformations, which are nonlinear from the outset. For cylindrical shells it is shown in [1 - 3] that the influence of the prebuckling rotations lead to a considerable reduction of the critical buckling loads, compared with the results of a linear stability analysis.

In the excellent monograph [4] the stability equations for nonlinear prebuckling deformations are given for quasi-shallow shells according to the Donnell-Vlassow theory, whereas the associated boundary conditions are not considered. Using energy considerations the full set of stability equations including boundary conditions is derived in [5] for the nonlinear shell theory with moderate rotations according to Sanders ("moderately small rotations") [6] and Koiter ("small finite deflections") [7]. As is shown in [8] this theory is a special case of the general moderate rotation shell theory consistent within the first approximation of the shell strain energy function.

In this paper the nonlinear shell equations, defining the prebuckling state, together with the associated full set of stability equations are given for the consistent moderate rotation shell theory. The nonlinear coefficients of the stability equations depend on the membrane stresses and rotations of the nonlinear prebuckling state. It is proved that even for rotation-free prebuckling deformations the stability equations lead to a nonlinear eigenvalue problem, which is reduced to a linear eigenvalue problem for the special cases of the simplified moderate rotation shell theory of Sanders and Koiter or for the quasi-shallow shell theory of Donnell-Mushtari-Vlassow.

In a forthcoming paper [9] the stability equations will be given for the first approximation shell theory with unrestricted rotations. It is shown that the stability equations depend nonlinearly not only on the prebuckling membrane forces and rotations, but also on the prebuckling couple stresses.

Nonlinear Governing Shell Equations

In this paper we are dealing with shells of constant thickness h with $(h/R) \ll 1$, where R is the smallest principal radius of curvature of the undeformed shell middle surface. We assume that the shell is elastic, homogeneous, isotropic and that the shell deformation is such that $(h/L)^2 \ll 1$, where L is the smallest wave length of deformation pattern of the shell middle surface. Furthermore it is presumed that the strains are small everywhere in the shell with $\eta \ll 1$, where η is the largest eigenvalue of the Green strain tensor.

The deformation of the shell middle surface can be specified by the middle surface strain tensor $\gamma_{\alpha\beta}$ and the change of curvature tensor $\kappa_{\alpha\beta}$:

$$\gamma_{\alpha\beta} = \frac{1}{2}(\bar{a}_{\alpha\beta} - a_{\alpha\beta}) \quad ; \quad \kappa_{\alpha\beta} = -(\bar{b}_{\alpha\beta} - b_{\alpha\beta}) \; , \tag{1}$$

where quantities indicated by a bar are referred to the deformed shell middle surface.

For the consistent first approximation shell theory the strain energy per undeformed unit area of the shell middle surface is given by the following quadratic form:

$$f = \frac{1}{2} H^{\alpha\beta\lambda\mu}(\gamma_{\alpha\beta}\gamma_{\lambda\mu} + \frac{h^2}{12}\kappa_{\alpha\beta}\kappa_{\lambda\mu}) \qquad (2)_1$$

with the modified tensor of elastic moduli

$$H^{\alpha\beta\lambda\mu} = \frac{Eh}{2(1+\nu)}(a^{\alpha\lambda}a^{\beta\mu} + a^{\alpha\mu}a^{\beta\lambda} + \frac{2\nu}{1-\nu}a^{\alpha\beta}a^{\lambda\mu}) \; , \qquad (2)_2$$

where E is Young's modulus and ν Poisson's ratio.

Using the following linearized strain and rotation components

$$\begin{aligned} \theta_{\alpha\beta} &= \frac{1}{2}(u_{\alpha|\beta} + u_{\beta|\alpha}) - b_{\alpha\beta}u_3 & \omega_{\alpha\beta} &= \frac{1}{2}(u_{\beta|\alpha} - u_{\alpha|\beta}) = \epsilon_{\alpha\beta}\varphi_3 \\ \varphi_\alpha &= u_{3,\alpha} + b^\lambda_\alpha u_\lambda & \varphi_3 &= \frac{1}{2}\epsilon^{\alpha\beta}u_{\beta|\alpha} \end{aligned} \tag{3}$$

the strain measures $\gamma_{\alpha\beta}$ and $\kappa_{\alpha\beta}$ for the first approximation shell theory with moderate rotations are given by the following expressions [8]:

$$\begin{aligned} \gamma_{\alpha\beta} &= \theta_{\alpha\beta} + \frac{1}{2}\varphi_\alpha\varphi_\beta + \frac{1}{2}a_{\alpha\beta}\varphi_3^2 - \frac{1}{2}(\theta_\alpha^\lambda\omega_{\lambda\beta} + \theta_\beta^\lambda\omega_{\lambda\alpha}) \\ \kappa_{\alpha\beta} &= \frac{1}{2}[\varphi_{\alpha|\beta} + \varphi_{\beta|\alpha} + b_\alpha^\lambda(\theta_{\lambda\beta} - \omega_{\lambda\beta}) + b_\beta^\lambda(\theta_{\lambda\alpha} - \omega_{\lambda\alpha})] \ . \end{aligned} \tag{4}$$

Assuming that distributed surface loads per unit area of the undeformed middle surface

$$p = p^\alpha a_\alpha + p_3 n \tag{5}$$

are given, that geometrical quantities are prescribed along the part C_1 of the boundary curve

$$u_\nu = u_\nu^* \quad ; \quad u_t = u_t^* \quad ; \quad u_3 = u_3^* \quad ; \quad \beta_\nu = \beta_\nu^* \tag{6}$$

with $\beta_\nu = -\varphi_\alpha \nu^\alpha$ and that statical quantities are prescribed along the part $C_2 = C \ominus C_1$

$$P^\nu = P^{\nu *} \quad ; \quad P^t = P^{t*} \quad ; \quad P_3 = P_3^* \quad ; \quad M^{\nu\nu} = M^{\nu\nu *} \tag*{(7)$_1$}$$

with the corner conditions at corner points $j = 1,\dots$ of C_2

$$F_j = F_j^* \ . \tag*{(7)$_2$}$$

All forces are assumed to be of dead-load type. Asterisks indicate given values on the boundary. If geometrical and statical quantities are prescribed on the same part of the boundary line they must be complementary to each other.

For all geometrically admissible deformations satisfying (6) the functional of total potential energy of the shell [10,11]

$$J_p(u) = \iint_S [f(\gamma_{\alpha\beta}(u), \kappa_{\alpha\beta}(u)) - p\cdot u]dS - \int_{C_2}(P^*\cdot u + M^{\nu\nu *}\beta_\nu)ds - \sum_{j=1}^{N} F_j^* u_{3j} \tag{8}$$

attains a stationary value for the solution $\bar{u}$. The vanishing of the first differential of (8) leads to the equilibrium equations

$$\left.\begin{aligned}&[\bar{N}^{\alpha\beta} - b^{\alpha}_{\lambda}\bar{M}^{\lambda\beta} + \frac{1}{2}\,\epsilon^{\beta\alpha}\bar{N}^{\lambda}_{\lambda}\bar{\varphi}_3 - \frac{1}{2}\,(\bar{N}^{\alpha}_{\lambda}\bar{\theta}^{\beta\lambda} - \bar{N}^{\beta}_{\lambda}\bar{\theta}^{\alpha\lambda}) - \frac{1}{2}\,(\bar{N}^{\alpha}_{\lambda}\bar{\omega}^{\beta\lambda} + \bar{N}^{\beta}_{\lambda}\bar{\omega}^{\alpha\lambda})]|_{\beta} -\\ &-b^{\alpha}_{\lambda}(\bar{M}^{\lambda\beta}|_{\beta} + \bar{N}^{\lambda\beta}\bar{\varphi}_{\beta}) + p^{\alpha} = 0\\ &(\bar{M}^{\alpha\beta}|_{\alpha} + \bar{N}^{\alpha\beta}\bar{\varphi}_{\alpha})|_{\beta} + b_{\alpha\beta}[\bar{N}^{\alpha\beta} - b^{\alpha}_{\lambda}\bar{M}^{\lambda\beta} - \frac{1}{2}\,(\bar{N}^{\alpha}_{\lambda}\bar{\omega}^{\beta\lambda} + \bar{N}^{\beta}_{\lambda}\bar{\omega}^{\alpha\lambda})] + p_3 = 0\end{aligned}\right\}\ \text{in}\ S \quad (9)$$

and to the static boundary conditions:

$$\left.\begin{aligned}&\nu_{\alpha}\nu_{\beta}[\bar{N}^{\alpha\beta} - b^{\alpha}_{\lambda}\bar{M}^{\lambda\beta} + \frac{1}{2}\,\epsilon^{\beta\alpha}\bar{N}^{\lambda}_{\lambda}\bar{\varphi}_3 - \frac{1}{2}\,(\bar{N}^{\alpha}_{\lambda}\bar{\theta}^{\beta\lambda} - \bar{N}^{\beta}_{\lambda}\bar{\theta}^{\alpha\lambda}) - \frac{1}{2}\,(\bar{N}^{\alpha}_{\lambda}\bar{\omega}^{\beta\lambda} + \bar{N}^{\beta}_{\lambda}\bar{\omega}^{\alpha\lambda})]-\\ &-b_{\nu t}\bar{M}^{t\nu} - P^{\nu *} = 0\\ &t_{\alpha}\nu_{\beta}[\bar{N}^{\alpha\beta} - b^{\alpha}_{\lambda}\bar{M}^{\lambda\beta} + \frac{1}{2}\,\epsilon^{\beta\alpha}\bar{N}^{\lambda}_{\lambda}\bar{\varphi}_3 - \frac{1}{2}\,(\bar{N}^{\alpha}_{\lambda}\bar{\theta}^{\beta\lambda} - \bar{N}^{\beta}_{\lambda}\bar{\theta}^{\alpha\lambda}) - \frac{1}{2}\,(\bar{N}^{\alpha}_{\lambda}\bar{\omega}^{\beta\lambda} + \bar{N}^{\beta}_{\lambda}\bar{\omega}^{\alpha\lambda})]-\\ &-b_{tt}\bar{M}^{t\nu} - P^{t*} = 0\\ &\nu_{\beta}(\bar{M}^{\alpha\beta}|_{\alpha} + \bar{N}^{\alpha\beta}\bar{\varphi}_{\alpha}) + \frac{d}{ds}\,\bar{M}^{t\nu} - P^{3*} = 0\\ &\bar{M}^{\nu\nu} - M^{\nu\nu *} = 0\\ &\bar{M}^{t\nu}(s_j + 0) - \bar{M}^{t\nu}(s_j - 0) - F^{*}_{j} = 0 \quad ; \quad j = 1,\ldots\end{aligned}\right\}\ \text{on}\ C_2 \quad (10)$$

Using the constitutive relations

$$\bar{N}^{\alpha\beta}(\bar{u}) = H^{\alpha\beta\lambda\mu}\bar{\gamma}_{\lambda\mu}(\bar{u}) \qquad \bar{M}^{\alpha\beta}(\bar{u}) = \frac{h^2}{12}\,H^{\alpha\beta\lambda\mu}\bar{\kappa}_{\lambda\mu}(\bar{u}) \tag{11}$$

the field quantities of (9) and (10) are considered as functions of the middle surface displacement field $\bar{u} = \bar{u}^{\alpha}a_{\alpha} + \bar{u}_3 n$. Together with the geometrical boundary conditions (6) the equations (10) and (11) form a set of inhomogeneous nonlinear partial differential equations.

In a load-displacement plot the equilibrium configurations form a fundamental equilibrium path, where relative minima and relative maxima and also branchings of secondary equilibrium paths can exist. The relative maxima correspond to limit loads, at which loss of stability and snap-through buckling of the structure occur. At a bifurcation or branching point the nonlinear shell equations have multiple solutions, one corresponding to each branch of the equilibrium path. A central aspect of the nonlinear shell analysis is the determination of the critical loads, at which loss of stability and loss of uniqueness, respectively, occur.

Consistent Stability Equations

To derive the stability equations in an appropriate Lagrangean description all field quantities of shell deformations will be referred to the known undeformed shell middle surface. With $\bar{u}$ we denote the displacement field of the shell middle surface from the undeformed reference configuration to the fundamental equilibrium configuration, while $\hat{u}$ is the displacement field from the fundamental configuration to an adjacent configuration:

$$u = \bar{u} + \hat{u} \tag{12}$$

yielding the linearized middle surface strain and rotation measures

$$\theta_{\alpha\beta} = \bar{\theta}_{\alpha\beta} + \hat{\theta}_{\alpha\beta} \quad ; \quad \varphi_i = \bar{\varphi}_i + \hat{\varphi}_i \quad ; \quad \omega_{\alpha\beta} = \bar{\omega}_{\alpha\beta} + \hat{\omega}_{\alpha\beta} \,, \tag{13$_1$}$$

with the definitions

$$\begin{aligned}
\bar{\theta}_{\alpha\beta} &= \frac{1}{2}(\bar{u}_{\alpha|\beta} + \bar{u}_{\beta|\alpha}) - b_{\alpha\beta}\bar{u}_3 & \hat{\theta}_{\alpha\beta} &= \frac{1}{2}(\hat{u}_{\alpha|\beta} + \hat{u}_{\beta|\alpha}) - b_{\alpha\beta}\hat{u}_3 \\
\bar{\varphi}_\alpha &= \bar{u}_{3,\alpha} + b^\lambda_\alpha \bar{u}_\lambda & \hat{\varphi}_\alpha &= \hat{u}_{3,\alpha} + b^\lambda_\alpha \hat{u}_\lambda \\
\bar{\varphi}_3 &= \frac{1}{2}\epsilon^{\alpha\beta}\bar{u}_{\beta|\alpha} & \hat{\varphi}_3 &= \frac{1}{2}\epsilon^{\alpha\beta}\hat{u}_{\beta|\alpha} \\
\bar{\omega}_{\alpha\beta} &= \epsilon_{\alpha\beta}\bar{\varphi}_3 & \hat{\omega}_{\alpha\beta} &= \epsilon_{\alpha\beta}\hat{\varphi}_3 \;.
\end{aligned} \tag{13$_2$}$$

Using (13) the strain measures $\gamma_{\alpha\beta}$ and $\kappa_{\alpha\beta}$ are obtained as:

$$\begin{aligned}
\gamma_{\alpha\beta} &= \bar{\gamma}_{\alpha\beta} + \frac{1}{2}(\bar{\varphi}_\alpha\hat{\varphi}_\beta + \hat{\varphi}_\alpha\bar{\varphi}_\beta) + a_{\alpha\beta}\bar{\varphi}_3\hat{\varphi}_3 - \frac{1}{2}(\bar{\theta}^\lambda_\alpha\hat{\omega}_{\lambda\beta} + \hat{\theta}^\lambda_\alpha\bar{\omega}_{\lambda\beta} + \bar{\theta}^\lambda_\beta\hat{\omega}_{\lambda\alpha} + \hat{\theta}^\lambda_\beta\bar{\omega}_{\lambda\alpha}) + \hat{\gamma}_{\alpha\beta} \\
\kappa_{\alpha\beta} &= \bar{\kappa}_{\alpha\beta} + \hat{\kappa}_{\alpha\beta} \,,
\end{aligned} \tag{13$_3$}$$

where $\bar{\gamma}_{\alpha\beta}$, $\hat{\gamma}_{\alpha\beta}$ and $\bar{\kappa}_{\alpha\beta}$, $\hat{\kappa}_{\alpha\beta}$, respectively, are given by (4) with corresponding indications.

Following the procedure outlined in [5] the increment in potential energy

$\Delta J_p(\bar{u};\hat{u})$ can be written in the following form

$$J_p(\bar{u};\hat{u}) = J_p(\bar{u}+\hat{u}) - J_p(\bar{u})$$

$$= \frac{1}{2} D^{(2)}J_p(\bar{u};\hat{u}) + \frac{1}{3!} D^{(3)}J_p(\bar{u};\hat{u}) + \frac{1}{4!} D^{(4)}J_p(\bar{u};\hat{u}) \tag{14}$$

where $D^{(n)}J_p(\bar{u};\hat{u})$ is the n-th Gâteaux differential of the total potential energy. The first Gâteaux differential $D^{(1)}J_p(\bar{u};\hat{u})$ vanishes, if $\bar{u}$ is assumed to represent a fundamental equilibrium configuration.

The most important contribution to the increment (14) is the second differential $D^{(2)}J_p(\bar{u};\hat{u})$, which we can split into two parts

$$D^{(2)}J_p(\bar{u};\hat{u}) = J_A(\hat{u},\hat{u}) - J_C(\bar{u};\hat{u},\hat{u}) \quad , \tag{15}$$

where the quadratic functional $J_A(\hat{u},\bar{u})$

$$J_A(\hat{u},\hat{u}) = \iint_S H^{\alpha\beta\lambda\mu}(\hat{\theta}_{\alpha\beta}\hat{\theta}_{\lambda\mu} + \frac{h^2}{12}\hat{\kappa}_{\alpha\beta}\hat{\kappa}_{\lambda\mu})dS \tag{16}$$

is independent of the deformation $\bar{u}$ and represents the double elastic strain energy for a small deformation $\hat{u}$. The functional $J_C(\bar{u};\hat{u},\hat{u})$

$$\begin{aligned} J_C(\bar{u};\hat{u},\hat{u}) = -\iint_S \{ & H^{\alpha\beta\lambda\mu}[2\hat{\theta}_{\alpha\beta}(\bar{\varphi}_\lambda\hat{\varphi}_\mu + a_{\lambda\mu}\bar{\varphi}_3\hat{\varphi}_3 - \bar{\theta}^\nu_\lambda\hat{\omega}_{\nu\mu} - \hat{\theta}^\nu_\lambda\bar{\omega}_{\nu\mu}) \\ & +(\bar{\varphi}_\alpha\hat{\varphi}_\beta + a_{\alpha\beta}\bar{\varphi}_3\hat{\varphi}_3 - \bar{\theta}^\nu_\alpha\hat{\omega}_{\nu\beta} - \hat{\theta}^\nu_\alpha\bar{\omega}_{\nu\beta})(\bar{\varphi}_\lambda\hat{\varphi}_\mu + a_{\lambda\mu}\bar{\varphi}_3\hat{\varphi}_3 - \bar{\theta}^\kappa_\lambda\hat{\omega}_{\kappa\mu} - \\ & - \hat{\theta}^\kappa_\lambda\bar{\omega}_{\kappa\mu})] + \bar{N}^{\alpha\beta}(\hat{\varphi}_\alpha\hat{\varphi}_\beta + a_{\alpha\beta}\hat{\varphi}_3^2 - 2\hat{\theta}^\nu_\alpha\hat{\omega}_{\nu\beta})\}dS \end{aligned} \tag{17}$$

depends on membrane stresses $\bar{N}^{\alpha\beta}$ and rotations $\bar{\varphi}_i$ of the fundamental state.

Functional (16) is defined on the set of elements $u \in \mathcal{D}_{J_A}$ satisfying sufficient differentiability conditions to assure the existence of (16) and satisfying the homogeneous geometric boundary conditions

$$u_\nu = u_t = u_3 = \beta_\nu = 0 \qquad \text{on } C_1 \; . \tag{18}$$

On the other hand functional (17) is defined on the set of elements $u \in \mathcal{D}_{J_C}$ sufficiently differentiable and satisfying the homogeneous geometric boundary conditions

$$u_\nu = u_t = u_3 = 0 \qquad \text{on } C_1 . \tag{19}$$

Comparing the domains $\mathcal{D}_{J_A}$ and $\mathcal{D}_{J_C}$ the inclusion

$$\mathcal{D}_{J_A} \subset \mathcal{D}_{J_C} \tag{20}$$

follows.

To derive the stability equations in the form of homogeneous partial differential equations we introduce the scalar product

$$(u,v) = \iint_S u \cdot v \, dS = \iint_S (u^\alpha v_\alpha + u^3 v_3) dS \tag{21}$$

and transform the functionals (16) and (17) by partial integration, yielding

$$J_A(\hat{u},\hat{u}) = (A\hat{u},\hat{u}) \quad ; \quad J_C(\bar{u};\hat{u},\hat{u}) = (C(\bar{u})\hat{u},\hat{u}) \quad , \tag{22}$$

where A is the operator of the linear shell theory defined by

$$Au = - \left\{ \begin{array}{l} [n^{\alpha\beta}(u) - b^\alpha_\lambda M^{\lambda\beta}(u)]|_\beta - b^\alpha_\lambda M^{\lambda\beta}(u)|_\beta \\ \\ M^{\alpha\beta}(u)|_{\alpha\beta} + b_{\alpha\beta}[n^{\alpha\beta}(u) - b^\alpha_\lambda M^{\lambda\beta}(u)] \end{array} \right\} \quad \text{on } S \tag{23}$$

on the set $\mathcal{D}_A$ of sufficiently differentiable vector-valued functions satisfying (18) on C_1 and the homogeneous static boundary conditions

$$\left. \begin{array}{l} \nu_\alpha \nu_\beta [n^{\alpha\beta}(u) - b^\alpha_\lambda M^{\lambda\beta}(u)] - b_{\nu t} M^{t\nu}(u) = 0 \\ \\ t_\alpha \nu_\beta [n^{\alpha\beta}(u) - b^\alpha_\lambda M^{\lambda\beta}(u)] - b_{tt} M^{t\nu}(u) = 0 \\ \\ \nu_\beta M^{\alpha\beta}(u)|_\alpha + \frac{d}{ds} M^{t\nu}(u) = 0 \\ \\ M^{\nu\nu}(u) = 0 \\ \\ F_j(u) = 0 \quad ; \quad j = 1,\dots \end{array} \right\} \quad \text{on } C_2 \tag{24}$$

with

$$n^{\alpha\beta}(u) = H^{\alpha\beta\lambda\mu}\theta_{\lambda\mu}(u) \quad . \tag{25}$$

The operator $C(\bar{u})$ depends nonlinearly on the fundamental displacement field $\bar{u}$ and is defined by

$$C(\bar{u})u = \left\{ \begin{array}{l} \frac{1}{2}\,[\phi^{\alpha\beta}(\bar{u},u) + \phi^{\beta\alpha}(\bar{u},u) + \epsilon^{\beta\alpha}T(\bar{u},u)]|_{\beta} - b^{\alpha}_{\lambda}\psi^{\lambda}(\bar{u},u) \\ \\ \psi^{\beta}(\bar{u},u)|_{\beta} + b_{\alpha\beta}\phi^{\alpha\beta}(\bar{u},u) \end{array} \right\} \tag{26}$$

on the set $\mathcal{D}_C$ of sufficiently differentiable vector-valued displacement fields satisfying (19) on C_1 and the homogeneous static boundary conditions

$$\left. \begin{array}{l} \nu_{\alpha}\nu_{\beta}\phi^{\alpha\beta}(\bar{u},u) = 0 \\ \\ t_{\alpha}\nu_{\beta}[\phi^{\alpha\beta}(\bar{u},u) + \phi^{\beta\alpha}(\bar{u},u) + \epsilon^{\beta\alpha}T(\bar{u},u)] = 0 \\ \\ \nu_{\beta}\psi^{\beta}(\bar{u},u) = 0 \end{array} \right\} \text{ on } C_2 \tag{27}$$

In (26) and (27) we introduced the shortenings:

$$\begin{aligned}
\Lambda^{\alpha\beta}(\bar{u},u) &= H^{\alpha\beta\lambda\mu}[\bar{\varphi}_{\lambda}(\bar{u})\varphi_{\mu}(u) + a_{\lambda\mu}\bar{\varphi}_3(\bar{u})\varphi_3(u) - \bar{\theta}^{\nu}_{\lambda}(\bar{u})\omega_{\nu\mu}(u) - \theta^{\nu}_{\lambda}(u)\bar{\omega}_{\nu\mu}(\bar{u})] \\
\phi^{\alpha\beta}(\bar{u},u) &= \Lambda^{\alpha\beta}(\bar{u},u) - [n^{\alpha\lambda}(u) + \Lambda^{\alpha\lambda}(\bar{u},u)]\bar{\omega}_{\lambda\cdot}{}^{\beta}(\bar{u}) - \bar{N}^{\alpha\lambda}(\bar{u})\omega_{\lambda\cdot}{}^{\beta}(u) \\
\psi^{\beta}(\bar{u},u) &= [n^{\alpha\beta}(u) + \Lambda^{\alpha\beta}(\bar{u},u)]\bar{\varphi}_{\alpha}(\bar{u}) - \bar{N}^{\alpha\beta}(\bar{u})\varphi_{\alpha}(u) \\
T(\bar{u},u) &= [n^{\lambda}_{\lambda}(u) + \Lambda^{\lambda}_{\lambda}(\bar{u},u)]\bar{\varphi}_3(\bar{u}) + \bar{N}^{\lambda}_{\lambda}(\bar{u})\varphi_3(u) + \epsilon_{\lambda\beta}[n^{\alpha\beta}(u)\bar{\theta}^{\lambda}_{\alpha}(\bar{u}) + \\
&\quad + \Lambda^{\alpha\beta}(\bar{u},u)\bar{\theta}^{\lambda}_{\alpha}(\bar{u}) + \bar{N}^{\alpha\beta}(\bar{u})\theta^{\lambda}_{\alpha}(u)] \; .
\end{aligned} \tag{28}$$

It can be shown that the operators A and $C(\bar{u})$ are symmetric:

$$(A\hat{u},\hat{\hat{u}}) = (\hat{u},A\hat{\hat{u}}) \quad \forall \hat{u},\hat{\hat{u}} \in \mathcal{D}_A \quad ; \quad (C(\bar{u})\hat{u},\hat{\hat{u}}) = (\hat{u},C(\bar{u})\hat{\hat{u}}) \quad \forall \hat{u},\hat{\hat{u}} \in \mathcal{D}_C \tag{29}$$

According to the energy criterion of stability the critical deformations are defined by the non-trivial solutions of a vanishing second energy differential. With

$$
\begin{aligned}
D^{(2)}J_p(\bar{u};\hat{u}) &= J_A(\hat{u},\hat{u}) - J_C(\bar{u};\hat{u},\hat{u}) && \hat{u} \in \mathcal{D}_{J_A} \\
D^{(2)}J_p(\bar{u};\hat{u}) &= (A\hat{u} - C(\bar{u})\hat{u},\hat{u}) && \hat{u} \in \mathcal{D}_A
\end{aligned}
\tag{30}
$$

we obtain the stability equations of the moderate rotation shell theory in the two alternative forms

$$
\begin{aligned}
&J_A(\hat{u},\hat{u}) - J_C(\bar{u};\hat{u},\hat{u}) = 0 && \hat{u} \in \mathcal{D}_{J_A} \quad ; \quad \hat{u} \neq 0 \\
&A\hat{u} - C(\bar{u})\hat{u} = 0 && \hat{u} \in \mathcal{D}_A \quad ; \quad \hat{u} \neq 0 \;,
\end{aligned}
\tag{31}
$$

where the functionals J_A and J_C are defined by (16) and (17) and the operators A and $C(\bar{u})$ by (23) - (28).

Stability Equations for Rotation-Free Prebuckling Deformations

For many stability problems it can be assumed that buckling occurs from rotation-free fundamental equilibrium configurations. In this case the stability equations can be given again in the form (31) with the same quadratic form J_A according to (16) and the same linear operator A according to (23)-(24), but the functional J_C and the operator $C(\bar{u})$ can be simplified essentially by omitting all terms depending on the prebuckling rotations $\bar{\varphi}_i$. From (17) we obtain

$$
\begin{aligned}
J_C(\bar{u};\hat{u},\hat{u}) = - \iint_S \{ & H^{\alpha\beta\lambda\mu}(-2\hat{\theta}_{\alpha\beta}\bar{\theta}^{\nu}_{\lambda}\hat{\omega}_{\nu\mu} + \bar{\theta}^{\nu}_{\alpha}\hat{\omega}_{\nu\beta}\bar{\theta}^{\kappa}_{\lambda}\hat{\omega}_{\kappa\mu}) + \\
& + \bar{n}^{\alpha\beta}(\hat{\varphi}_{\alpha}\hat{\varphi}_{\beta} + a_{\alpha\beta}\hat{\varphi}_3^2 - 2\hat{\theta}^{\nu}_{\alpha}\hat{\omega}_{\nu\beta})\}dS \;.
\end{aligned}
\tag{32}
$$

The operator $C(\bar{u})$ according to (26) - (27) yields for $\bar{\varphi}_i = 0$:

$$
C(\bar{u})u = \left\{
\begin{aligned}
& \frac{1}{2}\,[-H^{\alpha\beta\lambda\mu}\bar{\theta}^{\nu}_{\lambda}\omega_{\nu\mu} - \frac{1}{2}\,(\bar{n}^{\alpha\lambda}\omega_{\lambda.}{}^{\beta} + \bar{n}^{\beta\lambda}\omega_{\lambda.}{}^{\alpha}) + \epsilon^{\beta\alpha}\bar{n}^{\lambda}_{\lambda}\varphi_3 + \epsilon^{\beta\alpha}\epsilon_{\rho\kappa}(n^{\lambda\rho}\bar{\theta}^{\kappa}_{\lambda} - \\
& \qquad - H^{\pi\rho\lambda\mu}\bar{\theta}^{\nu}_{\lambda}\omega_{\nu\mu}\bar{\theta}^{\kappa}_{\pi} + \bar{n}^{\lambda\rho}\theta^{\kappa}_{\lambda})]|_{\beta} - b^{\alpha}_{\beta}\bar{n}^{\lambda\beta}\varphi_{\lambda} \\
& (\bar{n}^{\alpha\beta}\varphi_{\alpha})|_{\beta} - b_{\alpha\beta}(H^{\alpha\beta\lambda\mu}\bar{\theta}^{\nu}_{\lambda}\omega_{\nu\mu} + \bar{n}^{\alpha\lambda}\omega_{\lambda.}{}^{\beta})
\end{aligned}
\right\}
\tag{33}
$$

defined on the set $\mathcal{D}_C$ of sufficiently differentiable vector-valued functions satisfying the homogeneous geometric boundary conditions (19) and, corresponding to (27), the following static boundary conditions on C_2:

$$\nu_\alpha \nu_\beta [-H^{\alpha\beta\lambda\mu}\bar{\theta}^\nu_\lambda \omega_{\nu\mu} - \frac{1}{2}(\bar{n}^{\alpha\lambda}\omega_{\lambda\cdot}^{\ \beta} + \bar{n}^{\beta\lambda}\omega_{\lambda\cdot}^{\ \beta}) + \epsilon^{\beta\alpha}\bar{n}^\lambda_\lambda\varphi_3 + \epsilon^{\beta\alpha}\epsilon_{\rho\kappa}(n^{\lambda\rho}\bar{\theta}^\kappa_\lambda - H^{\pi\rho\lambda\mu}\bar{\theta}^\nu_\lambda\omega_{\nu\mu}\bar{\theta}^\kappa_\pi + \bar{n}^{\lambda\rho}\theta^\kappa_\lambda)] = 0$$

$$t_\alpha \nu_\beta [-H^{\alpha\beta\lambda\mu}\bar{\theta}^\nu_\lambda \omega_{\nu\mu} - \frac{1}{2}(\bar{n}^{\alpha\lambda}\omega_{\lambda\cdot}^{\ \beta} + \bar{n}^{\beta\lambda}\omega_{\lambda\cdot}^{\ \alpha}) + \epsilon^{\beta\alpha}\bar{n}^\lambda_\lambda\varphi_3 + \epsilon^{\beta\alpha}\epsilon_{\rho\kappa}(n^{\lambda\rho}\bar{\theta}^\kappa_\lambda - H^{\pi\rho\lambda\mu}\bar{\theta}^\nu_\lambda\omega_{\nu\mu}\bar{\theta}^\kappa_\pi + \bar{n}^{\lambda\rho}\theta^\kappa_\lambda)] = 0$$

$$\nu_\beta \bar{n}^{\alpha\beta}\varphi_\alpha = 0 \ . \tag{34}$$

With (32) - (34) it follows that even for rotation-free prebuckling deformations the stability equations (31) depend nonlinearly on the membrane stresses and rotations of the fundamental state.

Linearization near a Critical State

To solve the stability equations (31) depending nonlinearly on the prebuckling deformations the following approximating procedure can be used. If an equilibrium configuration $\bar{u}$ in the neighbourhood of a supposed critical state is calculated or estimated, the deformations in this neighbourhood can be linearized, which allows to introduce a load factor λ such that the linearized deformations can be represented by $\lambda\bar{u}$. Then the stability equations (31) can be represented by a quadratic eigenvalue problem of the form

$$J_A(\hat{u},\hat{u}) - \lambda J_{C_1}(\bar{u};\hat{u},\hat{u}) - \lambda^2 J_{C_2}(\bar{u};\hat{u},\hat{u}) = 0 \qquad \forall \hat{u} \in \mathcal{D}_{J_A} \ ; \ \hat{u} \neq 0$$

$$A\hat{u} - \lambda C_1(\bar{u})\hat{u} - \lambda^2 C_2(\bar{u})\hat{u} = 0 \qquad \forall \hat{u} \in \mathcal{D}_A \ ; \ \hat{u} \neq 0 \ . \tag{35}$$

The eigenvalues λ of (35) are the critical load factors, at which nonlinear bifurcation buckling or limit load buckling occurs. Because of lack of space the functionals J_{C_1}, J_{C_2} and the operators $C_1(\bar{u})$, $C_2(\bar{u})$ cannot be given here explicitly.

References

1. Almroth, B.O.: Influence of edge conditions on the stability of axially compressed cylindrical shells. AIAA Journ. 4/1 (1966) 134-140.

2. Yamaki, N.: Influence of prebuckling deformations on the buckling of circular cylindrical shells under external pressure. AIAA Journ. 7/4 (1969), 753-755.

3. Brush, D.O.: Prebuckling rotations and cylinder buckling analysis. J. Eng. Mech. Div., Proc. ASCE 106/2 (1980) 225-232.

4. Brush, D.O. and Almroth, B.O.: Buckling of bars, plates and shells, McGraw-Hill, New York 1975.

5. Stumpf, H.: On the linear and nonlinear stability analysis in the theory of thin elastic shells. XV th Int. Congr. Theor. Appl. Mech., Toronto 1980, Ing. Arch. 1981.

6. Sanders, J.L.: Nonlinear theories of thin shells. Quart. Appl. Math. 21 (1963) 21-36.

7. Koiter, W.T.: On the nonlinear theory of thin shells. Proc. Kon. Ned. Akad. Wet, Ser. B., 69 (1966) 1-54.

8. Pietraszkiewicz, W.: Introduction to the nonlinear theory of shells. Mitt. Inst. Mech. 10, Ruhr-Univ. Bochum 1977.

9. Nolte, L.-P.: Nonlinear stability analysis in the theory of thin elastic shells with unrestricted rotations, to appear: Mitt. Inst. Mech., Ruhr-Univ. Bochum.

10. Stumpf, H.: The derivation of dual extremum and complementary stationary principles in geometrical non-linear shell theory. Ing. Arch. 48 (1979) 221-237.

11. Schmidt, R. and Pietraszkiewicz, W.: Variational principles in the geometrically non-linear theory of shells undergoing moderate rotations. Ing. Arch. 50 (1981) 187-201.

12. Labisch, F.K.: Duality, variational principles and error bounds for the case of von Kārmān plate. IV Symp. Trends in Appl. of Pure Math. to Mech., Bratislava 1981, to appear in: Proc. Pitman Publ., London - San Franzisko - Melbourne 1982.

The Concept of Reduced-Membrane-Buckling Determining Bounds of Buckling Loads and Imperfection-Sensitivity of Shells

U. WITTEK

Institut für Konstruktiven Ingenieurbau
Lehrstuhl III
Ruhr-Universität Bochum, Germany

Summary

A complete buckling and post-buckling theory of shells is derived. In view of the dominating significance of the linear membrane energy passing from the fundamental to the adjacent state, linear membrane reduction factors are introduced. By means of a reduced-membrane-buckling concept all dominating buckling phenomenons may be explained and imperfection-sensitivity, knockdown-factors and worst imperfection parameters may be computed by a linear reduced bifurcation analysis.

1. Nonlinear basic shell equations

To derive the basic ideas of the concept of reduced-membrane-buckling I refer to the simplest discription of nonlinear shell equations of the type of Donnell-Marguerre. (For non shallow buckling modes see [1]).

The equations of equilibrium are

$$n^{(\alpha\beta)}|_{\alpha} + \qquad p^{\beta} = 0 \tag{1.1}$$

$$n^{(\alpha\beta)}[B_{\alpha\beta} + u_3|_{\alpha\beta} + u_{03}|_{\alpha\beta}] + m^{(\alpha\beta)}|_{\alpha\beta} + p^3 = 0 \,, \tag{1.2}$$

where $(\dots)/_{\alpha}$ stands for the covariant differentiation with respect to the coordinates Θ^{α} of the shell middle surface, $n^{(\alpha\beta)}$ the symmetric components of the membrane stress tensor, $m^{(\alpha\beta)}$ of the bending tensor, $B_{\alpha\beta}$ the curvature tensor, p^i the components of the load vector und u_i the components of the deformation vector. A geometric imperfection is discribed by the component u_{o3}.

The kinematic relations combine the first and second strain tensor $\alpha_{(\alpha\beta)}$ and $\omega_{(\alpha\beta)}$ with the displacements u_i of the shell middle surface

$$\alpha_{(\alpha\beta)} = \frac{1}{2}(u_\alpha|_\beta + u_\beta|_\alpha - 2u_3 B_{\alpha\beta} + u_3|_\alpha u_3|_\beta$$

$$+ u_{03}|_\alpha u_3|_\beta + u_3|_\alpha u_{03}|_\beta) \tag{1.3}$$

$$\omega_{(\alpha\beta)} = -u_3|_{\alpha\beta} . \tag{1.4}$$

The elastic constitutive equations are

$$n^{(\alpha\beta)} = D H^{\alpha\beta\varrho\lambda} \alpha_{(\varrho\alpha)} \tag{1.5}$$

$$m^{(\alpha\beta)} = B H^{\alpha\beta\varrho\lambda} \omega_{(\varrho\lambda)} , \tag{1.6}$$

with the elasticity tensor

$$H^{\alpha\beta\varrho\lambda} = \frac{1}{2}[A^{\alpha\varrho} A^{\beta\lambda} + A^{\alpha\lambda} A^{\beta\varrho} + \nu(\varepsilon^{\alpha\varrho}\varepsilon^{\beta\lambda} + \varepsilon^{\alpha\lambda}\varepsilon^{\beta\varrho})], \tag{1.7}$$

the membrane D and bending stiffness B

$$D = \frac{Eh}{1-\nu^2} \quad ; \quad B = \frac{Eh^3}{12(1-\nu^2)} . \tag{1.9}$$

In the above equations $A^{\alpha\beta}$ represent the contravariant components of the metric tensor, $\varepsilon^{\alpha\beta}$ the permutation tensor, E the elasticity modulus, ν the Poisson's ratio and h the wall thickness.

2. Reduced membrane bifurcation buckling

To derive the stability equations for neutral equilibrium of the perfect shell ($u_{o3} \equiv 0$) it is necessary to start with the nonlinear equilibrium equations (1.1) and (1.2) and the linear kinematic relations (1.3) and (1.4). The nonlinear equilibrium state is identified as an adjacent state arising from a first variation of the displacement field of the fundamental state. Thus the displacement field u_i and all other field variables are to split up into the two parts of the fundamental state $(\bar{\ldots})$ and the incremental infinitesimal adjacent state $(\overset{+}{\ldots})$:

$$\begin{bmatrix} u_i \\ \alpha_{(\alpha\beta)} \\ n^{(\alpha\beta)} \\ \omega_{(\alpha\beta)} \\ m^{(\alpha\beta)} \end{bmatrix} = \lambda \begin{bmatrix} \bar{u}_i \\ \bar{\alpha}_{(\alpha\beta)} \\ \bar{n}^{(\alpha\beta)} \\ \bar{\omega}_{(\alpha\beta)} \\ \bar{m}^{(\alpha\beta)} \end{bmatrix} + \varepsilon \begin{bmatrix} \overset{+}{u}_i \\ \overset{+}{\alpha}_{(\alpha\beta)red} \\ \overset{+}{n}^{(\alpha\beta)}_{red} \\ \overset{+}{\omega}_{(\alpha\beta)} \\ \overset{+}{m}^{(\alpha\beta)} \end{bmatrix} \tag{2.1}$$

where the fundamental state is assumed as a linear stress state depending on a normalized load factor λ. ε represents a perturbation parameter with the attitude $\lim\limits_{\varepsilon\to 0}\lambda = \lambda_{cr}$.

To derive the different buckling behavior of the mentioned parts of strain energies the incremental membrane effects are introduced in a reduced manner

$$\dot{\alpha}_{(\alpha\beta)red} = r\ \dot{\alpha}_{(\alpha\beta)} \tag{2.2}$$

$$\dot{n}^{(\alpha\beta)}{}_{red} = r\cdot\dot{n}^{(\alpha\beta)} = r\cdot D\cdot H^{\alpha\beta\varrho\lambda}\ \dot{\alpha}_{(\varrho\lambda)}\ , \tag{2.3}$$

where r represents a linear membrane factor

$$0 \leq r \leq 1\ . \tag{2.4}$$

Introduction of (2.1) in (1.1) and (1.2) delivers two sets of equations: the terms without the factor ε represent the equilibrium equations of the fundamental state, the remaining terms with the factor ε are the linearized stability equations for neutral equilibrium. If we assume that the fundamental state is a linear membrane stress state ($\bar{m}^{(\alpha\beta)} \equiv 0$), loading conditions are conservative ($\overset{+}{p}{}^{i} \equiv 0$) and the influence of the deformation of the fundamental state is omitted ($\bar{u}_i = 0$) the following membrane-reduced stability equations are obtained

$$r\cdot\dot{n}^{(\alpha\beta)}|_{\alpha} = 0 \tag{2.5}$$

$$\underbrace{r\cdot\dot{n}^{(\alpha\beta)}B_{\alpha\beta}}_{} + \underbrace{\dot{m}^{\alpha\beta}|_{\alpha\beta}}_{} + \underbrace{\lambda\cdot\bar{n}^{(\alpha\beta)}\dot{u}_3|_{\alpha\beta}}_{} = 0\ . \tag{2.6}$$

$$[\ \ r\cdot\underline{K}_{eM} + \underline{K}_{eB} + \lambda\cdot\underline{K}_g\ \ ]\ \dot{\underline{v}} = 0$$

Due to the curvature ($B_{\alpha\beta}$ in 2.6, 1.3) these differential equations are coupled in all three displacement components $\overset{+}{u}_i$. For the case of a flat plate ($B_{\alpha\beta} = 0$) there is no such coupling and the critical load factor λ_{cr} is then only determined by the incremental bending effects in (2.6). Thus we notice that in general for a curved shell - and especially depending on the curvature - there exists a stability load λ_{cr} which consists of two parts

$$\lambda_{cr} = \lambda_B + r\cdot\lambda_M\ , \tag{2.7}$$

where λ_B corresponds to the incremental linear bending and λ_M to the incremental linear membrane effects.

As to the membrane reduction factor r three different kinds of bifurcation exist

1. <u>r = 1 : classical bifurcation</u>

$$\lambda_{cr} = \lambda_B + \lambda_M \qquad \text{with}\ \ \dot{n}^{(\alpha\beta)}\ \text{and}\ \dot{m}^{(\alpha\beta)} \tag{2.8}$$

2. <u>r = 0 : inextensional (isometric) bifurcation</u>

$$\lambda_{cr} = \lambda_B \qquad \text{with}\ \ \dot{n}^{(\alpha\beta)} = \dot{\alpha}_{(\alpha\beta)} \equiv 0 \tag{2.9}$$

3. 0 < r < 1 : membrane-reduced-bifurcation

$$\lambda_{cr} = \lambda_B + r\cdot\lambda_M \qquad \text{with} \quad r\cdot \overset{+}{n}^{(\alpha\beta)} \text{ and } r\cdot\overset{+}{\alpha}_{(\alpha\beta)} \tag{2.10}$$

3. Reduced membrane postbuckling

To discuss the behavior of λ_B and λ_M after passing the point of bifurcation the concept of reduced membrane buckling is carried over to the initial postbuckling theory [2,4].

In addition to the series development (2.1) an additional displacement field $\overset{++}{u}_i$ which should be orthogonal to the buckling mode $\overset{+}{u}_i$ [3] must be taken into consideration. Beyond it the kinematic relations (1.3) of the perfect shell must be applied in their complete nonlinear form. For the introduced variables we then get

$$\begin{bmatrix} u_i \\ \alpha_{(\alpha\beta)} \\ n^{(\alpha\beta)} \\ \omega_{(\alpha\beta)} \\ m^{(\alpha\beta)} \end{bmatrix} = \lambda \begin{bmatrix} \bar{u}_i \\ \bar{\alpha}_{(\alpha\beta)} \\ \bar{n}^{(\alpha\beta)} \\ \bar{\omega}_{(\alpha\beta)} \\ \bar{m}^{(\alpha\beta)} \end{bmatrix} + \varepsilon \begin{bmatrix} \overset{+}{u}_i \\ \overset{+}{\alpha}_{(\alpha\beta)red} \\ \overset{+}{n}^{(\alpha\beta)}_{red} \\ \overset{+}{\omega}_{(\alpha\beta)} \\ \overset{+}{m}^{(\alpha\beta)} \end{bmatrix} + \varepsilon^2 \begin{bmatrix} \overset{++}{u}_i \\ \overset{++}{\alpha}_{(\alpha\beta)red} \\ \overset{++}{n}^{(\alpha\beta)}_{red} \\ \overset{++}{\omega}_{(\alpha\beta)} \\ \overset{++}{m}^{(\alpha\beta)} \end{bmatrix} + \dots \tag{3.1}$$

and for the development of the load factor itself

$$\lambda/\lambda_{cr}(r) = 1 + \lambda_1(r)\cdot\varepsilon + \lambda_2(r)\cdot\varepsilon^2 , \tag{3.2}$$

where the first row on the right hand side in (3.1) represents the known solution of the fundamental state, the second row the known solution of the bifurcation point, whereas the variables within the third row as well as the postbuckling parameters λ_2(r) and λ_1(r) must still be determined.

The membrane effects have again been introduced in a reduced manner where the separation between linear and nonlinear parts (1.3) now must be considered. For the point of bifurcation only linear effects are acting so that

$$\overset{+}{\alpha}_{(\alpha\beta)red} = r\cdot\overset{+}{\alpha}^L_{(\alpha\beta)} \tag{3.3}$$

$$\overset{+}{n}^{(\alpha\beta)}_{red} = r\cdot D\cdot H^{\alpha\beta\varrho\lambda}\, \overset{+}{\alpha}^L_{(\varrho\lambda)} . \tag{3.4}$$

If we introduce (3.1a) into the kinematic relation (1.3) for the perfect shell we find that on the level of the membrane strains there exist a coupling between the displacement fields $\overset{+}{u}_i$ and $\overset{++}{u}_i$. As only the linear membrane effects shall be attached with the linear reduction factor r we get

$$\overset{++}{\alpha}_{(\alpha\beta)red} = r\cdot\overset{++}{\alpha}^{L}_{(\alpha\beta)} + \overset{+}{\alpha}^{NL}_{(\alpha\beta)} \; . \tag{3.5}$$

The additional membrane strains $\overset{++}{\alpha}_{(\alpha\beta)}$ consist of the linear contribution of the additional displacement field $\overset{++}{u}_i$ which should be reduced

$$r\cdot\overset{++}{\alpha}^{L}_{(\alpha\beta)} = r\cdot\frac{1}{2}[\overset{++}{u}_{\alpha}|_{\beta} + \overset{++}{u}_{\beta}|_{\alpha} - 2\overset{++}{u}_3 B_{\alpha\beta}] \tag{3.6}$$

and the nonlinear contribution of the buckling displacement field $\overset{+}{u}_i$ which should not be reduced

$$\alpha^{NL}_{(\alpha\beta)} = \frac{1}{2} u_3|_{\alpha} u_3|_{\beta} \quad . \tag{3.7}$$

For the constitutive relation (1.5) we then find

$$\overset{++}{n}{}^{(\alpha\beta)}_{red} = r\cdot D H^{\alpha\beta\varrho\lambda}\overset{++}{\alpha}^{L}_{(\varrho\lambda)} + \frac{D}{2}H^{\alpha\beta\varrho\lambda}\overset{+}{u}_3|_{\varrho}\overset{+}{u}_3|_{\lambda} \; . \tag{3.8}$$

The starting point for the determination of $\lambda_1(r)$ and $\lambda_2(r)$ is the nonlinear principle of virtual work under the assumption of a linear membrane fundamental state. Applying the series development (3.1) to this principle, using the orthogonality conditions between the higher displacement fields $\overset{++}{u}_i$ with the buckling mode $\overset{+}{u}_i$ finally leads [1,4] to the destination of the reduced postbuckling coefficients

$$\lambda_1(r) = -\frac{3}{2\lambda_{cr}(r)}\,\frac{\iint\limits_F [\overset{+}{n}{}^{(\alpha\beta)}_{red}\,\overset{+}{u}_3|_{\alpha}\,\overset{+}{u}_3|_{\beta}]\,dF}{\iint\limits_F [\bar{n}^{(\alpha\beta)}\,\overset{+}{u}_3|_{\alpha}\,\overset{+}{u}_3|_{\beta}]\,dF} \tag{3.9}$$

$$\lambda_2(r) = -\frac{1}{\lambda_{cr}(r)}\,\frac{\iint\limits_F [2\overset{+}{n}{}^{(\alpha\beta)}_{red}\,\overset{++}{u}_3|_{\alpha}\,\overset{+}{u}_3|_{\beta} + \overset{++}{n}{}^{(\alpha\beta)}_{red}\,\overset{+}{u}_3|_{\alpha}\,\overset{+}{u}_3|_{\beta}]\,dF}{\iint\limits_F [\bar{n}^{(\alpha\beta)}\,\overset{+}{u}_3|_{\alpha}\,\overset{+}{u}_3|_{\beta}]\,dF} \; . \tag{3.10}$$

4. Significance of the linear membrane reduction for the point of bifurcation, the postbuckling behavior and the imperfection sensitivity

The dominating significance of the linear-membrane-reduction-concept to shell buckling phenomenons canbe summarized as follows (see Fig. 1 and 2):

1. In general the total amount of bifurcation load λ_{cr} (2.6) is the sum of a linear bending λ_B and a linear membrane λ_M contribution. Especially λ_M is strongly depending on the curvature of the shell and will always exist except for those boundary

conditions which allow an a-priori inextensional buckling pattern. By use of the reduction factor r reduced bifurcation loads $\lambda_{cr}(r)$ can be determined.

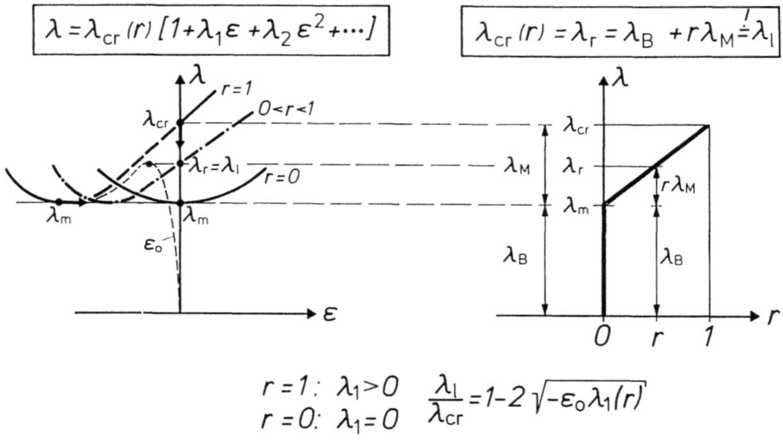

Fig. 1. Buckling concept with reduced membrane energy

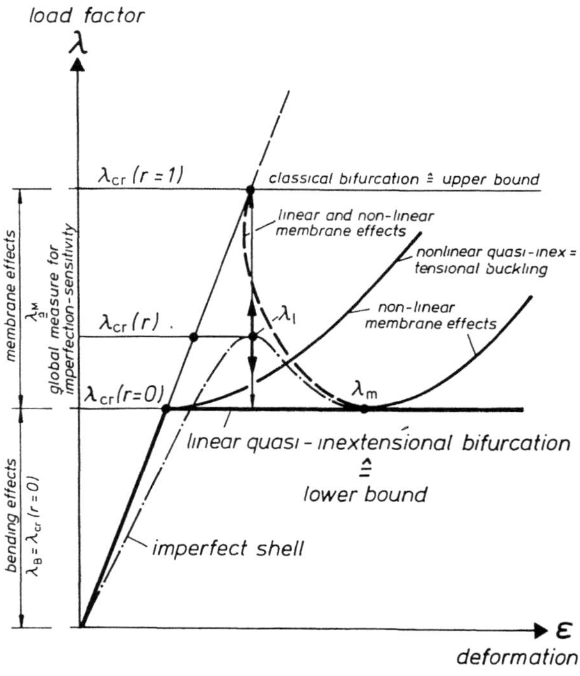

Fig. 2. Buckling mechanism of imperfection-sensitive shells and concept of reduced-membrane-buckling

2. In the asymmetric case of bifurcation ($\lambda_1 \neq 0$) the postbuckling area is approximated by $\lambda / \lambda_{cr}(r) = 1 + \lambda_1(r) \cdot \varepsilon$. As can be seen from (3.9) and Fig. 1 the reduction factor r of the linear membrane contribution markedly influences the initial slope of the postbuckling load deflection curve. If we reduce the reduction factor the critical load $\lambda_{cr}(r)$ and the initial slope $\lambda_1(r)$ are reduced too. Thus λ_M looses its stabilizing effect in the postbuckling range.

3. For the extreme case r = 0 (inextensional buckling) $\lambda_1(r)$ becomes zero and the asymmetric bifurcation changes into symmetric bifurcation with $\lambda/\lambda_{cr}(r) = 1 + \lambda_2(r) \cdot \varepsilon^2$. Thus the amount of linear membrane buckling λ_M decides about symmetric of asymmetric buckling behavior.

4. From (3.10) in connection with (3.8) we see that in the case of r = 0 λ_2 becomes positive due to the purely nonlinear membrane effects. Thus as can be seen from Fig. 2 the arising path of the load deflection curve depends only on the nonlinear membrane strains.

5. In an extreme estimation of the deep postbuckling range (Fig. 1 and 2) the deminishing process might be marked to such a degree that the whole stabilizing effect of λ_M will be deminished. In this case the minimum buckling load λ_m corresponds to the buckling load λ_B of the linear bending effects. Not until passing λ_m the purely nonlinear membrane strains will lead as well as for the nonlinear quasi-inextensional buckling theory to a higher load bearing capacity.

The above mentioned five topics characterized the buckling and postbuckling behavior of the perfect shell. In the case of geometric imperfections the limit load λ_l or the knockdown-factors λ_l/λ_{cr} are [3, 4]

$$\lambda_l/\lambda_{cr} \approx 1 - 2\sqrt{-\varepsilon_0 \lambda_1} \qquad (4.1)$$

for asymmetric buckling with $\varepsilon_0 \lambda_1 < 0$ and

$$\lambda_l/\lambda_{cr} \approx 1 - \frac{3}{2}\sqrt[3]{-2\varepsilon_0^2 \lambda_2} \qquad (4.2)$$

for symmetric buckling with $\lambda_1 = 0$ and $\lambda_2 < 0$. In (4.1, 4.2) the imperfection parameter ε_0 is the normalized amount of a geometric imperfection amplitude u_{o3} whose pattern is identical with the buckling mode $\overset{+}{u}_3$.

6. As can be seen from (3.9, 3.10) λ_1 and λ_2 depend strongly on the linear membrane contribution at the point of bifurcation. A comparison with (4.1, 4.2) yields that the knockdown-factor depends on ε_0 and λ_1 (λ_2) and thus on the amount of the reduction factor r.

7. The mechanical effect of the geometric imperfections consists in an a-priori reduction of the membrane bifurcation part λ_M of the perfect shell equivalent to the amplitude and pattern of the geometric imperfections. Thus the limit load λ_1 of the imperfect shell may be substituted by the membrane reduced bifurcation load of the perfect shell (see Fig. 1 and 2)

$$\lambda_l = \lambda_{cr}(r) = \lambda_B + r(\varepsilon_0)\cdot\lambda_M \,, \tag{4.3}$$

where the reduction factor r is now depending on the imperfection parameter ε_o.

8. The strong connection between linear membrane energy and geometric imperfections delivers a heuristic criterion for the most unfavourable imperfections. Those imperfections, as to their amplitude and their distribution, are worst, who will minimize the linear part of the imperfect extensional strain energy [6].

9. Geometric imperfections lead to limit loads (Fig. 2) which will scatter between λ_{cr} (no imperfections) and λ_B (worst imperfections) of the perfect shell.

Two bounds can be given

a) an upper bound is formed by the classical linear bifurcation load $\lambda_{cr}(r = 1)$ of the perfect shell,

b) a lower bound is formed by the quasi-inextensional bifurcation load $\lambda_{cr}(r = 0) = \lambda_B$, again for the perfect shell.

Thus the linear mambrane buckling contribution λ_M becomes a measurement for the global imperfection-sensitivity and forms the ultimate knockdown-factor.

10. In the case when the geometric imperfections are known, the reduced bifurcation load $\lambda_{cr}(r)$ corresponding (4.3) delivers a measurement for the specific imperfection-sensitivity. The correlation between the reduction factor r and the imperfection parameter ε_o can be achieved either by a limitation of the result of the initial postbuckling analysis by λ_B or by determining the worst geometric imperfection corresponding to the above mentioned topic 8. If a single trustful experimental buckling result with exactly known imperfection amplitude and pattern is known this also might serve together with λ_B as a correlation possibility.

5. Numerical aspects of the reduced eigenvalue-problem

The computation of reduced buckling loads $\lambda_{cr}(r)$ of arbitrary shells will be preferably based on the finite-element-method. In this case the energetic bifurcation criterion is used. If unreduced classical bifurcation (r = 1) is regarded this criterion leads to the eigenvalue problem

$$[\underline{K}_e + \lambda_{cr} \cdot \underline{K}_g] \cdot \overset{+}{\underline{v}}_{cr}(r=1) = 0 \tag{5.1}$$

with $\underline{K}_e$ the elastic stiffness matrix of the adjacent state, $\underline{K}_g$ the geometric stiffness matrix and $\overset{+}{\underline{V}}_{cr}$ the buckling mode in terms of the nodal displacements. When the linear membrane reduction is introduced a reduced eigenvalue problem must be solved with the specific attribute that the eigenfunctions $\overset{+}{\underline{V}}_{cr}$ (r = 1) of the unreduced solution are employed (2.6) (2.5)

$$[\underline{K}_{ered} + \lambda_{cr}(r) \cdot \underline{K}_g] \cdot \overset{+}{\underline{v}}_{cr}(r=1) = 0 \tag{5.2}$$

with

$$\underline{K}_{ered} = r \cdot \underline{K}_{eM} + \underline{K}_{eB} \ . \tag{5.3}$$

The solution of (5.2) may be avoided if the total amount of elastic strain energy $\overset{+}{\Pi}$ at the point of bifurcation is computated for the case of unreduced and reduced membrane effects

$$\overset{+}{\Pi} = \frac{1}{2} \lambda_{cr} \cdot \overset{+}{\underline{v}}{}_{cr}^T \underline{K}_g \overset{+}{\underline{v}}_{cr} + \frac{1}{2} \overset{+}{\underline{v}}{}_{cr}^T \underline{K}_e \overset{+}{\underline{v}}_{cr} \tag{5.4}$$

$$\overset{+}{\Pi} = \lambda_{cr} \cdot \overset{+}{\Pi}_g + \overset{+}{\Pi}_e \tag{5.5}$$

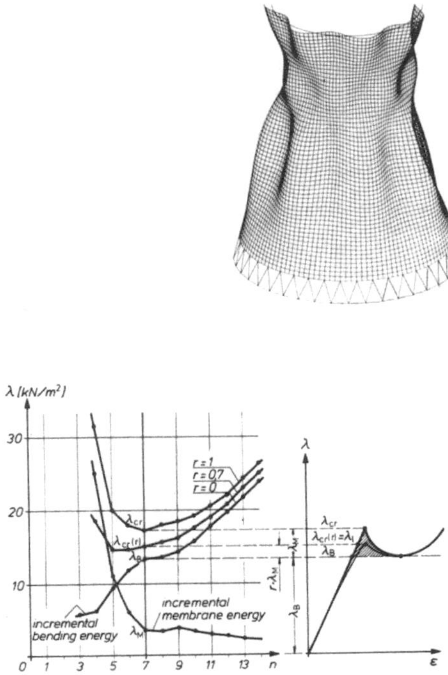

Fig. 3. Membrane-reduced-buckling for a hyperboloidal shell

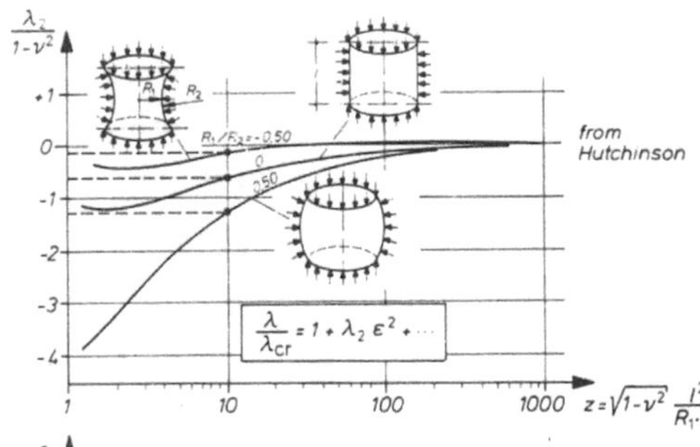

type	$\frac{R_1}{R_2}$	λ_{cr}	$\frac{\lambda_B}{\lambda_{cr}}$	λ_2	n	$\varepsilon_{o\,max}$
hyperboloid	- 0,50	$0{,}643118 \cdot 10^3$	0,928	- 0,14	11	0,0214
circular cylinder	0	$1{,}00289 \cdot 10^3$	0,816	- 0,62	13	0,0473
ellipsoid	+ 0,50	$1{,}77869 \cdot 10^3$	0,622	- 1,27	14	0,1276

$E = 3 \cdot 10^7\ MN/m^2$
$\nu = 0$
$l = 10{,}00\,m$
$R_1 = 31{,}60\,m$
$h = 0{,}316\,m$

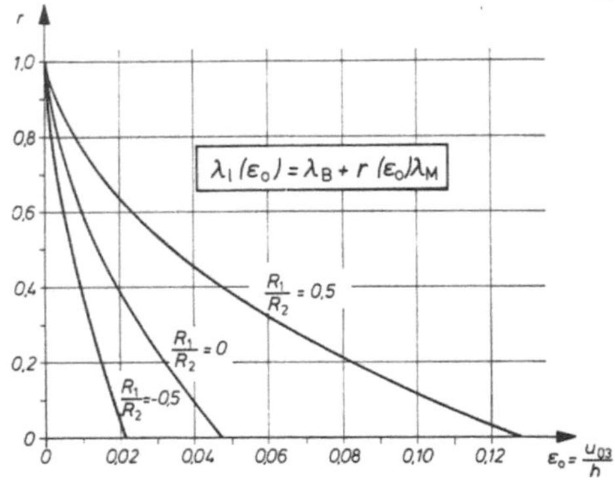

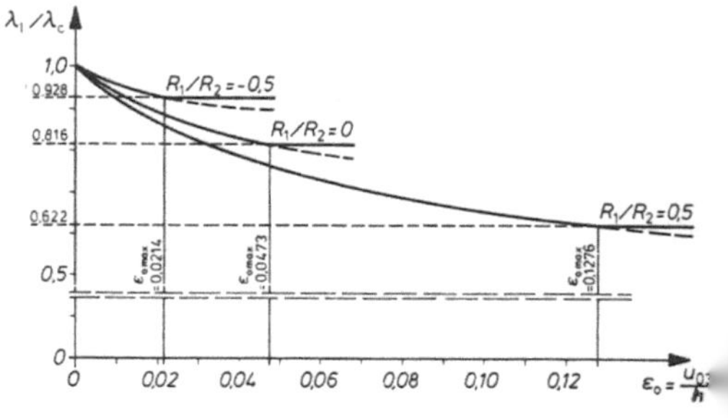

Fig. 4. Correlation between initial postbuckling and membrane-reduced-buckling

and

$$\dot{\Pi}_{red} = \frac{1}{2}\,\lambda_{cr}(r)\cdot\dot{\underline{v}}_{cr}^{T}\,\underline{K}_g\,\dot{\underline{v}}_{cr} + \frac{1}{2}\,\dot{\underline{v}}_{cr}^{T}\,\underline{K}_{e\,red}\,\dot{\underline{v}}_{cr} \tag{5.6}$$

$$\dot{\Pi}_{red} = \dot{\underline{v}}_{cr}(r)\cdot\dot{\Pi}_g + \dot{\Pi}_{e\,red}\ . \tag{5.7}$$

As easily can be seen the reduced critical bifurcation load will be determined by

$$\lambda_{cr}(r) = \lambda_{cr}\,\frac{\dot{\Pi}_{e\,red}}{\dot{\Pi}_e}\ . \tag{5.8}$$

6. Characteristic results

A hyperboloidal shell under lateral compression is regarded in Fig. 3. The classical bifurcation analysis with r = 1 yields a bifurcation load of λ_{cr} = 17,1 kN/m^2 and a pattern with n = 7 waves in the circumferential direction. The corresponding bifurcation loads of the reduced theory ensue form the ratio of the reduced to the complete strain energy of the buckled state. The significance of the reduced bifurcation concept for the post-buckling range and the imperfection sensitivity is plotted qualitatively on the right hand side of the picture. Geometric imperfections may in this case reduce the actual buckling load to 78 % of the classical bifurcation load.

The correlation between the reduction factor r and the imperfection parameter ε_o is shown on Fig. 4, where the buckling results of the initial postbuckling theory [5] are compared with the results of the reduced membrane concept. Due to the significance of the inextensional buckling load λ_B equation (4.2) may serve to determine the worst imperfection parameter $\varepsilon_{o\,max}$ with $\lambda_1 = \lambda_B$ and λ_2 from [5]. Using the imperfection law (4.2) and the substitution (4.3) the correlation $r(\varepsilon_o)$ is found and plotted on the lower part of Fig. 4. Thus limit loads of imperfect shells λ_1, knockdown-factor λ_B/λ_{cr} and worst geometric imperfection parameter $\varepsilon_{o\,max}$ may be computed by a linear procedure: the mambrane-reduced-buckling-concept.

References

1. Wittek, U.: Beitrag zum Tragverhalten der Strukturen bei endlichen Verformungen unter besonderer Beachtung des Nachbeulmechanismus dünner Flächentragwerke. Techn.-wiss. Mittlg. Nr. 80-1, Inst. f. Konstr. Ingenieurbau, Ruhr-Universität Bochum 1980.

2. Koiter, W.T.: Over de stabiliteit van het elastisch evenwicht.Dissertation Delft 1945, NASA TT F-10, 833,1967.

3. Koiter, W.T.: General equations of elastic stability for thin shells. Proceedings - Symposium on the theory of shells to honor Lloyd Hamilton Donnell University of Houston 1967.

4. Budiansky, B.: Post-buckling behavior of cylinders in torsion. IUTAM-Symposium Theory of Thin Shells, 1967. Springer Verlag Berlin Heidelberg New York, 1969.

5. Hutchinson, J.W.: Initial Post-Buckling Behavior of Toroidal Shell Segments. Int. J. Solids Struct., Vol. 3, 1967.

6. Wittek, U.: Kriterium für ungünstigste geometrische Imperfektionen der Schalenbeulung. ZAMM, Band 61, T 71-73, 1981.

7. Eckstein, U.; Krätzig, W.B.; Wittek, U.: Finite-Element-Berechnungen zur Grenztragfähigkeit der Rotationsschalen. Techn.-wiss. Mittlg. Nr. 80-4, Inst. f. Konstr. Ingenieurbau, Ruhr-Universität Bochum 1980.

Axially-Loaded, Imperfect, Cylindrical Shells

G. J. SIMITSES and I. SHEINMAN

School of Engineering Science and Mechanics,
Georgia Institute of Technology, Atlanta

Summary

The paper deals with buckling of axially-loaded, imperfect, circular, cylindrical shells. First, a short historical sketch of the problem is presented, and then a solution methodology is described through which the complete analysis of the system is achieved, including post-limit point behavior. The solution scheme is then employed to study the effect of the radius to thickness and length to radius ratios on the response characteristics of an imperfect, unstiffened, simply-supported (classical), circular, cylindrical shell. Moreover, the effect of the imperfection amplitude is studied for a fixed geometry shell configuration.

Introduction

Because of the multitude of the reported theoretical and experimental investigations on the problem, a very short historical sketch is presented, herein. Excellent reviews with details are presented in Refs. 1-4.

The first theoretical investigations on the subject were performed by Lorenz [5,7], Timoshenko [6,8], and Southwell [9]. The first experimental studies are those of Lilly (see Ref. 10), Robertson [11], Flügge [12], Wilson and Newmark [13], Lundquist [10] and Donnell (his experimental results are reported in Ref. 10). The initial theoretical investigations were based on many simplifying assumptions and they reduced the mathematical model to a linear eigen-boundary-value problem (classical bifurcation approach). Comparison between theoretical predictions (critical loads) and experimental results (buckling load) revealed discrepancy of unacceptable magnitude. A tremendous effort was made in order to explain the discrepancy both analytically and experimentally.

From the analytical point of view, the initial simplifying assumptions were re-evaluated and removed. This led to studies which attempted to attribute the discrepancy to (a) effect of prebuckling deformations [12,14-16], (b) effect of, primarily, in-plane boundary conditions [17-20] and (c) initial geometric imperfections.

Initially, the imperfection sensitivity of the system was established through strict postbuckling analyses of the perfect geometry configuration [21-26]. In addition, some of these investigators explained that the minimum postbuckling equilibrium load is a measure of the load carrying capacity of the system. This latter thinking came to an end when Hoff, Madsen, and Mayers [27] concluded from their calculations that the minimum postbuckling load tended towards zero with increasing number of terms for the transverse displacement component and with diminishing thickness. In this limiting case the Yoshimura buckle pattern [28] can be achieved.

Another approach for imperfection sensitivity studies is to deal directly with the imperfect configuration and employ nonlinear kinematic relations. The first attempt along these lines is due to Donnell [29]. Efforts reported in Refs. 30-35 fall in this category, but with varied success. Moreover, Koiter [36] was the first to question the use of the minimal postbuckling load as a measure of the load carrying capacity. He also dealt directly with the imperfect configuration, but his theory is limited to the neighborhood of the classical bifurcation load (immediate postbuckling) and therefore to small initial imperfections. Many researchers adopted this approach and most of their investigations can be found in Ref. 4.

The single and most important conclusion of all the theoretical investigations is that the primary reason for the discrepancy between (linear) theoretical critical loads and buckling loads is that the system is extremely sensitive to initial geometric imperfections. Imperfections of the inward type have more detrimental effect.

In parallel to the above analytical investigations, many experimental studies are performed with the same objective in mind (explain the discrepancy). While the old buckling loads fell in the range of 15% - 50% of the classical critical load, the new ones [37-41] with carefully manufactured specimens fall in the range of 40% - 90% of the classical critical load. Note that Ref. 37 obtained also postbuckling curves for both axially and pressure loaded cylinders. Moreover, this work was extended by Thielemann and Esslinger [42,43] to include theoretical postbuckled state calculations on the basis of observed experimental results. The theoretical predictions are based on a Galerkin procedure with the emphasis on post-limit point equilibrium positions. The theoretical study of these two papers belongs to the group of postbuckling analyses of perfect configurations.

The present paper is analytical in nature and belongs in the group of those investigations that deal with the imperfect system. It is also an extension and improvement of the one presented by the same authors [35] in 1977. The

1977 paper describes a solution methodology capable of predicting limit-point loads only. Incidentally, all such investigations, with the exception of those for which the response of the shell is characterized by a small number of degrees of freedom, are only concerned with finding limit point loads (critical loads) and not post-limit point behavior. The only exception to this is the newest version of the computer code STAGS [44], which has not been released yet. The methodology of Ref. 35 has been modified in order to obtain the post-limit point behavior of the imperfect shell. The description of the solution scheme is given under separate heading.

Solution Methodology

The method of solution is an improved modification of the one described in [35]. The same steps and nomenclature are used, herein, and these are not repeated for the sake of brevity. Only a broad description is presented in this paper.

For finding pre-limit point equilibrium positions, the applied load level, $\bar{N}_{xx}$, is taken as known, the linear ($n\equiv0$) solution is taken to be the approximate solution, and the small corrections (in W_i's and f_i's) are obtained through the solution of the linearized (with respect to the corrections) differential equations. Note that, in this range, the stiffness matrix is positive definite.

For finding post-limit point equilibrium positions (in a range of negative stiffness matrix), the numerical scheme is modified. The load parameter, $\bar{N}_{xx}$, is taken to be unknown, and one of the displacement parameters W_i replaces it as a known parameter. Great care must be exercised in choosing this W_i. This is done by observing how the various W_i's change with $\bar{N}_{xx}$ changes in the pre-limit point range, and choosing a W_i that tends to increase in a smooth and continuous manner, but most importantly is one of the most dominant displacement terms. In this post-limit point range, the linear solution cannot be taken as the initial estimate for the needed iterations. Therefore, the last converged, pre-limit point solution is used as an initial estimate for finding the first post-limit point solution. From there on, in this same range, the previous solution is utilized as an initial estimate. Needless to say that this latter procedure may also be used in the pre-limit point range, starting near the undeformed position. Unfortunately, this procedure is not very economical with regard to computer time and, therefore, it is very inefficient in this range.

In generating data, in order to investigate pre- and post-limit point behavioral response of axially loaded cylindrical shells, the solution procedure goes as follows: first, the system of equations is solved for a small level of the applied load, $\overline{N}_{xx}$, (taken as known). Then, solutions are sought for step increases in $\overline{N}_{xx}$, until the process fails to converge. The load level at which the solution fails to converge is a measure of the limit point or critical load. As explained in Ref. 35, when approaching the critical load, the increment in the applied load must be small and the sign of the determinant of the coefficients of the response must be checked. If convergence fails, the load level is over the limit point. But if convergence does not fail and the sign of the determinant changes from what it was at the previous load level, then the load level is also over the limit point. Desired accuracy can be achieved by taking smaller and smaller increments in $\overline{N}_{xx}$. Note that a cost penalty must be paid for improving the accuracy in $\overline{N}_{xx}$ by this approach. It is also observed that by employing this procedure (algorithm in which $\overline{N}_{xx}$ is known and the response, W_i, f_i, is unknown), no solution can be obtained past the limit point. Because of this, the new algorithm is employed at this point of the solution procedure. The new algorithm, as already explained, simply changes the role of one of the displacement terms with that of the applied load $\overline{N}_{xx}$. While the first procedure is followed, the most dominant displacement term is identified (or a group of terms). At some level before the limit point, the procedure is switched and a solution is formed that corresponds to an increment in the chosen dominant displacement parameter. To this end, the previous solution is used as an initial solution. The procedure is continued until the entire post-limit point response is obtained. During this phase of the solution procedure, some convergence failures can also occur. These failures can be attributed to one of two reasons: (a) either the increment in the dominant displacement parameter is too large or (b) the NP (number of mesh points) is too small for an accurate description of the response. Both of these can easily be corrected. In this second phase, large increments are purposely used in order to save computer time. If the solution fails to converge, then the increment is automatically reduced.

Numerical integration is used to find the total potential and end shortening. By this solution procedure, the entire load-displacement or load-end shortening curves can be obtained for a given imperfection and each wave number n.

Numerical Results and Discussion

Numerical results are presented for the geometry described below:

$R = 4$ in., $t = 0.004$ in., 0.008 in., 0.016 in., 0.050 in.;

$L = 4$ in., 12 in., 20 in., 40 in.; $E = 10.5 \times 10^6$ psi; $\nu = 0.3$;

with $W^o(x,y) = t\xi\left[-\cos\frac{2\pi x}{L} + 0.1\sin\frac{\pi x}{L}\cos\frac{ny}{R}\right]$

and SS-3 Boundary Conditions (classical) where ξ is the imperfection amplitude and it is taken to be equal to one, except for R/t = 1,000 and L/R = 1, in which case ξ = 0.5, 1.0, and 4.0.

Before discussing the results, a few more clarifying remarks about the geometry are needed. The geometry is taken from Refs.35 (Example 3) and 34. Note that in these references only the critical load is given and not the complete behavior. This geometry employs, virtually, an axisymmetric imperfection. Note that the non-axisymmetric amplitude is 10% of the axisymmetric amplitude. A smaller value was tried (1% for the non-axisymmetric amplitude) and the response (see Fig. 1; R/t = 500) is, for all practical purposes, the same. The only difference is in the value for $\bar{N}_{xx_{cr}}$ (limit-point load). This difference only reflects the effect of imperfection amplitude, i.e., for ξ = 1, $\bar{N}_{xx_{cr}}$ = 39.40 lbs/in. for 1% non-axisymmetric amplitude, while $\bar{N}_{xx_{cr}}$ = 35.00 lbs/in. for 10% non-axisymmetric amplitude. Note that, in the former case, the maximum imperfection amplitude is 1.01t while in the latter it is 1.10t. The classical load for this case is 101.81 lbs/in. The reason that the 10% amplitude is used in the numerical results obtained is that the higher the non-axisymmetric amplitude, the faster the solution. Finally, results are generated for several values of n (number of circumferential full waves). This is needed in order to obtain a clear picture of the complete response.

Some results are presented in graphical and tabular form. In the ensuing discussion, including conclusions, the statements are based on all generated data. Table 1 summarizes the most important results of the study, including limit point loads, $\bar{N}^{\ell}_{xx}$, minimum post-limit point loads, $\bar{N}^{m}_{xx}$, their ratios to the classical load, λ^{ℓ} and λ^{m}, and the range of n-values (number of circumferential full waves) for which data are generated. Note that when ξ = 0.5 the geometry corresponds to that reported in Ref. 35, and the limit-point load is exactly the same as that reported in that paper.

The data are also presented in graphical form on Figs. 2-4.

Fig. 2 is a plot of $\bar{N}_{xx}$ versus average end shortening for ξ = 1. These data are generated for several values of full waves, n, around the circumference. From this figure, it is clear that, as the system is loaded quasi-statically from zero, the load-deflection curve is the same and independent of n. The limit-point load, $\bar{N}_{xx_{cr}}$, is definitely n-dependent. It is observed that the value of the total potential corresponding to the lowest limit load (and associated n) is the smallest of all values corresponding to the same load

TABLE 1. Summery of Results for Axially-Loaded Unstiffened Shells

R/t	L/R	ξ	classical $\bar{N}_{xx}$ lbs/in	$\bar{N}^{\ell}_{xx}$ lbs/in	$\bar{N}^{m}_{xx}$ lbs/in	n	λ^{ℓ}	λ^{m}	n^{ℓ}	n^{m}
250	1	1	407.23	101.10	42.89	6-8	0.248	0.1053	8	6
250	3	1	407.23	242.40	46.00	4-5	0.595	0.1130	5	5
250	5	1	407.23	292.80	47.79	3-5	0.719	0.1174	4	4
250	8	1	407.23	355.80	132.90	3-4	0.873	0.326	3	3
250	10	1	407.23	356.30	159.60	3-5	0.875	0.3919	3	3
1000	1	0.5	25.42	16.61	0.5244	8-13	0.653	0.0206	13	9
1000	1	1	25.42	11.34	0.5649	8-13	0.446	0.0222	13	9
500	1	1	101.81	35.00	7.80	7-11	0.344	0.0766	10	8
250	1	1	407.23	101.10	42.890	6-8	0.248	0.105	8	6
80	1	1	3977.00	623.08	599.700	4-6	0.157	0.151	5	5

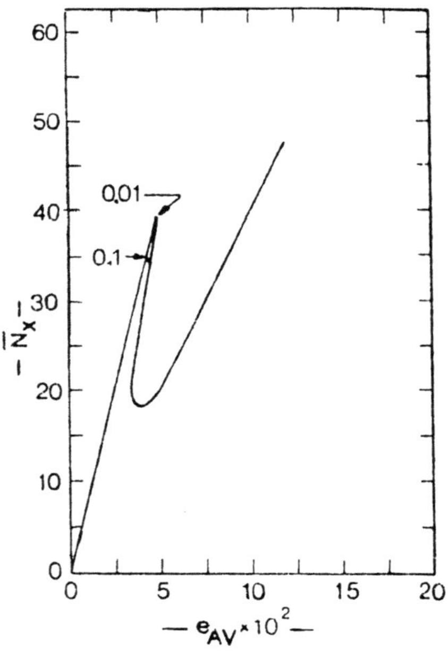

Fig. 1. Effect of the Asymmetric Imperfection Amplitude (R/t = 500; L/R = 1; n = 10)

position). For this value of ξ (which corresponds to w^o_{max} = 1.10h), the limit point occurs at $\bar{N}_{xx}$ = 11.34 lbs/in. [λ^{ℓ} = 0.446]. In the post-limit point region, the unstable branch shows several changes from n = 13 to n = 12 to n = 11. These changes occur at the unstable portion of the curve. The change from n = 11 to n = 10, etc., to n = 8, occur at the stable portion of the curves. This implies that if one can transverse the post-limit point branches, he would move along the n = 13 (with decreasing load) curve, then along the n = 12 and n = 11 curves (with decreasing load). Then, along the n = 11 curve, the system moves with increasing load until it reaches the n = 10 curve. Then it moves along the n = 10 curve until it intersects the n = 9 curve, etc. In reality, though, under dead weight loading, the system reaches the limit point, and then it snaps-through (violent buckling) towards far stable equilibrium positions. During the snapping process, it is clear from this figure that the shell experiences changes in the circumferential mode, corresponding to various n-values. As R/t decreases to 500, 250, and 80 the results are similar to those corresponding to R/t = 1,000. The composite response is shown on Fig. 3. Next, the effect of L/R is examined through several geometries for which R/t = 250 and $\xi = 1$. The results shown on Fig. 4, which is a plot of λ ($= \bar{N}_{xx}/\bar{N}_{xx_{cl}}$). It is observed from the generated data that, as L/R increases the entire static equilibrium response corresponds to one wave number (n = 4 for L/R = 5 and n = 3 for L/R = 8 and 10). Note also that the n-value decreases as L/R increases. One may question the validity of the results corresponding to n = 3 (because of the use of Donnell-type of equations [42]) but they do seem reasonable. Note also from Fig. 4 that the imperfection sensitivity of the shell decreases as L/R increases.

Finally, for R/t = 1,000 and L/R = 1, the effect of the imperfection amplitude parameter ξ on the limit-point load and minimum post-limit point load is shown on Fig. 5.

In summary, one may list the following observations:

(1) Relatively short and thin imperfect cylinders are subject to violent buckling, followed by a change in the wave number during the buckling process.

(2) Imperfection sensitivity decreases with both increasing R/t and L/R values. This definitely means that imperfection sensitivity decreases with increasing values of the Batdorf curvature parameter, $Z = (L^2/Rt)\cdot(1-\nu^2)^{1/2}$.

(3) The minimum post-limit point load, for a given imperfection, increases with increasing values of L/R.

References

1. Fung, Y. C.; Sechler, E. E.: Instability of thin elastic shells. Structural Mechanics. Proceedings of the First Symposium on Naval Structural Mechanics. Editors: J. N. Goodier and N. J. Hoff. Pergamon Press: New York (1958) 115.
2. Hoff, N. J.: The perplexing behavior of thin circular cylindrical shells in axial compression. Israel J. Techn. 4, 1 (1966) 1-28.
3. Stein, M.: Recent advances in shell buckling. AIAA Paper No. 68-103. AIAA 6th Aerospace Sciences Meeting, New York, N.Y., 1968.
4. Hutchinson, J. W.; Koiter, W. T.: Postbuckling theory. Appl. Mech. Reviews. 12 (1970) 1353-1366.
5. Lorenz, R.: Achsensymmetrische verzerrungen in dünnwandigen Hohlzylinder. Z.V.D.I., 52 (1908) 1706-1713.
6. Timoshenko, S. P.: Einige stabilitätsprobleme der elastizitatstheorie. Z. Math, Physik, 58 (1910) 337 (Abstract of paper published in 1907 in the Bulletin of the Polytechnic Institute of Kier).
7. Lorenz, R.: Die nicht achsensymmetrische knickung dünnwandiger hohlzylinder. Physik. Zeits. Leipzig., Jahrg. 12 (1911) 241-260.

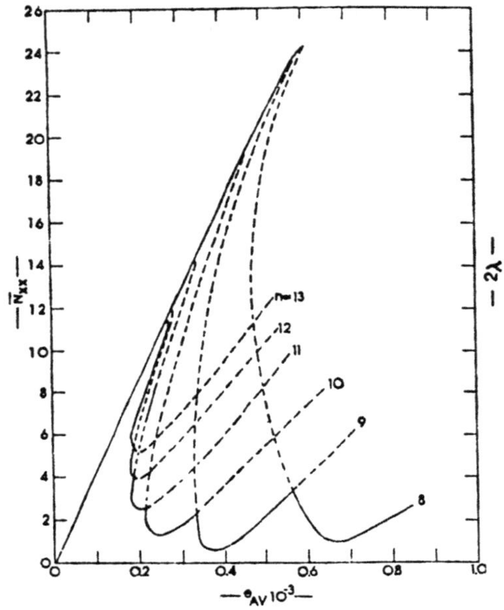

Fig. 2. Response Characteristics
(ξ=1; R/h = 1,000; L/R =1.0)

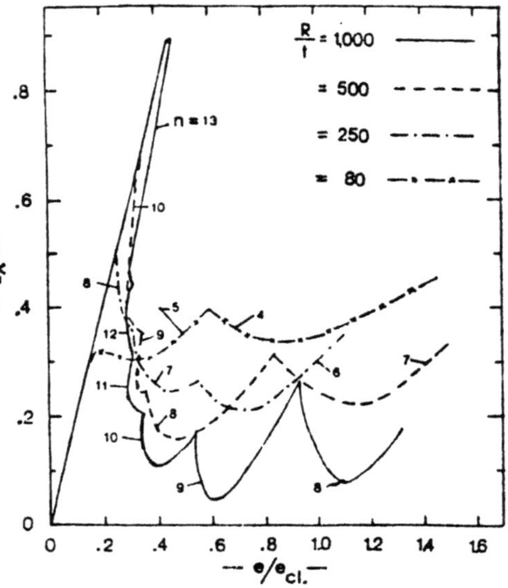

Fig. 3. Effect of R/t on Shell Response
(L/R = 1; ξ = 1.0)

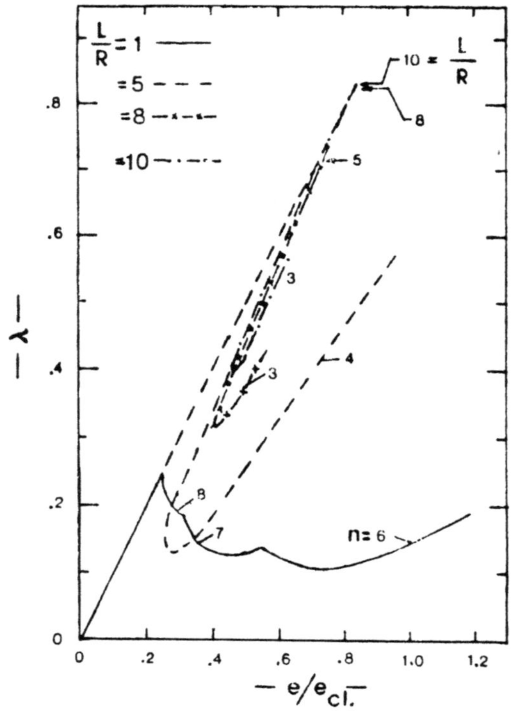

Fig. 4. Effect of L/R on Shell Response
(R/t = 250; ξ = 1.0)

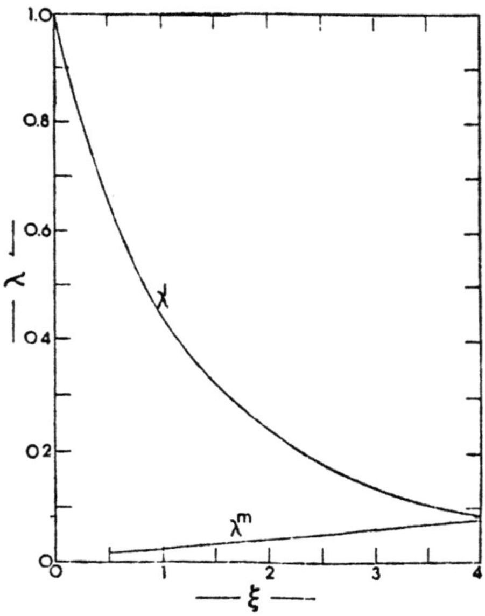

Fig. 5. Effect of Imperfection Amplitude (R/L = 1; R/t = 1,000)

8. Timoshenko, S. P.: Buckling of cylinders (exact title unknown). Bulletin Electrotechnical Institute, St. Petersburg, XI (1914).
9. Southwell, R. V.: On the general theory of elastic stability. Phil. Trans. Roy. Soc. (London), 213, series A (1914) 187-244.
10. Lundquist, E. E.: Strength tests of thin-walled duralumin cylinders in compression. NACA TR-473 (1933).
11. Robertson, A.: The strength of tubular struts. ARC Rep. and Mem. No. 1185 (1929).
12. Flügge, W.: Die stabilität der kreiszylinderschale. Ingenieur-Archiv. Band III, 5 (1932) 463-506.
13. Wilson, W. M.; Newmark, N. M.: The strength of thin cylindrical shells as columns. Bulletin No. 255, Eng. Exp. Station, Univ. of Illinois (1933).
14. Stein, M.: The effect on the buckling of perfect cylinders of prebuckling deformations and stresses induced by edge support. Collected Papers on Instability of Shell Structures. NASA TN D-1510 (1962) 217.
15. Fisher, G.: Über den einfluss der gelenkigen lagerung auf die stabilität dünnwandiger kreiszylinderschalen unter axiallast und innendruck. Z Flugwissenschaften, Jahrg. 11, Heft 3 (1963) 111.
16. Stein, M.: The influence of prebuckling deformations and stresses on the buckling of perfect cylinders. NASA TR R-190 (1964).
17. Ohira, H.: Local buckling theory of axially compressed cylinders. Proceedings of the Eleventh Japan Nat'l. Congress of Appl. Mech. (1961) 37.
18. Nachbar, W.; Hoff, N. J.: On edge buckling of axially compressed circular cylindrical shells. Quarterly of Appl. Math., 20 (1962) 267-271.
19. Hoff, N. J.; Soong, T. C.: Buckling of circular cylindrical shells in axial compression. Int'l. J. of Mech. Sci., 7 (1965) 489.
20. Hoff, N. J.; Rehfield, L. W.: Buckling of axially compressed circular cylindrical shells at stresses smaller than the classical critical value. J. Appl. Mech., 32 (1965) 533.
21. Von Karman, Th.; Tsien, H. S.: The buckling of thin cylindrical shells under axial compression. J. Aero. Sci., 8, 8 (1941) 303.
22. Leggett, D. M. A.; Jones, R. P. N.: The behavior of a cylindrical shell under axial compression when the buckling load has been exceeded. A.R.C. Rep. and Mem., No. 2190 (1942).
23. Michielsen, H. F.: The behavior of thin cylindrical shells after buckling under axial compression. J. Aero. Sci., 15, 12 (1948) 739.
24. Kempner, J.: Postbuckling behavior of this cylindrical shells. J. Aero. Sci., 21, 5 (1954) 329.
25. Almroth, B. O.: Postbuckling behavior of axially compressed circular cylinders. AIAA J., 1, 3 (1963) 630-633.
26. Sobey, A. J.: The buckling strength of a uniform circular cylinder loaded in axial compression. A.R.C. Rep. and Mem. No. 3366 (1964).
27. Hoff, N. J.; Madson, W. R.; Mayers, J.: The postbuckling equilibrium of axially compressed circular cylindrical shells. AIAA Journal, 4 (1966) 126-133.
28. Yoshimura, Y.: On the mechanism of buckling of a circular cylindrical shell under axial compression. NACA TM 1390 (1955).
29. Donnell, L. H.: A new theory for the buckling of this cylinders under axial compression. Trans of ASME, 56 (1934) 795.
30. Donnell, L. H.; Wan, C. C.: Effects of imperfections on buckling of thin cylinders and columns under axial compression. Trans of ASME, 72, 1 (1950) 73.
31. Loo, T. T.: Effects of large deflections and imperfections on the elastic buckling of cylinders under torsion and axial compression. Proceedings of 2nd U.S. Congress of Appl. Mech. (1950) 345.

32. Lee, L. H. N.: Effects of modes of initial imperfections on the stability of cylindrical shells under axial compression. Collecter Papers on Stability of Shell Structures. NASA TN D-1510 (1962) 143.
33. Narasimhan, K. Y.; Hoff, N. J.: Snapping of imperfect thin-walled circular cylindrical shells of finite length. J. Appl. Mech., 38, 1 (1971) 162-171.
34. Arbocz, J.; Sechler, E. E.: On the buckling of axially compressed imperfect cylindrical shells. J. Appl. Mech., 41, 3 (1974) 737-743.
35. Sheinman, I.; Simitses, G. J.: Buckling analysis of geometrically imperfect stiffened cylinder under axial compression. AIAA J., 15, 3 (1977) 374-382.
36. Koiter, W. T.: On the stability of elastic equilibrium. Thesis, Polytechnic Institute of Delft, H. J. Paris, Amsterdam (1945) (in Dutch).
37. Thielemann, W. F.: On the postbuckling behavior of thin cylindrical shells. Collected Papers on Instability of Shell Structures, NASA TN D-1510 (1962) 203.
38. Babcock, C. D.; Sechler, E. E.: The effect of initial imperfections on the buckling stress of cylindrical shells. NASA TN D-2005 (1963).
39. Tennyson, R. C.: A note on the classical buckling load of circular cylindrical shells under axial compression. AIAA J., 1, 2 (1963) 202.
40. Almroth, B. O.; Holmes, A. M. C.; Brush, D. O.: An experimental study of the buckling of cylinders under axial compression. Exper. Mech., 4, 9 (1964) 263.
41. Horton, W. H.; Durham, S. C.: Imperfections, a main contributor to scatter in experimental values of buckling load. Int'l. J. Solids and Structures, 1 (1965) 59.
42. Thielemann, W. F.; Esslinger, M. E.: On the postbuckling behavior of thin-walled axially compressed circular cylinders of finite length. Proceedings of 70th Anniversary Symposium on the Theory of Shells to Honor H. L. Donnell, Ed. by D. Muster, Univ. of Houston (1967) 433-479.
43. Thielemann, W. F.; Esslinger, M. E.: On the postbuckling equilibrium and stability of thin-walled circular cylinders under axial compression. Proceedings of Second IUTAM Symposium in the Theory of Thin Shells, Ed. by F. I. Niordson, Springer-Verlag, Berlin (1967) 264-293.
44. Almroth, B.O. of Lockheed Missile and Space Company, Palo Alto Research Laboratory: Private Communications, March 1981.

Stability of Cylindrical Shells with Piecewise Constant Thickness under Combined Loading

S. D. KISLIAKOV

Higher Institute of Civil Engineering and Architecture (VIAS),
Sofia, Bulgaria

Summary

Stability of vertical storage tanks for liquid products is investigated. The tank is considered to be thin elastic shell with piecewise constant thickness and with or without stiffening rings. Instability may occur as a result of three basic loadings: vertical load from a stationary roof, pressure due to vacuum, and wind loading. Correct stability criteria for this problem have been established. Then each of the basic loading states has been investigated. Possibility of buckling as a result of shear stresses due to wind load has been found. It has been further established that the first eigenvalue of the hydrostatic pressure stability problem under some conditions may prove to be located in the complex domain.

Introduction. Stability Criteria

The stability problem for cylindrical shell with piecewise constant thickness, with or without ring stiffeners, under combined loading, is of considerable interest. This structure is a good model for vertical storage tanks for liquid products. Buckling of such tanks may be caused by three possible loadings - alone or in combination with each other (Fig. 1): a) axial pressure from the roof and from the weight of the wall itself; b) vacuum or underpressure due to cooling or drawing-off the liquid; c) wind loading. The b-loading means lateral uniformly distributed pressure. The wind loading is considerably more complicated - in circumferential direction it is distributed as shown in Fig. 2 and besides, heightwise variation is possible too. We are not going to discuss here all the solutions of this problem given in the literature. Usually some kind of thickness averaging is performed, or different courses are considered as separate cylindrical shells almost independent from each other [1,2]. What is more essential is that the wind load is replaced by an "equivalent" transverse uniformly distributed pressure, and then the well known stability

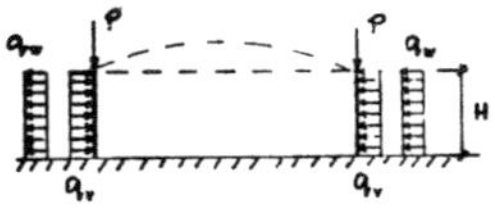

Fig.1. Vertical cylindrical storage tank under combined loading

criterion

$$\frac{\sigma_1}{\sigma_{1\,cr}} + \frac{\sigma_2}{\sigma_{2\,cr}} \leqq 1 \qquad (1)$$

is applied. Here the indices 1,2 stand for the basic states - axial compression and summarized transverse uniformly distributed pressure respectively; "cr" means the corresponding critical value of the stress. In this way, state "2" unites the above mentioned loadings b) and c).

Taking into account the influence of the stiffening rings runs into considerable difficulties. With few exceptions this problem has not been investigated in the literature known to the author.

But the most essential shortcomings of the existing methods are connected with the wind loading. Reducing it to uniformly distributed pressure through no matter what a coefficient is wrong in principle because this pressure causes only compression all along the shell, and the real wind loading causes in some regions compression, in other ones tension, and in third ones (around the"zero line" of the cross section) considerable tangential stresses. This means that in the stability analysis the wind loading is to be considered as a separate (third) basic state in which the shell is subjected to bending and shear. At that, either the normal or the tangential stresses may prove to be decisive. That is why the stability analysis should be based on the criterion

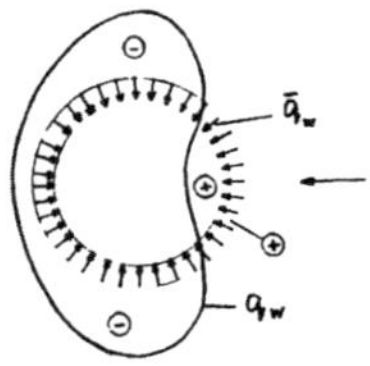

Fig.2. Distribution of the wind loading in circumferential direction

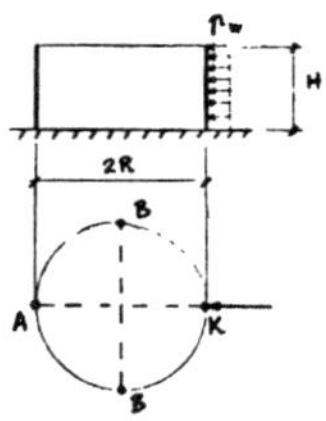

Fig.3. A-, B-, and K- regions of instability

$$\frac{\sigma_1}{\sigma_{1\,cr}} + \frac{\sigma_2}{\sigma_{2\,cr}} + \frac{\sigma_3}{\sigma_{3\,cr}} \leqq 1 \qquad (2)$$

when the influence of the normal stresses due to wind loading is checked, and correspondingly on the criterion

$$\frac{\sigma_1}{\sigma_{1cr}} + \frac{\sigma_2}{\sigma_{2cr}} + \left(\frac{\tau}{\tau_{cr}}\right)^2 \leqq 1 \tag{3}$$

when the influence of the shear stresses is involved. Inequalities (2), (3) have been suggested in [3] as stability criteria for closed cylindrical shells under combined loading. They are based on the general Papkovich criteria for stability under sophisticated loading.

Let us analyze the third state - bending and shear due to wind. We will present this load as uniformly distributed along the height, and applied on one generatrix of the cylinder (Fig.3). In this case three types of instability are possible, as pointed out by Darevskiy [4] : 1) in the vicinity of the A-generatrix (region A) as a result of axial compressive stresses; 2) in the vicinity of the B-generatrix (region B) as a result of shear stresses; 3) in the vicinity of the K-generatrix (region K) as a result of high local pressure. In each of the three zones mentioned the buckling takes place with the well known wave pattern.

Both criteria (2),(3) are to be applied separately to each course considered as a medium-long shell. The first criterion for A-region of the i-th course has the form

$$\frac{\sigma_{1i}}{\sigma_{1i}^{cr}} + \frac{\lambda}{\lambda_{cr}} + \frac{\hat{p}_{1i}}{\hat{p}_{1cr}^{i}} \leqq 1, \tag{4}$$

where λ_{cr} is the critical parameter of the transverse uniformly distributed load, and λ is its actual value. Let us note that λ_{cr} must be determined for the whole shell taking into account the thicknesses of all courses under the one considered and above it. By $\hat{p}_{1i}$ is noted the dimensionless value of the maximum compressive stress along the contour of i-th course (compression domain at bending). For its calculation one can make use of the formula (see [3])

$$\hat{p}_{1i} = \frac{p_{1i} \cdot R}{E h_i} = \frac{M_i}{\pi R^2 h_i} \cdot \frac{R}{E h_i} \,. \tag{5}$$

Here $M_i = \frac{1}{2} p_w H_i^2$, where H_i is the distance between the lower edge of i-th course to the upper edge of the tank. For $\hat{p}_{1cr}^{i}$ may be accepted the value $1{,}25\,\sigma_{1i}^{cr}$ i.e. a 25%-increase for the critical stress at axial compression. The reasons for such an increase are known.

Finally, criterion (4) takes the form

$$\frac{1}{\sigma_{1i}^{cr}}\left[\sigma_{1i}+\frac{p_w \cdot H_i^2}{2{,}5\pi E R h_i^2}\right]+\frac{\lambda}{\lambda_{cr}} \leqq 1 . \tag{6}$$

Now let us consider the stability criterion for the B-region. The analysis of the stability problem under torsion shows [3] , that tangential stresses are responsible for instability at $H < 4R$, and this condition is fulfilled in all practically encountered tanks. If the tank is considered as a cantilever shell under uniformly distributed load the latter's critical value at constant thickness h may be found from Darevskiy's formula

$$Q_{cr} = 1{,}12\pi E h^2 \sqrt[4]{\frac{Rh}{H^2}} . \tag{7}$$

For a tank with piecewise constant thickness it may be approximately accepted that for the i-th course

$$Q_{cr,i} = 1{,}12\pi E h_{i,av}^2 \sqrt[4]{\frac{R h_{i,av}}{H_i^2}} , \tag{8}$$

where $h_{i,av}$ is the average thickness of all the courses above the i-th (including itself). Since

$$\tau/\tau_{cr} = Q/Q_{cr} , \tag{9}$$

for the criterion (3) we finally obtain

$$\frac{\sigma_{1i}}{\sigma_{1i}^{cr}}+\frac{\lambda}{\lambda_{cr}}+\frac{(p_w H_i)^2 H_i}{12{,}39 E^2 h_{i,av}^4 \sqrt{R h_{i,av}}} . \tag{10}$$

The estimation of buckling danger for K-domain may be brought to a criterion of the type (1).

Stability under Uniformly Distributed Pressure

Now we have to analyze separately the basic three loadings in order to determine their critical values. Most interesting appears to be determination of the parameter λ_{cr} because the critical values of the other two loads can be determined approximately, more or less clearly. Moreover, the influence of the ring stiffeners is most essential exactly in this state, where the load is also the highest.

Among the solutions of the stability problem for cylindrical shell with piecewise constant thickness and stiffening rings under hydrostatic load the most general one is given by Darevskiy. It leads to the set of differential equations

$$\varepsilon_i\left(\frac{\partial^8 w_i}{\partial\varphi^8}+2\,\frac{\partial^6 w_i}{\partial\varphi^6}+\frac{\partial^4 w_i}{\partial\varphi^4}\right)+\frac{\partial^4 w_i}{\partial\xi^4}+$$

$$+\frac{\lambda q_{0i}R}{E_i h_i}\left(\frac{\partial^6 w_i}{\partial\varphi^6}+\frac{\partial^4 w_i}{\partial\varphi^4}\right)=0 \qquad (i=1,2,\dots,m)$$

$$\Big\{\Big[\frac{E_i^0 J_i}{R^2}\Big(\frac{\partial^2}{\partial\varphi^2}+1\Big)+\lambda k_i a_i R q_i^0\Big]\frac{\partial^4}{\partial\varphi^4}\Big(\frac{\partial^2}{\partial\varphi^2}+1\Big)w_i-$$

$$-\varepsilon_i h_i R E_i\Big[(2-\nu_i)\frac{\partial^2}{\partial\varphi^2}+1\Big]\frac{\partial^3}{\partial\xi\partial\varphi^2}\Big(\frac{\partial^2}{\partial\varphi^2}+1\Big)w_i-$$

$$-E_i h_i R\,\frac{\partial^3 w_i}{\partial\xi^3}\Big\}_{\xi=\frac{1}{2}\rho_i}+\Big\{\varepsilon_{i+1}h_{i+1}RE_{i+1}\Big[(2-$$

$$-\nu_{i+1})\frac{\partial^2}{\partial\varphi^2}+1\Big]\frac{\partial^3}{\partial\xi\partial\varphi^2}\Big(\frac{\partial^2}{\partial\varphi^2}+1\Big)w_{i+1}+$$

$$+E_{i+1}h_{i+1}R\,\frac{\partial^3 w_{i+1}}{\partial\xi^3}\Big\}_{\xi=-\frac{1}{2}\rho_{i+1}}.$$

$$(i=1,2,\dots,m-1) \tag{11}$$

These equations, with the notations, are given in [5]. Only the quantities γ_i are not given there, and they have been found in [6] to be

$$\gamma_i=\sqrt[4]{3(1-\nu_i^2)}\,\sqrt{R/h_i}\;. \tag{12}$$

Now about the solution of Eqs. (11). In [5] the additional normal displacements w_i are sought in the form

$$w_i=C_i\cdot\cos(\mu_i\xi+\alpha_i)\sin n\varphi,\quad (i=1,2,\dots,m), \tag{13}$$

where C_i, $\mu_i>0$, α_i and n are unknown constants, and $-\frac{\rho_i}{2}\leqq\xi\leqq+\frac{\rho_i}{2}$. Free supported ends of the shell are assumed, and this may be accepted for a cantilever shell with end stiffening ring under hydrostatic pressure. Along the boundaries between each two intervals continuity of the normal displacement w is assured. So Eqs. (11) yield the set

$$\lambda=\frac{E_i h_i}{q_{0i}R}\Big[\varepsilon_i(n^2-1)+\frac{\mu_i^4}{n^4(n^2-1)}\Big],\quad (i=1,2,\dots,m) \tag{14}$$

$$\lambda = \frac{E_i^0 J_i}{k_i a_i q_i^0 R^3}(n^2-1) - \frac{E_i h_i}{k_i a_i q_i^0}\mu_i\Big[\varepsilon_i(2-\nu_i-n^{-2}) +$$

$$+ \frac{\mu_i^2}{n^4(n^2-1)}\Big]\operatorname{tg}\left(\frac{1}{2}\mu_i\rho_i+\alpha_i\right) - \frac{E_{i+1}\cdot h_{i+1}}{k_i a_i q_i^0}\mu_{i+1}\times$$

$$\times\Big[\varepsilon_{i+1}(2-\nu_{i+1}-n^{-2}) + \frac{\mu_{i+1}^2}{n^4(n^2-1)}\Big]\operatorname{tg}\left(\frac{1}{2}\mu_{i+1}\rho_{i+1}-\alpha_{i+1}\right)$$

$$(i=1,2,\dots,m-1) \tag{15}$$

For any taken integer n these equations give two values of λ which eventually must be equal to each other. For different n is obtained $\lambda(n)$, and $\lambda_{min} = \lambda_{cr}$.

Now it is interesting to follow the variation of the critical value λ_{cr} as a result of variation of some parameters. For the sake of simplicity we will consider a shell without rings, and besides, we will assume the same Young's modulus and Poisson's ratio for all the courses. In this case it is convenient to equate the two values of λ obtained from the equations

$$\lambda = \frac{Eh_1}{Rq_{01}}\left[\varepsilon_1(n^2-1) - \frac{\mu_1^4}{n^4(n^2-1)}\right] \tag{16}$$

$$\lambda = \frac{Eh_1}{Rq_{01}}A_1(n) + \frac{Eh_1}{k_1 a_1 q_1^0}B_1(\mu_1,n)\operatorname{ctg}\mu_1\rho_1 -$$

$$- \frac{Eh_2}{k_1 a_1 q_1^0}B_2(\mu_2,n)\frac{\operatorname{tg}\mu_2\rho_2 - \operatorname{tg}\left(\frac{1}{2}\mu_2\rho_2+\alpha_2\right)}{1+\operatorname{tg}\mu_2\rho_2\cdot\operatorname{tg}\left(\frac{1}{2}\mu_2\rho_2+\alpha_2\right)}, \tag{17}$$

where

$$A_i(n) = \frac{J_i q_{0i}(n^2-1)}{R^2 h_i k_i a_i q_i^0} \tag{18}$$

$$B_i(n, \mu_i) = \mu_i \left[\left(2 - \nu - \frac{1}{n^2} \right) \varepsilon_i + \frac{\mu_i^2}{n^4(n^2-1)} \right], \qquad (19)$$

and the μ_i -coefficients are calculated from the formula

$$\mu_i^4 = \frac{h_1 q_{0i}}{h_i q_{01}} \mu_1^4 + n^4 (n^2-1)^2 \left(\frac{h_1 q_{0i}}{h_i q_{01}} \varepsilon_1 - \varepsilon_i \right). \qquad (20)$$

Some values for μ_1 and n (integer!) are chosen. From Eqs. (16),(17) $\lambda_I(\mu_1, n)$ and $\lambda_{II}(\mu_1, n)$ can be determined. These two quantities form two curves in the $(\lambda - \mu_1)$-plane, and their intersection point gives the sought value μ_1 . The procedure is performed again for other integers n, until $\lambda_{cr} = \lambda_{min}(n)$ be obtained.

The detailed investigations carried out by the author of this paper and a group of his assistants (including Dr. T. Bobev, Dr. Sht. Dukova and Dr. Khr. Kunchev) have shown some peculiarities. In some cases the increasing thickness of one of the courses lead to decreasing critical value, and in some other cases a small increase of the thickness caused drastical increase of this value. At the same time n also changed. A more thorough investigation of this phenomenon was performed. In Fig. 4 are shown the curves $\lambda_I(\mu_1)$ and $\lambda_{II}(\mu_1)$ at different wave numbers n, for particular shells. The values of $\lambda_I(\mu_1)$ in the whole allowable interval for μ_1 are incomparably smaller than these of $\lambda_{II}(\mu_1)$, and for the chosen scale the first curve practically coincides with μ_1 axis. Hence, the values of μ_1 for which the two curves intersect each other practically coincide with the zeros of the function $\lambda_{II}(\mu_1)$.

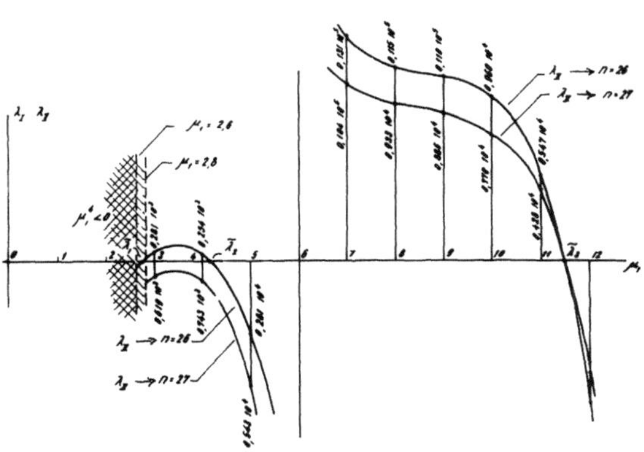

Fig.4. Shifting of the λ_{II} -curves for different wave numbers

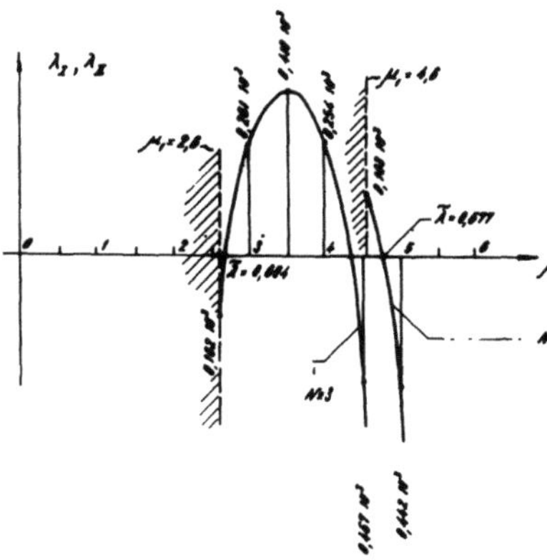

Fig.5. Initial parts of the λ_{II} -curves for different shells

As seen from the figure, the curves $\lambda_{II}(\mu_1)$, which correspond to different wave numbers (n=26 and n=27), have the same shape but with increasing n these curves are shifting downwards, and their origin (from which on the condition $\mu_i^4 > 0$ is fulfilled) is shifting rightwards. This situation takes place also at the curves for n=28,29,... which are not shown in the figure.

This gradual change of the curves λ_{II} which, if considered separately, seems natural leads to sudden loss of some roots $\tilde{\lambda}$ at definite values of n, which cannot be accepted as normal. For instance, at n=26 the least root is $\tilde{\lambda}_1 \simeq 0{,}604$. At n=27 however $\tilde{\lambda}_1$ and $\tilde{\lambda}_2$ do not exist any more, and the lowest root is $\tilde{\lambda}_3 \simeq 3{,}50$.

In Fig.5 are given with more details the initial intervals of the graphs constructed for Shell Nr 3 (S3) at n=26, and for Shell Nr 4 (S4) at n=27. The two shells differ from each other only by the thickness of the first course which for S4 is thinner than for S3. It should be expected that $\lambda_{cr}|S4$ is smaller than $\lambda_{cr}|S3$. But the carefully performed and checked digital results show just the contrary which obviously is not true. The explanation is that the decreasing thickness of the first course causes considerable shifting of the imaginary value limit for μ_1. At that, the interval from the $\lambda_{II}(\mu_1)$- graph which contains the first root vanishes.

In some cases loss of roots $\tilde{\lambda}$ may be observed as a result of both causes - shifting of the λ_{II} -curves downwards and shifting of the μ_1 limit rightwards. This is displayed in Fig. 6.

Further a cylindrical shell with only two courses was thoroughly investigated in order to clarify if an average thickness might be involved without essential loss of accuracy. In each course the equation

$$\frac{D_i}{h_i} \nabla^8 w_i + \frac{E}{R^2} \frac{\partial^4 w_i}{\partial x^4} + q \frac{R}{h_i} \nabla^4 \left(\frac{\partial^2 w_i}{\partial y^2} \right) = 0 \qquad (21)$$

is valid, where D_i stands for cylindrical rigidity. For the normal displacements the following functions have been accepted:

$$\left.\begin{aligned} w_1(x,y) &= f_1 \cdot \sin k_1 x \cdot \sin \beta y \\ w_2(x,y) &= f_2 \cdot \sin k_2 (l_2 - x) \cdot \sin \beta y . \end{aligned}\right\} \qquad (22)$$

Here $\beta = n/R$, and k_1, k_2 are additional parameters which enable us to shift the maximum depth of the displacement towards the weaker course.

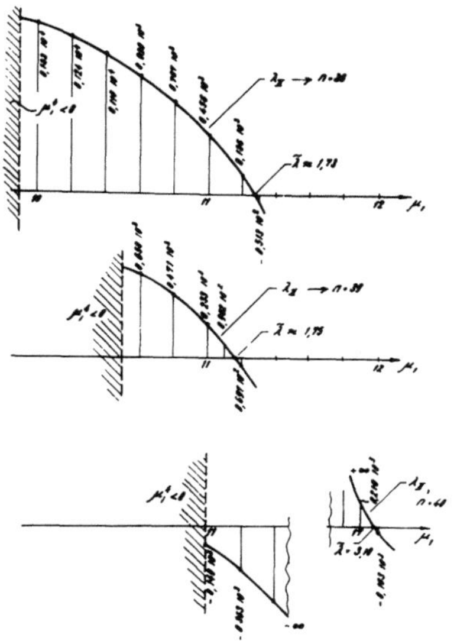

Fig.6. Simultaneous shifting of λ_{II}- curve downwards and of μ_1- limit rightwards

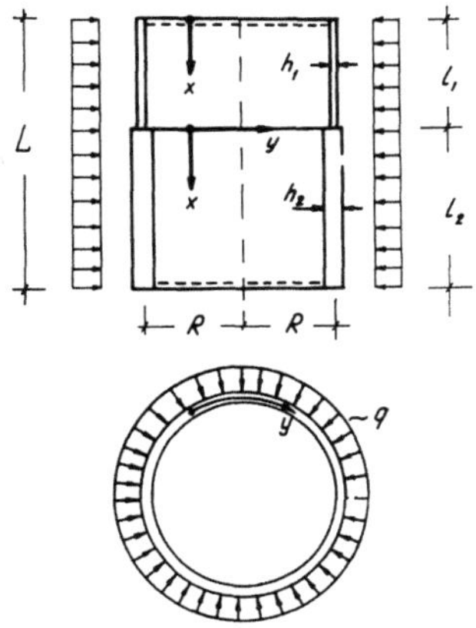

Fig.7. Iwo course cylindrical shell under hydrostatic pressure

Eqs. (22) satisfy rigorousl**y** following boundary conditions (see Fig.7):

$$\left.\begin{aligned} &w_1(0,y)=0 \qquad && \left(\frac{\partial^2 w_1}{\partial x^2}\right)_{x=0}=0 \\ &w_2(l_2,y)=0 \qquad && \left(\frac{\partial^2 w_2}{\partial x^2}\right)_{x=l_2}=0 \end{aligned}\right\} \tag{23}$$

$$\left.\begin{aligned} &w_1(l_1,y)=w_2(0,y)\neq 0 \\ &\left(\frac{\partial w_1}{\partial x}\right)_{x=l_1}=\left(\frac{\partial w_2}{\partial x}\right)_{x=0}\neq 0 . \end{aligned}\right\} \tag{24}$$

Inserting (22) into (21) and solving the equations obtained with respect to q yield following result:

$$\bar{q}_V = F_1(\Psi_1) + \varphi_1(\Psi_1) \tag{25}$$

$$\bar{q}_V = \nu_1^3 F_2(\Psi_2) + \nu_1 \varphi_2(\Psi_2), \tag{26}$$

where $\nu_1 = h_2/h_1$, $\Psi_{1,2} = R k_{1,2}/\pi$, and $F_{1,2}$ and $\varphi_{1,2}$ are some explicit functions of $\Psi_{1,2}$.

The problem to find the critical load is now reduced to following: given n, ν_1, and $\lambda_1 = h_1/R$, to determine values of Ψ_1 and Ψ_2 such that the values for $\bar{q}=10^8 q/E$ from (25) and (26) would be equal to each other, and besides, to satisfy the coupling condition

$$\frac{\operatorname{tg} k_2 l_2}{k_2} = -\frac{\operatorname{tg} k_1 l_1}{k_1}. \tag{27}$$

The results of the calculations show that the critical load for the shell with average thickness is close to the "exact" value and, moreover, on the conservative side provided $\nu_1 \leqq 1{,}5$ and $R/L \leqq 2$, $(l_1 = l_2)$. Because of the lack of space the other two basic states will not be considered in this paper.

References

1. Berezin, V.L.; Shutov, V.E.: Prochnost' i ustoychivost' rezervuarov i truboprovodov.(in Russian). Moscow, Nedra 1973.
2. Vertical steel welded storage tanks with butt-welded shells for the petroleum industry. BS 2654. 1973.
3. Vol'mir, A.S.: Ustoychivost' deformiruemikh sistem (in Russian). Moscow, Nauka 1967.
4. Darevskiy, V.M.: Ustoychivost' konsol'noy tsilindricheskoy obolochki pri izgibe poperechnoy siloy s krucheniem i vnutrennim davleniem (in Russian). Sborn. Rascheti prostr. konstr., 5 (1959), Moscow, Gosstroyizdat.
5. Darevskiy, V.M.: Ustoychivost' podkreplennoy tsilindricheskoy obolochki peremennoy tolshchini pri peremennom vneshnem davlenii (in Russian). Mekhanika tverdogo tela, 2 (1975).
6. Kisliakov, S.D.: On the stability of cylindrical storage tanks for oil products (in Bulgarian). Yearly of Higher Institutes, Technical mechanics, vol. XIII,1 (1979).

Energy-Consistent Linear and Nonlinear Stability Equations for Hyperelastic Shells

Y. Başar / W.B. Krätzig

Institut für Konstruktiven Ingenieurbau,
Ruhr-Universität Bochum, Germany

Summary

Shell structures may respond to external loadings by highly nonlinear primary and secondary equilibrium paths, possibly intersecting each other. The paper presents an energy-consistent nonlinear shell theory for moderate rotations in a formulation, which is similar to that one of the linear theory. It derives the fundamental equations for linear and nonlinear bifurcation analysis in operator form.

1. Description of the shell space

Fig. 1 shows the middle surface F of an arbitrary shell continuum in its deformed state. If points P of F are described by a set of convected Gaussian coordinates Θ^α

$$\underset{\sim}{r} = \underset{\sim}{r}\,(\Theta^1, \Theta^2)\,, \tag{1.1}$$

the deformed frame of reference $\underset{\sim}{a}_\alpha$, $\underset{\sim}{a}_3$ of P can be introduced by:

$$\underset{\sim}{a}_\alpha = \underset{\sim}{r}_{,\alpha}\,; \quad \underset{\sim}{a}_3 = \frac{\underset{\sim}{a}_1 \times \underset{\sim}{a}_2}{\sqrt{a}} \quad . \tag{1.2}$$

Thus we find for the metric- and curvature tensor of F:

$$a_{\alpha\beta} = \underset{\sim}{a}_\alpha \cdot \underset{\sim}{a}_\beta\,; \; b_{\alpha\beta} = -\frac{1}{2}\,(\underset{\sim}{a}_\alpha \cdot \underset{\sim}{a}_{3,\beta} + \underset{\sim}{a}_\beta \cdot \underset{\sim}{a}_{3,\alpha})\,. \tag{1.3}$$

Introducing the Christoffel - symbols by

$$\Gamma^\lambda_{\alpha\beta} = \underset{\sim}{a}^\lambda \cdot \underset{\sim}{a}_{\alpha,\beta}\,, \quad \Gamma^\lambda_{\lambda\alpha} = \frac{(\sqrt{a})_{,\alpha}}{\sqrt{a}}\,, \tag{1.4}$$

the covariant derivatives $(\ \ )||_\alpha$ with respect to the deformed middle surface can be defined in the usual manner [5].

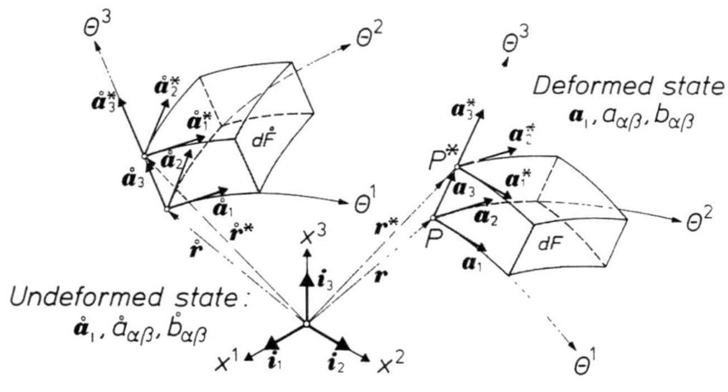

Fig. 1: Undeformed and deformed state

Furthermore we denote the distance between any arbitrary point P^* of the shell outside F and its orthogonal projection P by Θ^3. As given in fig. 1 the position vector $\underset{\sim}{r}^*$ of P^* then is represented by

$$\underset{\sim}{r}^* = \underset{\sim}{r} + \Theta^3 \underset{\sim}{a}_3 . \qquad (1.5)$$

Consequently we find for the frame of reference of P^*:

$$\underset{\sim}{a}^*_\alpha = \underset{\sim}{r}^*_{,\alpha} = \underset{\sim}{a}_\alpha + \Theta^3 \underset{\sim}{a}_{3,\alpha} = (\delta^\rho_\alpha - \Theta^3 b^\rho_\alpha)\, \underset{\sim}{a}_\rho = \mu^\rho_\alpha \underset{\sim}{a}_\rho , \qquad (1.6)$$

$$\underset{\sim}{a}^*_3 = \underset{\sim}{r}^*_{,3} = \underset{\sim}{a}_3 . \qquad (1.7)$$

In the following all functions related to a point P^* outside F will be starred.

Analogous relations will be also established for all geometrical elements of the undeformed middle surface $\overset{\circ}{F}$ and denoted by a circle as superscript $(\overset{\circ}{\ })$, e.g. $\overset{\circ}{\underset{\sim}{a}}_\alpha$, $\overset{\circ}{\underset{\sim}{a}}_3$, $\overset{\circ}{a}_{\alpha\beta}$, ... Covariant derivatives with respect to $\overset{\circ}{F}$ will be characterized by a single stroke $(\ )|_\alpha$.

2. State of deformation

Fig. 2 demonstrates the state of deformation of the middle surface of the shell space. Under the Kirchhoff-Love-Hypothesis the undeformed unit normal $\overset{\circ}{\underset{\sim}{a}}_3$ deforms into a vector $\underset{\sim}{a}_3$, preserving its normality and unit length. Thus the displacement vector $\underset{\sim}{v}^*$ of an arbitrary point P^* of the shell space can be represented by the linear form

$$\underset{\sim}{v}^* = \underset{\sim}{r}^* - \overset{\circ}{\underset{\sim}{r}}{}^* = (\underset{\sim}{r} - \overset{\circ}{\underset{\sim}{r}}) + \Theta^3 (\underset{\sim}{a}_3 - \overset{\circ}{\underset{\sim}{a}}_3) = \underset{\sim}{v} + \Theta^3 \underset{\sim}{w},$$

which defines the displacement vector $\underset{\sim}{v}$ and the associated difference vector $\underset{\sim}{w}$ of the middle surface $\overset{\circ}{F}$:

$$\underset{\sim}{v} = \underset{\sim}{r} - \overset{\circ}{\underset{\sim}{r}} = v_\alpha \overset{\circ}{\underset{\sim}{a}}{}^\alpha + v_3 \overset{\circ}{\underset{\sim}{a}}{}^3 = v^\alpha \overset{\circ}{\underset{\sim}{a}}_\alpha + v^3 \overset{\circ}{\underset{\sim}{a}}_3, \tag{2.1}$$

$$\underset{\sim}{w} = \underset{\sim}{a}_3 - \overset{\circ}{\underset{\sim}{a}}_3 = w_\alpha \overset{\circ}{\underset{\sim}{a}}{}^\alpha + w_3 \overset{\circ}{\underset{\sim}{a}}{}^3 = w^\alpha \overset{\circ}{\underset{\sim}{a}}_\alpha + w^3 \overset{\circ}{\underset{\sim}{a}}_3. \tag{2.2}$$

Both vectors will be decomposed with respect to the undeformed frame of reference. Defining further the displacement gradients

$$\varphi_{\alpha\rho} = (v_\rho|_\alpha - v_3 \overset{\circ}{b}_{\rho\alpha}), \quad \varphi_{\alpha 3} = v_{3,\alpha} + v_\lambda \overset{\circ}{b}{}^\lambda_\alpha,$$
$$\varphi_{\alpha\cdot}^{\ \beta} = \varphi_{\alpha\rho} \overset{\circ}{a}{}^{\rho\beta}, \quad \varphi^{\rho}_{\cdot 3} = \varphi_{\beta 3} \overset{\circ}{a}{}^{\beta\rho} \tag{2.3}$$

by the well-known identity

$$\underset{\sim}{v}_{,\alpha} = (v_\rho|_\alpha - v_3 \overset{\circ}{b}_{\rho\alpha}) \overset{\circ}{\underset{\sim}{a}}{}^\rho + (v_{3,\alpha} + v_\lambda \overset{\circ}{b}{}^\lambda_\alpha) \overset{\circ}{\underset{\sim}{a}}_3 = \varphi_{\alpha\rho} \overset{\circ}{\underset{\sim}{a}}{}^\rho + \varphi_{\alpha 3} \overset{\circ}{\underset{\sim}{a}}{}^3 \tag{2.4}$$

we find for the transformations of the undeformed frame of reference into the deformed one:

$$\underset{\sim}{a}_\alpha = (\delta^\rho_\alpha + \varphi_{\alpha\cdot}^{\ \rho}) \overset{\circ}{\underset{\sim}{a}}_\rho + \varphi_{\alpha 3} \overset{\circ}{\underset{\sim}{a}}_3; \quad \underset{\sim}{a}_3 = w^\alpha \overset{\circ}{\underset{\sim}{a}}_\alpha + (1 + w_3) \overset{\circ}{\underset{\sim}{a}}_3. \tag{2.5}$$

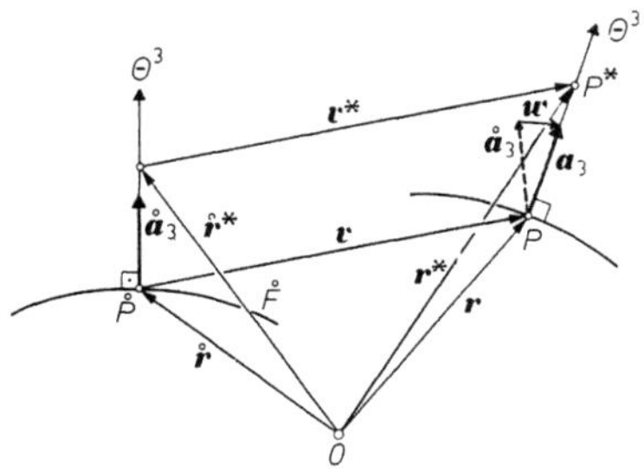

Fig. 2: Definition of displacement variables

Specifying now by virtue of (2.5) the Kirchhoff-Love-Hypothesis

$$\underset{\sim}{a}_\alpha \cdot \underset{\sim}{a}_3 = 0, \quad \underset{\sim}{a}_3 \cdot \underset{\sim}{a}_3 - \overset{\circ}{\underset{\sim}{a}}_3 \cdot \overset{\circ}{\underset{\sim}{a}}_3 = 0$$

we find the following equations:

$$w_\alpha + \varphi_{\alpha 3} + \varphi_{\alpha\rho} w^\rho + \varphi_{\alpha 3} w_3 = 0, \tag{2.6}$$

$$2w_3 + w_\alpha w^\alpha + w_3 w_3 = 0. \tag{2.7}$$

They are satisfied, up to terms of third and forth order, by the approximations [1]:

$$w_\alpha = -\varphi_{\alpha 3} + \varphi_{\alpha .}^{\ \lambda} \varphi_{\lambda 3}, \; w_3 = -\frac{1}{2} \varphi_{\alpha 3} \varphi_{\cdot 3}^{\alpha} \;. \tag{2.8}$$

By means of the same approximation, we obtain from (2.6) and (2.7):

$$w_\alpha + \varphi_{\alpha 3} + \varphi_{\alpha\rho} w^\rho = 0, \; w_3 = -\frac{1}{2} w_\alpha w^\alpha. \tag{2.9}$$

Finally we introduce the rotation vector $\underset{\sim}{\omega}$ by

$$\underset{\sim}{\omega} = \overset{\circ}{\underset{\sim}{a}}_3 \times \underset{\sim}{w} = \omega_\alpha \overset{\circ}{\underset{\sim}{a}}{}^\alpha = \omega^\alpha \overset{\circ}{\underset{\sim}{a}}_\alpha, \tag{2.10}$$

which stands orthogonal to $\overset{\circ}{\underset{\sim}{a}}_3$ and $\underset{\sim}{w}$. Its length $|\underset{\sim}{\omega}|$ corresponds with the sine of the angle of rotation $(\overset{\circ}{\underset{\sim}{a}}_3, \underset{\sim}{a}_3)$. According to (2.2), (2.10) we find:

$$\omega^\rho = \overset{\circ}{\varepsilon}{}^{\beta\rho} w_\beta, \; w_\beta = \overset{\circ}{\varepsilon}_{\beta\alpha} \omega^\alpha. \tag{2.11}$$

To define 2-dimensional strain measures we apply Green's strain tensor to the shell space:

$$\gamma_{(ij)} = \frac{1}{2} (\overset{*}{a}_{ij} - \overset{\circ *}{a}_{ij}) = \frac{1}{2} (\overset{*}{\underset{\sim}{a}}_i \cdot \overset{*}{\underset{\sim}{a}}_j - \overset{\circ *}{\underset{\sim}{a}}_i \cdot \overset{\circ *}{\underset{\sim}{a}}_j). \tag{2.12}$$

Considering in it (1.3), (1.6) as well as (2.3), (2.5) $\gamma_{(\alpha\beta)}$ may be approximated by the linear expression

$$\gamma_{(\alpha\beta)} = \alpha_{(\alpha\beta)} + \Theta^3 \omega_{(\alpha\beta)}, \tag{2.13}$$

which defines the 1. and 2. strain tensor of the middle surface:

$$\alpha_{(\alpha\beta)} = \frac{1}{2} (a_{\alpha\beta} - \overset{\circ}{a}_{\alpha\beta}) = \frac{1}{2} (\varphi_{\alpha\beta} + \varphi_{\beta\alpha} + \varphi_{\alpha\lambda} \varphi_{\beta .}^{\ \lambda} + \varphi_{\alpha 3} \varphi_{\beta 3}), \tag{2.14}$$

$$\begin{aligned} \omega_{(\alpha\beta)} = -(b_{\alpha\beta} - \overset{\circ}{b}_{\alpha\beta}) = \frac{1}{2} \Big[& (w_{\alpha|\beta} + w_{\beta|\alpha} - \overset{\circ}{b}{}_\alpha^\lambda \varphi_{\beta\lambda} - \overset{\circ}{b}{}_\beta^\lambda \varphi_{\alpha\lambda}) \\ & + (\overset{\circ}{b}_{\alpha\beta} w_\rho w^\rho + w^\lambda{}_{|\alpha} \varphi_{\beta\lambda} + w^\lambda{}_{|\beta} \varphi_{\alpha\lambda} + \overset{\circ}{b}{}_\alpha^\lambda w_\lambda \varphi_{\beta 3} + \overset{\circ}{b}{}_\beta^\lambda w_\lambda \varphi_{\alpha 3}) \Big] \end{aligned} \tag{2.15}$$

Additionally we have expressed w_3 in $\omega_{(\alpha\beta)}$ according to (2.9) and neglected terms of third order. With (2.9) $\omega_{(\alpha\beta)}$ also can be expressed in terms of the displacement gradients:

$$\omega_{(\alpha\beta)} = -\frac{1}{2}\Big[\varphi_{\alpha3}|_\beta+\varphi_{\beta3}|_\alpha+\overset{\circ}{b}{}^\lambda_\beta\varphi_{\alpha\lambda}+\overset{\circ}{b}{}^\lambda_\alpha\varphi_{\beta\lambda}-\varphi^\lambda_{.3}(\varphi_{\alpha\lambda}|_\beta+\varphi_{\beta\lambda}|_\alpha$$
$$+\overset{\circ}{b}_{\alpha\beta}\varphi_{\lambda3}-\overset{\circ}{b}{}^\lambda_\beta\varphi_{\alpha3}-\overset{\circ}{b}{}^\lambda_\alpha\varphi_{\beta3})\Big]. \tag{2.16}$$

Finally some further geometrical elements of the deformed middle surface shall be presented, in which quadratic deformations compared to geometrical quantities, are neglected:

the contravariant metric tensor:

$$a^{\alpha\beta} \approx \overset{\circ}{a}{}^{\alpha\beta}-2\alpha^{\alpha\beta} \approx \overset{\circ}{a}{}^{\alpha\beta}-\varphi^{\alpha\beta}-\varphi^{\beta\alpha} \tag{2.17}$$

$$\sqrt{\frac{a}{\overset{\circ}{a}}} \approx 1+\alpha^\lambda_\lambda \approx 1+\varphi_\lambda{}^\lambda_{\cdot} \tag{2.18}$$

the mixed curvature tensor: $b^\alpha_\beta = b_{\alpha\rho}a^{\rho\beta} = \overset{\circ}{b}{}^\beta_\alpha - L^\beta_\alpha$

$$L^\beta_\alpha = w^\beta|_\alpha+\overset{\circ}{b}{}^\lambda_\alpha\varphi_\lambda{}^\beta_{\cdot} \approx -\varphi^\beta_{.3}|_\alpha+\overset{\circ}{b}{}^\lambda_\alpha\varphi_\lambda{}^\beta_{\cdot} \tag{2.19}$$

the difference-Christoffelsymbol: $L^\lambda_{\beta\gamma} = \Gamma^\lambda_{\beta\gamma}-\overset{\circ}{\Gamma}{}^\lambda_{\beta\gamma}$

$$L^\alpha_{\beta\gamma} \approx \varphi_\beta{}^\alpha_{\cdot}|_\gamma-\varphi_{\beta3}\overset{\circ}{b}{}^\alpha_\gamma+\varphi^\alpha_{.3}\overset{\circ}{b}_{\beta\gamma} \tag{2.20}$$

$$\sqrt{\frac{a}{\overset{\circ}{a}}}L^\lambda_{\lambda\alpha} = \left(\sqrt{\frac{a}{\overset{\circ}{a}}}\right)_{,\alpha} = \left(\sqrt{\frac{a}{\overset{\circ}{a}}}\right)\Big|_\alpha = \left(\sqrt{\frac{a}{\overset{\circ}{a}}}\right)\Big\|_\alpha \tag{2.21}$$

3. Nonlinear conditions of equilibrium

The load vector $\underset{\sim}{p}$, representing the external surface loads acting on the deformed middle surface, is related always to elements of F. It may be decomposed, after multiplication by $\sqrt{\frac{a}{\overset{\circ}{a}}}$ in two different ways:

$$\sqrt{\frac{a}{\overset{\circ}{a}}}\,\underset{\sim}{p} = p^\alpha\overset{\circ}{\underset{\sim}{a}}_\alpha+p^3\overset{\circ}{\underset{\sim}{a}}_3 = \hat{p}^\alpha\underset{\sim}{a}_\alpha+\hat{p}^3\underset{\sim}{a}_3 \tag{3.1}$$

connected by

$$p^\alpha = \hat{p}^\rho(\delta^\alpha_\rho+\varphi_\rho{}^\beta_{\cdot}) + \hat{p}^3w^\alpha,\quad p^3 = \hat{p}^\alpha\varphi_{\alpha3}+\hat{p}^3(1+w_3). \tag{3.2}$$

The vector $\sqrt{\frac{a}{\overset{\circ}{a}}}\,\underset{\sim}{p}$ then is related to surface elements of the undeformed middle surface $\overset{\circ}{F}$; its components p^i are - for conservative loads - independent of the deformations.

The internal forces and moments along Θ^{α}=const. are described by two vectors $\underset{\sim}{n}^{\alpha}$ and $\underset{\sim}{m}^{\alpha}$, the decompositions of which

$$\underset{\sim}{n}^{\alpha} = n^{\alpha\beta}\underset{\sim}{a}_{\beta}+q^{\alpha}\underset{\sim}{a}_{3}, \quad \underset{\sim}{m}^{\alpha} = m^{(\alpha\rho)}\underset{\sim}{a}_{3}x\underset{\sim}{a}_{\rho} = m^{(\alpha\rho)}\varepsilon_{\rho\beta}\underset{\sim}{a}^{\beta} \tag{3.3}$$

define the in-plane stress tensor $n^{\alpha\beta}$, the transverse shear vector q^{α} and the moment tensor $m^{(\alpha\beta)}$. In addition the symmetric pseudo-stress tensor $\tilde{n}^{(\alpha\beta)}$ is introduced

$$\tilde{n}^{(\alpha\beta)} = n^{\alpha\beta}+m^{(\alpha\rho)}b^{\beta}_{\rho} \approx n^{\alpha\beta}+m^{(\alpha\rho)}(\overset{\circ}{b}{}^{\beta}_{\rho}-w^{\beta}|_{\rho}-\overset{\circ}{b}{}^{\lambda}_{\rho}\varphi_{\lambda\cdot}^{\ \beta}) \tag{3.4}$$

as a consequence of the equation of moment equilibrium around the $\underset{\sim}{a}_{3}$-axis.

The equilibrium of the deformed middle surface is described by the following vector equations [1, 4]:

$$\underset{\sim}{n}^{\alpha}||_{\alpha} + \underset{\sim}{p} = 0, \quad \underset{\sim}{m}^{\alpha}||_{\alpha} + \underset{\sim}{a}_{\alpha}x\underset{\sim}{n}^{\alpha} = 0. \tag{3.5}$$

Transformation of their covariant derivatives to the undeformed middle surface $\overset{\circ}{F}$ leads to

$$\left(\sqrt{\frac{a}{\overset{\circ}{a}}}\,\underset{\sim}{n}^{\alpha}\right)\Big|_{\alpha} + \sqrt{\frac{a}{\overset{\circ}{a}}}\,\underset{\sim}{p} = 0, \left(\sqrt{\frac{a}{\overset{\circ}{a}}}\,\underset{\sim}{m}^{\alpha}\right)\Big|_{\alpha} + \underset{\sim}{a}_{\alpha}x\left(\sqrt{\frac{a}{\overset{\circ}{a}}}\,\underset{\sim}{n}^{\alpha}\right) = 0. \tag{3.6}$$

We finally obtain, by using (2.5), (2.8), (3.1), (3.3), (3.4) and neglecting quadratic deformation terms:

$$N^{\alpha\beta}|_{\alpha} - Q^{\alpha}\overset{\circ}{b}{}^{\beta}_{\alpha} + p^{\beta} = 0, \quad N^{\alpha\beta}\overset{\circ}{b}_{\alpha\beta} + Q^{\alpha}|_{\alpha} + p^{3} = 0 \tag{3.7}$$

$$\begin{aligned} M^{\alpha\beta}|_{\alpha} - \left(\sqrt{\frac{a}{\overset{\circ}{a}}}\,q^{\alpha}\right)\left(\delta^{\beta}_{\alpha} + \varphi_{\alpha\cdot}^{\ \beta}\right) - \overset{\circ}{b}_{\alpha\rho}\Big[\left(\sqrt{\frac{a}{\overset{\circ}{a}}}\,m^{(\alpha\beta)}\right)\varphi^{\rho}_{\cdot 3} \\ - \left(\sqrt{\frac{a}{\overset{\circ}{a}}}\,m^{(\alpha\rho)}\right)\varphi^{\beta}_{\cdot 3}\Big] = 0, \end{aligned} \tag{3.8}$$

with nonlinear stress components:

$$\begin{aligned} N^{\alpha\beta} &= \left(\sqrt{\frac{a}{\overset{\circ}{a}}}\,n^{\alpha\rho}\right)\left(\delta^{\beta}_{\rho} + \varphi_{\rho\cdot}^{\ \beta}\right) + \left(\sqrt{\frac{a}{\overset{\circ}{a}}}\,q^{\alpha}\right)w^{\beta}, \\ &\approx \left(\sqrt{\frac{a}{\overset{\circ}{a}}}\,\tilde{n}^{(\alpha\rho)}\right)\left(\delta^{\beta}_{\rho} + \varphi_{\rho\cdot}^{\ \beta}\right) - \left(\sqrt{\frac{a}{\overset{\circ}{a}}}\,m^{(\alpha\rho)}\right)\left(\overset{\circ}{b}{}^{\beta}_{\rho} - w^{\beta}|_{\rho}\right) \\ &\quad + q^{\alpha}w^{\beta}, \end{aligned} \tag{3.9}$$

$$Q^{\alpha} = \left(\sqrt{\frac{a}{\overset{\circ}{a}}}\, q^{\alpha}\right) + \left(\sqrt{\frac{a}{\overset{\circ}{a}}}\, n^{\alpha\beta}\right) \varphi_{\beta 3},$$

$$\approx \left(\sqrt{\frac{a}{\overset{\circ}{a}}}\, q^{\alpha}\right) + \left(\sqrt{\frac{a}{\overset{\circ}{a}}}\, \tilde{n}^{(\alpha\beta)}\right) \varphi_{\beta 3} - \overset{\circ}{b}{}^{\lambda}_{\rho} \left(\sqrt{\frac{a}{\overset{\circ}{a}}}\, m^{(\alpha\rho)}\right) \varphi_{\lambda 3}, \quad (3.10)$$

$$M^{\alpha\beta} = \left(\sqrt{\frac{a}{\overset{\circ}{a}}}\, m^{(\alpha\rho)}\right) \left(\delta^{\beta}_{\rho} + \varphi_{\rho\cdot}^{\ \beta}\right) . \quad (3.11)$$

The equation of moment equilibrium around the $\underset{\sim}{a}_3$-axis, which is equivalent to (3.4), has not been included. By means of (2.8) we can express the abbreviation (3.9) solely in terms of the deformation gradients $\varphi_{\alpha\beta}$, $\varphi_{\alpha 3}$ and the stress components $\tilde{n}^{(\alpha\beta)}$, $m^{(\alpha\beta)}$, q^{α}.

4. Constitutive equations

Constitutive equations of hyperelastic materials are characterized by [4, 7]

$$\sqrt{\frac{a}{\overset{\circ}{a}}}\, \tilde{n}^{(\alpha\beta)} = \frac{1}{2} \left(\frac{\partial \pi_i}{\partial \alpha_{(\alpha\beta)}} + \frac{\partial \pi_i}{\partial \alpha_{(\beta\alpha)}} \right) ,$$

$$\sqrt{\frac{a}{\overset{\circ}{a}}}\, m^{(\alpha\beta)} = \frac{1}{2} \left(\frac{\partial \pi_i}{\partial \omega_{(\alpha\beta)}} + \frac{\partial \pi_i}{\partial \omega_{(\beta\alpha)}} \right) , \quad (4.1)$$

where π_i denotes the specific deformation energy density of the undeformed middle surface. By neglecting the stresses τ^{33} which are orthogonally to F and approximating the geometry of the undeformed continuum through

$$\overset{\circ}{a}{}^{*}_{\alpha\beta} \approx \overset{\circ}{a}_{\alpha\beta}, \ \overset{\circ}{a}{}^{*} \approx \overset{\circ}{a}, \quad (4.2)$$

we obtain for a H o o k e a n material

$$\pi_i = \frac{1}{2} \left(D H^{\alpha\beta\rho\lambda} \alpha_{(\alpha\beta)} \alpha_{(\rho\lambda)} + B H^{\alpha\beta\rho\lambda} \omega_{(\alpha\beta)} \omega_{(\rho\lambda)} \right) \quad (4.3)$$

with $$H^{\alpha\beta\rho\lambda} = \frac{1-\nu}{2} \left(\overset{\circ}{a}{}^{\alpha\lambda} \overset{\circ}{a}{}^{\beta\rho} + \overset{\circ}{a}{}^{\alpha\rho} \overset{\circ}{a}{}^{\beta\lambda} + \frac{2\nu}{1-\nu} \overset{\circ}{a}{}^{\alpha\beta} \overset{\circ}{a}{}^{\rho\lambda} \right) \quad (4.4)$$

and $$D = \frac{Eh}{1-\nu^2} , \ B = \frac{Eh^3}{12(1-\nu^2)} . \quad (4.5)$$

E and ν are Young's modulus and Poisson's ratio respectively. By introducing (4.3) in (4.1) we finally obtain

$$\tilde{n}^{(\alpha\beta)} = D H^{\alpha\beta\rho\lambda} \alpha_{(\rho\lambda)}, \ m^{(\alpha\beta)} = B H^{\alpha\beta\rho\lambda} \omega_{(\rho\lambda)} .$$

5. Boundary variables

We describe the forces acting along a unit length of the boundary curve C of the deformed middle surface by vectors $\underset{\sim}{n}$ and $\underset{\sim}{m}$. These vectors can be decomposed with respect to the orthogonal reference system

$$\mathring{C} : \underset{\sim}{\mathring{u}} = \underset{\sim}{\mathring{t}} x \underset{\sim}{\mathring{a}}_3 = \mathring{u}_\alpha \underset{\sim}{\mathring{a}}^\alpha , \quad \underset{\sim}{\mathring{t}} = \frac{d\underset{\sim}{\mathring{r}}}{d\mathring{s}} = \mathring{t}_\alpha \underset{\sim}{\mathring{a}}^\alpha , \quad \underset{\sim}{\mathring{a}}_3 = \underset{\sim}{\mathring{a}}^3 \tag{5.1}$$

of the undeformed curve $\mathring{C}$:

$$\frac{ds}{d\mathring{s}} \underset{\sim}{n} = n_t \underset{\sim}{\mathring{t}} + n_u \underset{\sim}{\mathring{u}} + n_3 \underset{\sim}{\mathring{a}}_3 , \quad \frac{ds}{d\mathring{s}} \underset{\sim}{m} = m_t \underset{\sim}{\mathring{t}} + m_u \underset{\sim}{\mathring{u}} + m_3 \underset{\sim}{\mathring{a}}_3 . \tag{5.2}$$

Observing

$$\frac{ds}{d\mathring{s}} \underset{\sim}{n} = \sqrt{\frac{\mathring{a}}{a}}\, \underset{\sim}{n}^\alpha \mathring{u}_\alpha , \quad \frac{ds}{d\mathring{s}} \underset{\sim}{m} = \sqrt{\frac{\mathring{a}}{a}}\, \underset{\sim}{m}^\alpha \mathring{u}_\alpha \tag{5.3}$$

and also (2.5), (3.3) and (3.9) to (3.11), we finally obtain

$$\begin{aligned} n_t &= N^{\alpha\beta} \mathring{u}_\alpha \mathring{t}_\beta , \quad n_u = N^{\alpha\beta} \mathring{u}_\alpha \mathring{u}_\beta , \quad n_3 = Q^\alpha \mathring{u}_\alpha , \\ m_t &= M^{\alpha\beta} \mathring{u}_\alpha \mathring{u}_\beta , \quad m_u = -M^{\alpha\beta} \mathring{u}_\alpha \mathring{t}_\beta , \quad m_3 = -m^{\alpha\beta} \varphi^\rho_{\cdot 3} \mathring{\varepsilon}_{\rho\beta} \mathring{u}_\alpha . \end{aligned} \tag{5.4}$$

Analogous expressions for $\underset{\sim}{v}$ und $\underset{\sim}{w}$, namely

$$\underset{\sim}{v} = v_t \underset{\sim}{\mathring{t}} + v_u \underset{\sim}{\mathring{u}} + v_3 \underset{\sim}{\mathring{a}}_3 , \quad \underset{\sim}{\omega} = \omega_t \underset{\sim}{\mathring{t}} + \omega_u \underset{\sim}{\mathring{u}} , \tag{5.5}$$

can be used for defining dual boundary deformations to (5.4):

$$\begin{aligned} v_t &= v_\alpha \mathring{t}^\alpha , \quad v_u = v_\alpha \mathring{u}^\alpha , \quad v_3 = v^3 , \\ \omega_t &= \omega_\alpha \mathring{t}^\alpha , \quad \omega_u = \omega_\alpha \mathring{u}^\alpha . \end{aligned} \tag{5.6}$$

In order to define prescribed boundary variables, we formulate the virtual work done by stresses acting along the deformed boundary C and by virtual deformation $\delta v^* = \delta v + \Theta^3 \delta w$. We can eliminate the dependant rotational component ω_u by known methods [4] and obtain thus:

$$\begin{aligned} \tilde{n}_t &= N^{\alpha\beta} \mathring{u}_\alpha \mathring{t}_\beta + \mathring{b}^\lambda_\beta \mathring{t}^\beta \mathring{t}_\lambda m_u , & v_t &= v_\alpha \mathring{t}^\alpha , \\ \tilde{n}_u &= N^{\alpha\beta} \mathring{u}_\alpha \mathring{u}_\beta + \mathring{b}^\lambda_\beta \mathring{t}^\beta \mathring{u}_\lambda m_u , & v_u &= v_\alpha \mathring{u}^\alpha , \\ \tilde{n}_3 &= Q^\alpha \mathring{u}_\alpha - \frac{\partial m_u}{\partial \mathring{s}} , & v_3 &= v^3 , \\ m_t &= M^{\alpha\beta} \mathring{u}_\alpha \mathring{u}_\beta , & \omega_t &= \omega_\alpha \mathring{u}^\alpha = -w_\beta \mathring{t}^\beta . \end{aligned} \tag{5.7}$$

6. Principle of virtual displacements

Let $C=C_r+C_t$ be the closed and smooth shell boundary curve with boundary conditions:

$$\underset{\sim}{v} = \underset{\sim}{v}^o, \ \underset{\sim}{w} = \underset{\sim}{w}^o \text{ on } C_r; \ \underset{\sim}{n} = \overset{o}{\underset{\sim}{n}}, \ \underset{\sim}{m} = \underset{\sim}{m}^o \text{ on } C_t. \tag{6.1}$$

The principle of virtual displacements can be cast into the form:

$$\begin{aligned}
\delta^*\hat{A}_a + \delta^*\hat{A}_i = & \iint_{\overset{o}{F}} (p^\alpha \delta v_\alpha + p^3 \delta v_3)\, d\overset{o}{F} + \int_{\overset{o}{C}_t} (n_t^o \delta v_t + n_u^o \delta v_u \\
& + n_3^o \delta v_3 + \overset{o}{m}_t \delta\omega_t + \overset{o}{m}_u \delta\omega_u)\, d\overset{o}{F} \\
& - \iint_{\overset{o}{F}} (\tilde{n}^{(\alpha\beta)} \delta\alpha_{(\alpha\beta)} + m^{(\alpha\beta)} \delta\omega_{(\alpha\beta)}) d\overset{o}{F} \\
& - \iint_{\overset{o}{F}} q^\alpha \Big[\delta w_\alpha + (\delta v_{3,\alpha} + \overset{o}{b}{}^\lambda_\alpha \delta v_\lambda) \\
& + \delta w^\lambda (v_\lambda|_\alpha - v_3 \overset{o}{b}_{\lambda\alpha}) \\
& + w^\lambda (\delta v_\lambda|_\alpha - \delta v_3 \overset{o}{b}_{\alpha\beta}) \Big]\, d\overset{o}{F} = 0 .
\end{aligned} \tag{6.2}$$

In this principle v_i and w_α can be varied independently due to the orthogonality condition (2.9) contained in the third surface integral. It can be shown by well-known methods [4] that the E u l e r - equations of this principle correspond to the equilibrium equations (3.7), (3.8) and the natural boundary conditions to (6.1).

7. Stability equations

The field equations of a geometrically nonlinear theory can be expressed by the following operator equations, where the column vectors of fig. 3 have been introduced:

$$\overset{o}{\underline{p}} = (\underline{D}_{eL} + \underline{D}_{eN}(\underline{u}))\ \underline{\sigma} \ \rightarrow \text{ equations of equilibrium}$$

$$\underline{\sigma} = D\underline{E}\ \underline{\varepsilon} \ \rightarrow \text{ constitutive law}$$

$$\underline{\varepsilon} = (\underline{D}_{kL} + \underline{D}_{kN}(\underline{u}))\ \underline{u} \ \rightarrow \text{ kinematic relations} \tag{7.1}$$

External variables:

$$\mathring{\boldsymbol{p}} = \begin{bmatrix} \boldsymbol{p}^{\alpha} \\ \hline \boldsymbol{p}^{3} \end{bmatrix} = \begin{bmatrix} p^{1} \\ p^{2} \\ \hline p^{3} \end{bmatrix} \qquad \boldsymbol{u} = \begin{bmatrix} \boldsymbol{v}_{\alpha} \\ \hline \boldsymbol{v}_{3} \end{bmatrix} = \begin{bmatrix} v_{1} \\ v_{2} \\ \hline v_{3} \end{bmatrix}$$

Internal variables:

$$\boldsymbol{\sigma} = \begin{bmatrix} \tilde{\boldsymbol{n}}^{(\alpha\beta)} \\ \hline \boldsymbol{m}^{(\alpha\beta)} \end{bmatrix} \qquad \boldsymbol{\epsilon} = \begin{bmatrix} \boldsymbol{\alpha}_{(\alpha\beta)} \\ \hline \boldsymbol{\omega}_{(\alpha\beta)} \end{bmatrix}$$

force variables   *deformation variables*

Fig. 3: Grouping of variables

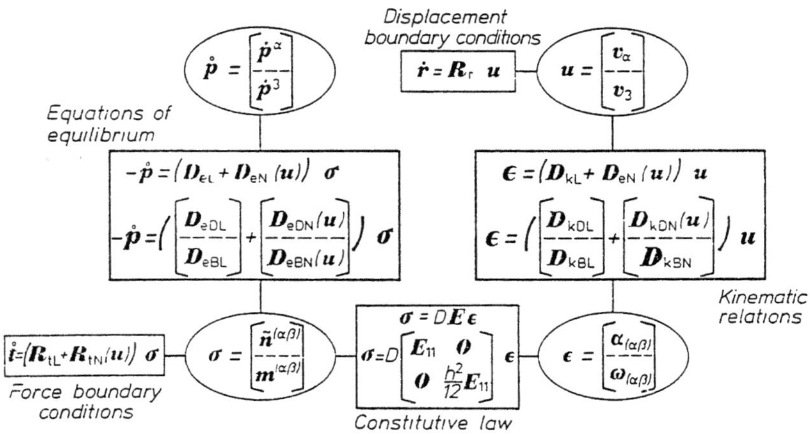

Fig. 4: Structure of the nonlinear shell theory

A survey of all field equations and boundary conditions of a nonlinear shell theory is offered by fig. 4. The operators $\underline{D}_{eL}$ $\underline{D}_{kL}$ and $\underline{E}$ of the linear theory are solely functions of the undeformed shell geometry, while $\underline{D}_{eN}$ and $\underline{D}_{kN}$ also contain linear terms of the displacements [3, 4]. In fig. 4 all operators are separated into their inplane (D) and bending (B) contributions.

In order to derive the stability equations we formulate the relations for a neighbouring state. Under the assumption of conservative loads (and boundary forces) we express the variables in the following form:

$$\begin{bmatrix} \bar{\underline{u}} \\ \bar{\underline{\varepsilon}} \\ \bar{\underline{p}} \\ \bar{\underline{\sigma}} \end{bmatrix} = \begin{bmatrix} \underline{u} \\ \underline{\varepsilon} \\ \mathring{\underline{p}} \\ \underline{\sigma} \end{bmatrix} + \varepsilon \begin{bmatrix} \overset{+}{\underline{u}} \\ \overset{+}{\underline{\varepsilon}} \\ \underline{O} \\ \overset{+}{\underline{\sigma}} \end{bmatrix}, \tag{7.2}$$

(first vector → neighbouring state; second vector → fundamental state)

in which ε is an arbitrary small number. After eliminating all equations of the fundamental state we obtain the nonlinear stability equations corresponding to (7.1):

$$\begin{aligned} \underline{O} &= \underline{D}_{eL}\overset{+}{\underline{\sigma}} + \underline{D}_{eN}(\overset{+}{\underline{u}})\ \underline{\sigma} + \underline{D}_{eN}(\underline{u})\ \overset{+}{\underline{\sigma}}, \\ \overset{+}{\underline{\sigma}} &= D\underline{E}\overset{+}{\underline{\varepsilon}}, \\ \overset{+}{\underline{\varepsilon}} &= \underline{D}_{kL}\overset{+}{\underline{u}} + \underline{D}_{kN}(\underline{u})\ \overset{+}{\underline{u}} + \underline{D}_{kN}(\overset{+}{\underline{u}})\ \underline{u}. \end{aligned} \tag{7.3}$$

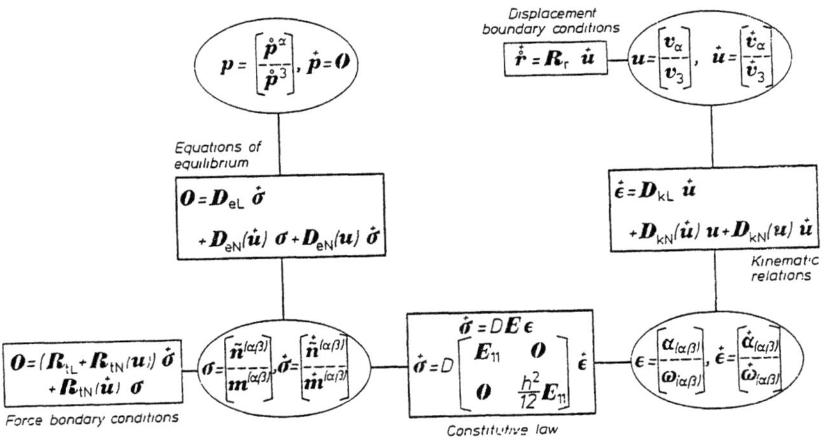

Fig. 5: Structure of the nonlinear stability equations in the theory of shells

The derived equations (7.3) and in fig. 5 are clearly nonlinear because of the nonlinear fundamental state described by fig. 4. Assuming a linear fundamental state by neglecting all displacement-depending suboperators

$$\begin{bmatrix} \underline{u} \\ \underline{\varepsilon} \\ \underline{\overset{\circ}{p}} \\ \underline{\sigma} \end{bmatrix} = \lambda \begin{bmatrix} \underline{\overset{\circ}{u}}^* \\ \underline{\overset{\circ}{\varepsilon}}^* \\ \underline{\overset{\circ}{p}}^* \\ \underline{\overset{\circ}{\sigma}}^* \end{bmatrix} \tag{7.4}$$

the linear stability equations can be derived:

$$\underline{O} = \underline{D}_{eL}\underline{\overset{+}{\sigma}} + \lambda(\underline{D}_{eN}(\underline{\overset{+}{u}})\ \underline{\overset{\circ}{\sigma}}^* + \underline{D}_{eN}(\underline{\overset{\circ}{u}}^*)\ \underline{\overset{+}{\sigma}}),$$

$$\underline{\overset{+}{\sigma}} = D\underline{E}\underline{\overset{+}{\varepsilon}},$$

$$\underline{\overset{+}{\varepsilon}} = D_{kL}\underline{\overset{+}{u}},$$

$$\underline{D}_{eL}D\underline{E}\ \underline{D}_{kL}\underline{\overset{+}{u}} + \lambda(\underline{D}_{eN}(\underline{\overset{+}{u}})\ \underline{\overset{\circ}{\sigma}}^* + \underline{D}_{eN}(\underline{\overset{\circ}{u}}^*)D\underline{E}\ \underline{D}_{kL}\underline{\overset{+}{u}}) = O\ . \tag{7.5}$$

Herein, λ may be considered as a load-uprating parameter. Its critical value can be determined by the linear Eigenvalue-problem (7.5). Finally by neglecting flexural effects in the fundamental state and also their deformations:

$$\underline{\overset{\circ}{\sigma}}^* = \begin{bmatrix} \underline{\tilde{n}}^{(\alpha\beta)} \\ \underline{m}^{(\alpha\beta)} \end{bmatrix} \approx \begin{bmatrix} \underline{\tilde{n}}^{(\alpha\beta)} \\ \underline{O} \end{bmatrix} = \underline{\overset{\circ}{\sigma}}^{**},\ \underline{\overset{\circ}{u}}^* \approx O,$$

we obtain the classical approximation of stability theory:

$$\underline{D}_{eL}\ D\underline{E}\ \underline{D}_{kL}\ \underline{\overset{+}{u}} + \lambda\underline{D}_{eN}(\underline{\overset{+}{u}})\ \underline{\overset{\circ}{\sigma}}^{**} = \underline{O}. \tag{7.6}$$

References

1. Başar, Y.: Eine Schalentheorie endlicher Verformungen und ihre Anwendung zur Herleitung der Stabilitätstheorie. ZAMM 52 (1972), pp. 197-211.

2. Başar, Y.: Zum allgemeinen Begriff der Stabilität und zur Stabilitätstheorie von Schalen. Konstruktiver Ingenieurbau Berichte, Vulkan-Verlag Essen 1973, pp. 5-24.

3. Başar, Y.; Krätzig, W.B.: Struktur konsistenter Grundgleichungen für das Beul- und Nachbeulverhalten allgemeiner Flächentragwerke. Der Stahlbau 46 (1977), pp. 138-144.

4. Başar, Y.; Krätzig, W.B.: Mechanik der Flächentragwerke, Theorie - Berechnungsmethoden - Anwendung, to appear in Vieweg Verlag.

5. Green, A.E.; Zerna, W.: Theoretical elasticity, 2nd ed., Clarendon Press, Oxford 1968.

6. Koiter, W.T.: On the nonlinear theory of thin shells. Proc. Kon. Ned. Akad. Wet., 1966, Ser. B, Vol. 69, No. 1, pp. 1-54.

7. Krätzig, W.B.: Optimale Schalengrundgleichungen und deren Leistungsfähigkeit. ZAMM 54 (1974), pp. 265-276.

8. Pietraszkiewicz, W.: Introduction to the non-linear theory of shells, Ruhr-Universität Bochum, Mitt. Inst. für Mechanik, Nr. 10, Mai 1977.

9. Pietraskiewicz, W.: Finite rotations and Lagrangean describtion on the non-linear theory of shells, Polish Scientific Publishers, Warszawa-Poznan 1979.

10. Sanders, J.L.: Non-linear theories for thin shells. Quart. Appl. Math., 1963, Vol. 21, pp. 21-36.

11. Simmonds, J.G.; Danielson, D.A.: Non-linear shell theory with finite rotation and stress-function vectors. J. Appl. Mech., Trans. ASME, Ser. E, 1972, Vol. 39, No. 4, pp. 1085-1090.

Inelastic Buckling of Thin, Prismatic Flat-Walled Structures

A. MARINETTI and G. OLIVETO

Istituto di Scienza delle Costruzioni
Università di Catania, Catania

Summary

A general sinusoidal stiffness matrix for the buckling analysis of thin, prismatic flat-walled elastic-plastic structures is derived. The constitutive laws of the flow-theory of plasticity associated with smooth yield surfaces are used. The use of this stiffness matrix to predict the buckling stresses and the associated modes is illustrated and some numerical results are used to show the range of applications which can be covered.

Introduction

A little more than a decade ago W. H. Wittrick, [1], derived a general sinusoidal stiffness matrix for the buckling and vibration analyses of elastic, thin flat-walled structures. Later, [2], he and F. W. Williams presented what they called an 'infallible' algorithm for finding any frequency of vibration or buckling load of a continuous (i.e. infinite dimensional) elastic structure. This algorithm was applied to frames, [3], and plate assemblies, [4]. The present paper is an attempt to extend some of the results in [1] and [4] in the elastic-plastic range. The incremental partial differential equations of equilibrium of a component plate are first presented in terms of forces. Then by using the constitutive laws of the flow theory of plasticity in the form given by Kachanov, [5], these equations are expressed in terms of displacement increments. In taking this step no allowance is made for unloading, which according to a generalized Shanley's analysis, though hardly justifiable, should provide the lower critical load, [5]. By retaining the fundamental hypothesis, made in [1] and [4], of sinusoidal modes of buckling, the partial differential equations of incremental equilibrium are reduced to ordinary ones. Solving these equations and applying the kinematic as well as the natural boundary conditions results in a trascendental stiffness matrix. The stiffness matrices of all the component plates may be assembled, according to standard rules of structural analysis, as was shown in [4], into the overall stiffness matrix of the assembly. The problem of finding the buckling stresses and the corresponding modes of a plate assembly reduces therefore to the solution of the non-linear eigenvalue problem

$$K\, \underline{d} = \underline{0} \tag{1}$$

in which $\underline{d}$ is the vector of the amplitudes of the sinusoidal nodal displacements and the entries of K are trascendental functions of the stresses and of the related properties of the material. The algorithm of Wittrick and Williams, [2], [4], may be applied successfully to the solution of problem (1) as far as the buckling mode (modes) corresponding to the required load does (do) not imply $\underline{d} = \underline{0}$. When this is the case a rather subtle but ingenious argument developed in [4], for the elastic problem, needs careful consideration in the present case which, as yet, has not been given. However in buckling problems it is the first buckling load which is of practical interest, and this cannot correspond to $\underline{d} = \underline{0}$.

Equilibrium equations

In what follows it will be assumed that the state of stress in the fundamental (i.e. unbuckled) configuration of equilibrium consists only of the constant (space independent) compressive generalized stress $\overline{N}_x$, fig. 1. The incremental equilibrium equations of a component plate may, therefore, be written as

$$\frac{\partial^2}{\partial x^2}(\delta M_x)+2\frac{\partial^2}{\partial x\partial y}(\delta M_{xy})+\frac{\partial^2}{\partial y^2}(\delta M_y)+\bar{N}_x\frac{\partial^2}{\partial x^2}(\delta w) = 0$$

$$\frac{\partial}{\partial x}(\delta N_x)+\frac{\partial}{\partial y}(\delta N_{xy})-\bar{N}_x\frac{\partial^2}{\partial x^2}(\delta u) = 0 \tag{2}$$

$$\frac{\partial}{\partial x}(\delta N_{xy})+\frac{\partial}{\partial y}(\delta N_y)-\bar{N}_x\frac{\partial^2}{\partial x^2}(\delta v) = 0$$

These are identical with those used in [4] and may be derived on the base of Novozhilov's non-linear theory.

Constitutive laws

The constitutive laws of an elastic-plastic hardening medium, based on the flow theory of plasticity, will be used in the form reported by Kachanov, [5],

$$\delta\sigma_x = \frac{E}{(1-\nu^2)\omega_1}(A_{yy}\delta\varepsilon_x - A_{xy}\delta\varepsilon_y)$$

$$\delta\sigma_y = \frac{E}{(1-\nu^2)\omega_1}(-A_{xy}\delta\varepsilon_x + A_{xx}\delta\varepsilon_y) \tag{3}$$

$$\delta\tau_{xy} = G\ \delta\gamma_{xy}$$

where

$$A_{xx} = E/E_t$$

$$A_{xy} = -\nu-(E/E_t-1)/2 \tag{4}$$

$$A_{yy} = 1+(E/E_t-1)/4$$

$$\omega_1 = 1+(E/E_t-1)(5-4\nu)/(4(1-\nu^2))$$

are given in terms of the Young's modulus E, of the tangent modulus E_t, and of the Poisson's ratio ν. The corresponding relations for unloading will not be needed here, but may be obtained from (3) and (4) by setting $E_t = E$.

Displacement's form of the equilibrium equations

When the strain increments $\delta\varepsilon_x$, $\delta\varepsilon_y$, $\delta\gamma_{xy}$ have been expressed in terms of displacement's increments δu, δv, δw, equations (3) may be used to derive through classical equilibrium considerations the following expressions for the increments of the generalized stresses.

$$\delta M_x = -\frac{D}{\omega_1}\left[A_{yy}\frac{\partial^2}{\partial x^2}(\delta w) - A_{xy}\frac{\partial^2}{\partial y^2}(\delta w)\right]$$

$$\delta M_y = -\frac{D}{\omega_1}\left[-A_{xy}\frac{\partial^2}{\partial x^2}(\delta w) + A_{xx}\frac{\partial^2}{\partial y^2}(\delta w)\right]$$

$$\delta M_{xy} = -D(1-\nu)\frac{\partial^2}{\partial x\partial y}(\delta w)$$

$$\delta N_x = \frac{Eh}{(1-\nu^2)\omega_1}\left[A_{yy}\frac{\partial}{\partial x}(\delta u) - A_{xy}\frac{\partial}{\partial y}(\delta v)\right] \tag{5}$$

$$\delta N_y = \frac{Eh}{(1-\nu^2)\omega_1}\left[-A_{xy}\frac{\partial}{\partial x}(\delta u) + A_{xx}\frac{\partial}{\partial y}(\delta v)\right]$$

$$\delta N_{xy} = Gh\left[\frac{\partial}{\partial y}(\delta u) + \frac{\partial}{\partial x}(\delta v)\right]$$

Substitution of (5) into (2) leads to the following system of partial differential equations in terms of displacements

$$\left[g\frac{\partial^4}{\partial x^4} + r\frac{\partial^4}{\partial x^2\partial y^2} + f\frac{\partial^4}{\partial y^4} + s\bar{N}_x\frac{\partial^2}{\partial x^2}\right](\delta w) = 0$$

$$d\frac{\partial^2}{\partial x^2}(\delta u) + b\frac{\partial^2}{\partial x\partial y}(\delta v) + a\frac{\partial^2}{\partial y^2}(\delta u) = 0 \tag{6}$$

$$m \frac{\partial^2}{\partial x^2}(\delta v)+b \frac{\partial^2}{\partial x \partial y}(\delta u)+f \frac{\partial^2}{\partial y^2}(\delta v) = 0$$

where

$$g = A_{yy} \quad ; \quad d = g-2(1+\nu)a\sigma/E$$

$$r = 2\left[(1-\nu)\omega_1 - A_{xy}\right] \quad ; \quad b = \omega_1(1-\nu)/2-A_{xy}$$

$$f = A_{xx} \quad ; \quad a = \omega_1(1-\nu)/2$$

$$s = \omega_1/D \quad ; \quad m = a\left[1-2(1+\nu)\sigma/E\right]$$

$$D = E\, h^3/(12(1-\nu^2)) \quad ;$$

Reduction to ordinary differential equations

The fundamental hypothesis of longitudinally sinusoidal modes of buckling made by Wittrick, [1], allows for the system of partial differential equations (6) to be reduced to an ordinary one. This hypothesis obviously restricts the class of structures which can be analysed but simplifies the analysis to such an extent that exact solutions may be sought. Accordingly we shall write

$$\{\delta u \; ; \; \delta v \; ; \; \delta w\} = \{U(y) \; ; \; V(y) \; ; \; W(y)\}\exp(i\pi x/\lambda) \qquad (7)$$

where λ is half of the wave length of the mode of buckling. Substitution of (7) into (6) provides

$$f \frac{d^4 W}{dy^4} -(\pi/\lambda)^2 r \frac{d^2 W}{dy^2} +(\pi/\lambda)^2 (g(\pi/\lambda)^2 - s\bar{N}_x)W = 0$$

$$a \frac{d^2 U_*}{dy^2} +(\pi/\lambda) b \frac{dV}{dy} -(\pi/\lambda)^2 d U_* = 0 \qquad (8)$$

$$f \frac{d^2 V}{dy^2} -(\pi/\lambda) b \frac{dU_*}{dy} -(\pi/\lambda)^2 m V = 0$$

having set $U_*(y) = -\sqrt{-1}\, U(y) = -iU(y)$.
These equations are uncoupled for out of plane and in plane displacements and may, therefore, be treated separately.

The sinusoidal stiffness matrix

This stiffness matrix relates the sinusoidal edge displacements, which arise when buckling occurs, to their force counterparts as shown in fig. 2. Edge forces and displacements may be expressed conveniently in the form

$$\{M_1 \,;\, P_{z1} \,;\, P_{y1} \,;\, P_{x1} \,;\, M_2 \,;\, P_{z2} \,;\, P_{y2} \,;\, P_{x2}\}\exp(i\pi x/\lambda)$$

$$\{\Psi_1 \,;\, W_1 \,;\, V_1 \,;\, U_1 \,;\, \Psi_2 \,;\, W_2 \,;\, V_2 \,;\, U_2\}\exp(i\pi x/\lambda)$$

with the understanding of taking the real or the imaginary part only. If we now define the vectors of the amplitudes of forces and displacements for edge j (j=1, 2) as

$$\underline{p}_j^t = \{M_j \,;\, P_{zj} \,;\, P_{yj} \,;\, iP_{xj}\}$$

$$\underline{d}_j^t = \{\Psi_j \,;\, W_j \,;\, V_j \,;\, iU_j\}$$

we may write these force-displacement relationships as

$$\underline{p}_1 = S_{11}\,\underline{d}_1 + S_{12}\,\underline{d}_2$$

$$\underline{p}_2 = S_{21}\,\underline{d}_1 + S_{22}\,\underline{d}_2$$

where S_{ij} are stiffness matrices which satisfy the reciprocity principle $S_{ij} = S_{ji}^t$. The immaginary unit i which is associated with P_{xj} and U_j accounts for a 90^0 phase difference between longitudinal and perpendicular components of forces and displacements. Owing to uncoupling between out-of-plane and in-plane terms, the above matrices are not full and take the form:

$$S_{11} = \begin{bmatrix} S_{MM} & -S_{MQ} & 0 & 0 \\ -S_{MQ} & S_{QQ} & 0 & 0 \\ 0 & 0 & S_{NN} & -S_{NT} \\ 0 & 0 & -S_{NT} & S_{TT} \end{bmatrix}$$

$$S_{22} = \begin{bmatrix} S_{MM} & S_{MQ} & 0 & 0 \\ S_{MQ} & S_{QQ} & 0 & 0 \\ 0 & 0 & S_{NN} & S_{NT} \\ 0 & 0 & S_{NT} & S_{TT} \end{bmatrix}$$

$$S_{12} = S_{21}^{t} = \begin{bmatrix} f_{MM} & f_{MQ} & 0 & 0 \\ -f_{MQ} & -f_{QQ} & 0 & 0 \\ 0 & 0 & -f_{NN} & -f_{NT} \\ 0 & 0 & f_{NT} & f_{TT} \end{bmatrix}$$

The entries of the above matrices may be evaluated by solving equations (8) and satisfying the boundary conditions on the longitudinal edges. The out-of-plane terms, after setting

$$\eta = \{1 + 4 \frac{f}{r^2} [(\lambda/\pi)^2 \text{sh}\sigma - g]\}^{1/2}$$

$$\alpha^2 = \frac{r}{2f} (1 + \eta) \quad ; \quad q_\alpha = \cosh \frac{\omega\alpha}{2}$$

$$\gamma^2 = -\gamma_*^2 = \frac{r}{2f} (1 - \eta) \quad ; \quad p_\alpha = \frac{1}{\alpha} \sinh \frac{\omega\alpha}{2}$$

$$\omega = \pi B/\lambda \quad ; \quad q_\gamma = \cosh \frac{\omega\gamma}{2} = \cos \frac{\omega\gamma_*}{2}$$

$$G_a = \frac{q_\alpha\ p_\gamma - p_\alpha\ q_\gamma}{\omega\eta\ (1 - \nu - \frac{A_{xy}}{\omega_1})} \quad ; \quad p_\gamma = \frac{1}{\gamma}\sinh\frac{\omega\gamma}{2} = \frac{1}{\gamma_*}\sin\frac{\omega\gamma_*}{2}$$

$$G_s = \frac{\alpha^2\ p_\alpha\ q_\gamma - \gamma^2\ q_\alpha\ p_\gamma}{\omega\eta\ (1 - \nu - \frac{A_{xy}}{\omega_1})}$$

may be written as

$$\frac{B}{D}\ (S_{MM} + f_{MM}) = 2\ \frac{p_\alpha\ p_\gamma}{G_a}$$

$$\frac{B}{D}\ (S_{MM} - f_{MM}) = 2\ \frac{q_\alpha\ q_\gamma}{G_s}$$

$$\frac{B^2}{D}\ (S_{MQ} + f_{MQ}) = \omega^2(1-\nu) - \omega\ \frac{p_\alpha\ q_\gamma + p_\gamma\ q_\alpha}{G_a}$$

$$\frac{B^2}{D}\ (S_{MQ} - f_{MQ}) = \omega^2(1-\nu) - \omega\ \frac{\alpha^2\ p_\alpha\ q_\gamma + \gamma^2\ q_\alpha\ p_\gamma}{G_s}$$

$$\frac{B^3}{D}\ (S_{QQ} + f_{QQ}) = 2\ \omega^2\ \frac{q_\alpha\ q_\gamma}{G_a}$$

$$\frac{B^3}{D}\ (S_{QQ} - f_{QQ}) = 2\ \omega^2(1-\eta^2)\ \frac{\omega_1^2}{A_{xx}^2}\ (1-\nu-\frac{A_{xy}}{\omega_1})^2\ \frac{p_\alpha\ p_\gamma}{G_s}$$

The in-plane terms may be written as

$$\frac{B}{Eh}\ (S_{NN} + f_{NN}) = \frac{q_\theta\ q_\phi}{H_s}$$

$$\frac{B}{Eh}\ (S_{NN} - f_{NN}) = \frac{f}{d}\ \theta^2\ \phi^2\ \frac{p_\theta\ p_\phi}{H_a}$$

$$\frac{B}{Eh}(S_{NT}+f_{NT}) = -\frac{\omega}{2(1+\nu)} + \frac{b}{f}\frac{1}{\phi^2-\theta^2}\frac{\phi^2\, p_\phi\, q_\theta - \theta^2\, p_\theta\, q_\phi}{H_s}$$

$$\frac{B}{Eh}(S_{NT}-f_{NT}) = -\frac{\omega}{2(1+\nu)} + \frac{b}{d}\frac{\theta^2\phi^2}{\theta^2-\phi^2}\frac{p_\phi\, q_\theta - p_\theta\, q_\phi}{H_a}$$

$$\frac{B}{Eh}(S_{TT}+f_{TT}) = \frac{d}{f}\frac{p_\phi\, p_\theta}{H_s}$$

$$\frac{B}{Eh}(S_{TT}-f_{TT}) = \frac{q_\theta\, q_\phi}{H_a}$$

having set

$$H_a = \frac{2(1+\nu)}{\omega d}\,\frac{(d-a\phi^2)\,\theta^2\, p_\theta\, q_\phi - (d-a\theta^2)\,\phi^2\, p_\phi\, q_\theta}{\theta^2-\phi^2}$$

$$H_s = \frac{2(1+\nu)}{\omega f}\,\frac{(d-a\phi^2)\, p_\phi\, q_\theta - (d-a\theta^2)\, p_\theta\, q_\phi}{\theta^2-\phi^2}$$

$$q_\theta = \cosh\frac{\omega\theta}{2} = \cos\frac{\omega\theta_*}{2} \quad ; \quad \theta^2 = -\theta_*^2 = \frac{1}{2}(S+T)$$

$$p_\theta = \frac{1}{\theta}\sinh\frac{\omega\theta}{2} = \frac{1}{\theta_*}\sin\frac{\omega\theta_*}{2} \quad ; \quad \phi^2 = -\phi_*^2 = \frac{1}{2}(S-T)$$

$$q_\phi = \cosh\frac{\omega\phi}{2} = \cos\frac{\omega\phi_*}{2} \quad ; \quad S = \frac{d}{a} + \frac{m}{f} - \frac{b^2}{af}$$

$$p_\phi = \frac{1}{\phi}\sinh\frac{\omega\phi}{2} = \frac{1}{\phi_*}\sin\frac{\omega\phi_*}{2} \quad ; \quad T^2 = -T_*^2 = S^2 - 4\frac{dm}{af}$$

It should be noticed that the above expressions for the coefficients of the stiffness matrix hold as far as η^2 and T^2 are positive. When η^2 becomes negative α^2 and γ^2 become a couple of conjugate complex numbers. Also (q_α, q_γ) and (p_α, p_γ) become conjugate couples and it may be easily checked that each out-of-plane term of the stiffness matrix remains real. The same argument may be used to show that also the in-plane terms remain real by considering the couples of conjugate complex numbers (θ^2, ϕ^2), (p_θ, p_ϕ), (q_θ, q_ϕ). For computational purposes it may be convenient to express the above matrices in terms of real variables. This has been done, [6], but the results are omitted here for sake of brevity.

Computational procedure

Once the stiffness matrices of each component plate have been evaluated and assembled into the overall stiffness matrix of the structure, the buckling problem becomes identical to the non-linear eigenvalue problem (1). The eigenvalues whose corresponding eigenmodes are characterized by $\underline{d} \neq \underline{0}$, may be obtained to any degree of accuracy by using a simplified version of the algorithm of Wittrick and Williams. Namely the stiffness matrix K, corresponding to a given value of the loading parameter and to a given number of half-wave lengths, is reduced to its upper triangular form by means of a Gaussian elimination without row interchanges. The number of negative entries in the main diagonal of this matrix will tell the number of the critical values of the load parameter which are smaller than the given one. In this way any number (not larger than the order of K) of critical values of the loading parameter corresponding to a given half-wave length may be efficiently separated and evaluated to the required degree of accuracy by means, for example, of a bisection procedure. Whole fractions of the length of the structure are assumed as successive half-wave lengths until these become sufficiently small and the critical values of the loading parameter sufficiently large. In this way a number of increasing eigenvalues may be selected. When only the first critical load is required, as often is the case, the procedure becomes much more simpler and efficient as only the lowest critical value will be required for any half-wave length. Once the required eigenvalue has been evaluated, the corresponding eigenmode (eigenmodes, if multiple) may be evaluated by standard methods.

Numerical applications

A general computer programme has been developed on the base of the theory outlined above. Any elastic-plastic hardening behaviour may be accomodated with little effort. The following types of tensile stress-strain curves have been incorporated:

1) Linear: $\delta\sigma=E(\delta\varepsilon)$

2) Bilinear: $\delta\sigma=E(\delta\varepsilon)$ for $\sigma<\sigma_y$ and $\delta\sigma=E_t(\delta\varepsilon)$ for $\sigma\geqq\sigma_y$
where $E_t=$ constant

3) Linear-hyperbolic:
$\delta\sigma=E(\delta\varepsilon)$ for $\sigma\leqq\sigma_p$ and $\delta\sigma=E_t(\delta\varepsilon)$ for $\sigma_p\leqq\sigma\leqq\sigma_y$
where σ_p is the proportional limit and

$$E_t = E\ \{1 - [(\sigma - \sigma_p)/(\sigma_y - \sigma_p)]^2\}$$

4) Ramberg-Osgood: $\varepsilon=(\sigma/E)+(\sigma/B)^n$ or $\delta\sigma=E_t(\delta\varepsilon)$

$$E_t = E\ /\ \{1 + nE(\sigma/B)^{n-1}\}$$

5) Parabolic: $\delta\sigma=E_t(\delta\varepsilon)$, $E_t=E(1 - \sigma/\sigma_y)^{1/2}$

In order to test the computer programme some numerical applications have been made with reference to relatively simple structures, as the cruciform and the box columns, fig. 3 a, b, which have been extensively studied in the past and for which are available trustworthy results. The former, which has been quite recently reconsidered by Hutchinson and Budiansky, [7], has the interesting feature of maintaining the critical load, in torsional modes, unaltered whether the linearly elastic behaviour is considered or the elastic-plastic one, namely $\sigma_c=G(h/B)^2$. Our numerical investigations have shown, fig. 4, that depending on the ratio L/B the first elastic critical load may occur either in a torsional or in a flexural mode, the latter occurring at higher values of L/B . Moreover the critical load corresponding to the torsional modes tends to the value given above for large values of L/B . Both curves relative to torsional and flexural modes are modified in the plastic range as is shown in fig. 5, relative to the linear-hyperbolic stress-strain relationship. However a rather more complete picture may be obtained from figs. 6, 7, relative to the linear-hyperbolic and the Ramberg-Osgood stress-strain relationships. As may be seen from fig. 6, where the broken lines represent what would be the results if the behaviour were linear-elastic, all curves corresponding to critical values larger than σ_y, in the elastic range, coalesce intc a single curve in the plastic range, whether the corresponding mode is flexural or not. For those curves corresponding to values of σ_c less than σ_y two points are worth noticing. One is that, even though the difference is negligible in the critical loads corresponding to torsional modes for elastic or elastic-plastic behaviour at large values of L/B , an appreciable variation is noticeable for small values of L/B , when the torsional mode is accompanied by flexural deformation of the cross-section. The second one is that for large values of L/B , though elastically the first mode would be torsional, in the plastic range it is not and the corresponding critical value is considerably smaller. The same trend may be observed in fig. 7, for the Ramberg-Osgood stress-strain relationship. The results concerning the box column are reported in figs. 8, 9, with reference to a linear-hyperbolic stress-strain relationship. The broken lines, as usual, refer to what would be the linear-elastic behaviour while the horizontal lines in fig. 8, represent the gross behaviour in the local buckling of the columns and were obtained by joining the points corresponding to whole values of L/B . A more detailed investigation has instead shown the behaviour reported in fig. 9, which is what should be expected for the problem at hand. It may be noticed that as the number of half-wave lengths in the local buckling increases, the actual behaviour approaches the gross behaviour reported in fig. 8. Besides this aspect most of the observations concerning the cruciform column remain valid in the present case, as may be gathered from fig.8. Further numerical investigations have been performed with reference to the panel of fig. 3c. The behaviour of this model is much more involved than the previous simple ones but some useful information is obtainable by plotting the critical stress σ_c against the half-wave length λ for a few lower modes, as done in fig. 10 using dimensionless variables. A linear-hyperbolic stress-strain relationship was used with plastic buckling occurring in the range $0.8 \leqq \sigma_c/\sigma_y \leqq 1$. The broken lines, as before, refer to an indefinite linear elastic behaviour. A rather interesting feature, also observed in [8], is that for small values of λ the modes tend to collect in bunches (of two modes in the present case). The line labeled 4-6 corresponds to one of those bunches and not, as it might appear, to a double eigenvalue, as may clearly be seen when it splits into two in the plastic range. The first four modes are shown for $\lambda/B = 10$ in fig. 11 where the broken lines refer to the linear elastic behaviour. It might also be interesting to notice that higher modes, though well separated elastically, tend to jam on the line $\sigma_c = \sigma_y$ in the plastic range. This being because of the particular stress-strain relationship which has been used in the calculations. If a Ramberg-Osgood law had been used instead, this would have not occurred to such an extent.

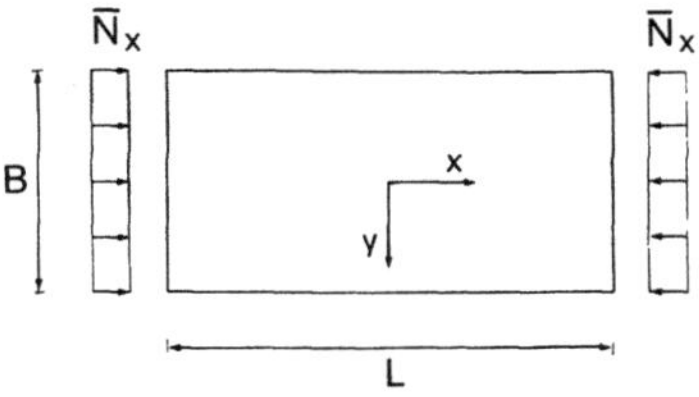

Fig.1

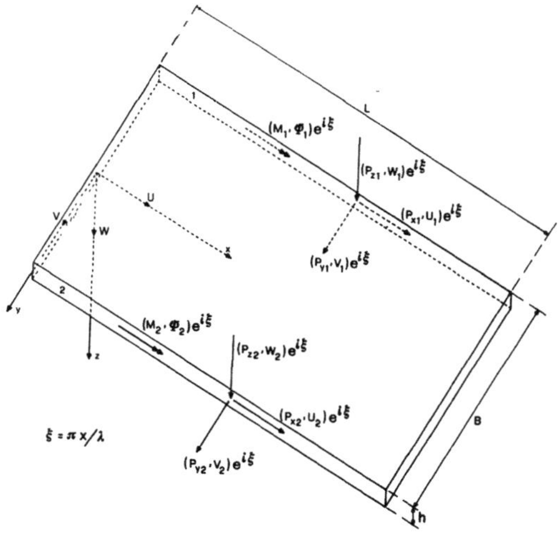

Fig.2

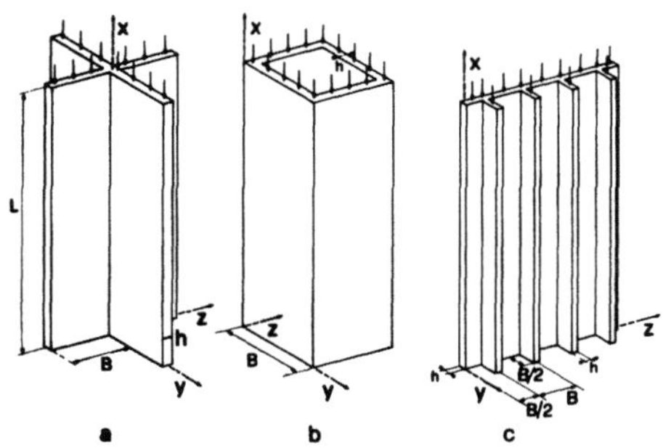

Fig.3

material	parameters
linear-elastic	E=206 KN/mm^2; ν=0.30 E=69 " ; ν=0.36
linear-hyperbolic	E=206 KN/mm^2; ν=0.30 σ_p=283 N/mm^2 σ_y=353 "
Ramberg-Osgood	E=69 KN/mm^2 ; ν=0.36 B=299 N/mm^2 n=10.24

Table 1. Numerical values of the mechanical parameters used in the calculations

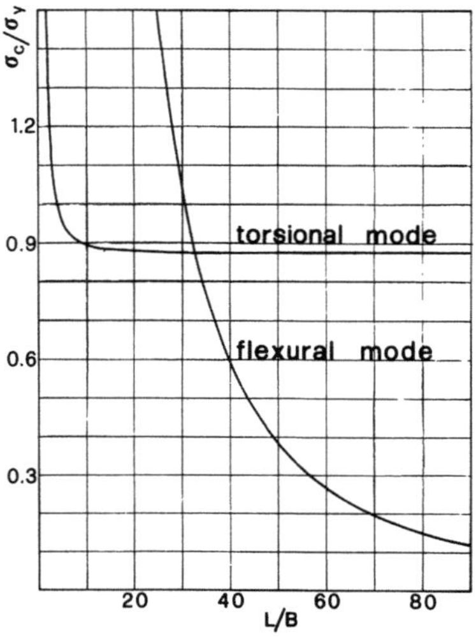

Fig.4.Cruciform column, elastic buckling. B/h = 16

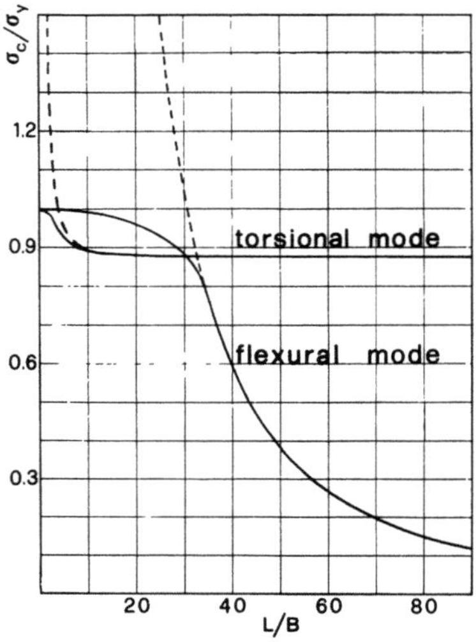

Fig.5.Cruciform column,elastic-plastic buckling. B/h = 16
----- linear
——— linear-hyperbolic

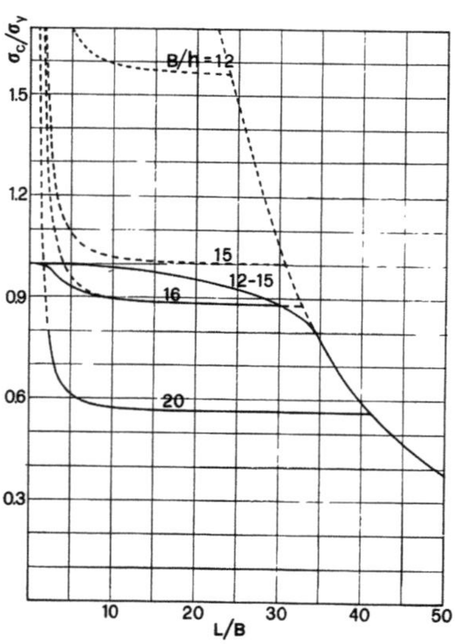

Fig.6.Cruciform column elastic-plastic buckling
----- linear
——— linear-hyperbolic

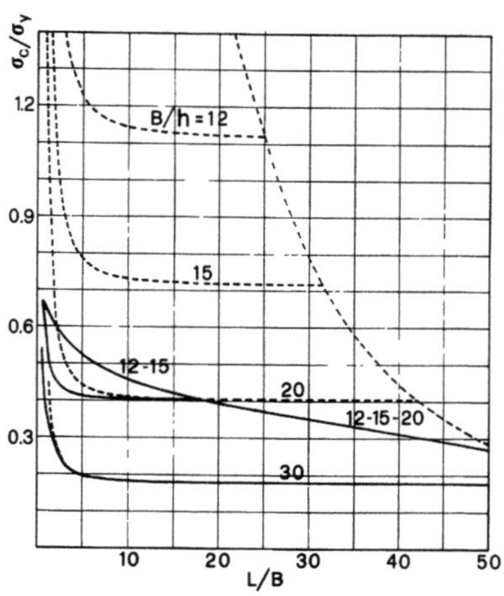

Fig.7.Cruciform column elastic-plastic buckling
----- linear
——— Ramberg-Osgood

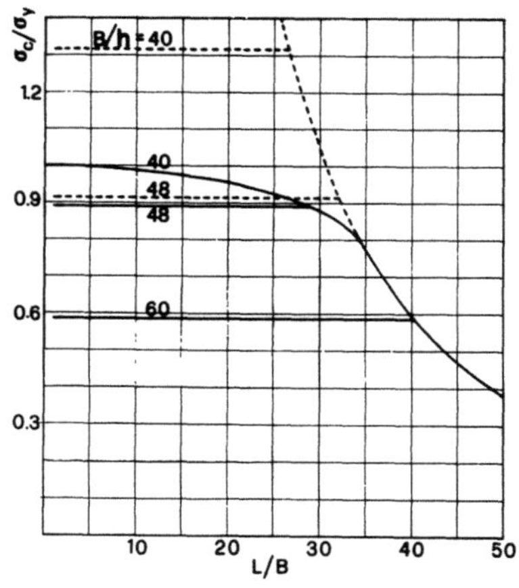

Fig.8.Box column
elastic-plastic buckling
----- linear
——— linear-hyperbolic

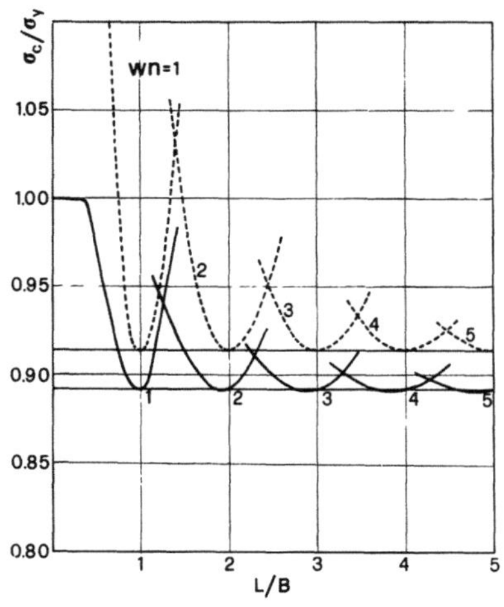

Fig.9.Box column
local buckling
----- linear
———linear-hyperbolic

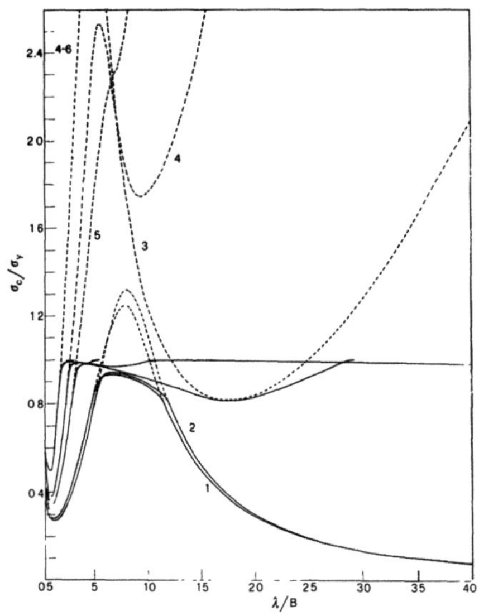

Fig.10.Stiffened panel. B/h=75
elastic-plastic buckling
----- linear
——— linear-hyperbolic

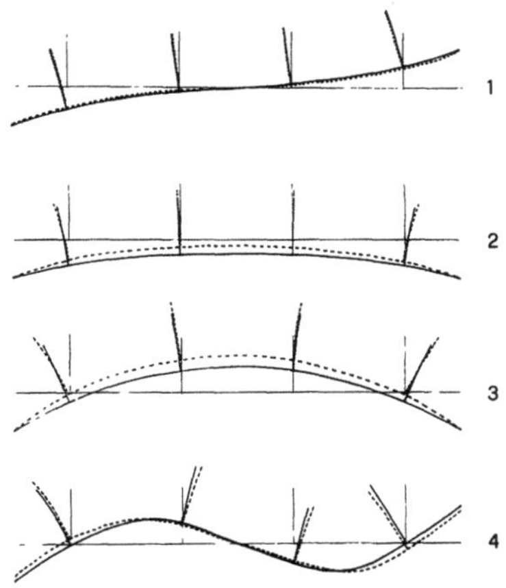

Fig.11.Stiffened panel. $\lambda/B=10$
buckling modes

Acknowledgments

This work has been supported by grants of the National Research Council of Italy (C.N.R.).

References

1. Wittrick, W.H.: General Sinusoidal Stiffness Matrices for Buckling and Vibration Analyses of thin flat-walled Structures. Int.J.mech.Sci. 10 (1968) 949-966.

2. Wittrick, W.H.; Williams, F.W.: A General Algorithm for Computing Natural Frequencies of Elastic Structures. Quart.Journ. Mech.and Applied Math., Vol.XXIV,Pt.3 (1971) 263-284.

3. Wittrick, W.H.; Williams, F.W.: An Algorithm for Computing Critical Buckling Loads of Elastic Structures. J.of Structural Mechanics 1 (1973) 497-518.

4. Wittrick, W.H.; Williams, F.W.: Natural Vibrations of thin, Prismatic Flat-Walled Structures. Proc.IUTAM SYMPOSIUM on high Speed Computing of Elastic Structures,Liège (1970) 563-588.

5. Kachanov, L.M.: Foundations of the theory of Plasticity. North-Holland (1971).

6. Marinetti, A.; Oliveto, G.: Instabilità in campo inelastico di Strutture a elementi piani di spessore sottile. Internal Report n.37, Istituto di Costruzioni Civili, Università di Catania (1981).

7. Hutchinson, J.W.; Budiansky, B.: Analytical and Numerical Study of the Effects of Initial Imperfections on the Inelastic Buckling of a Cruciform Column. Buckling of Structures. Budiansky, B.(ed) Springer-Verlag (1976) 98-105.

8. Williams, F.W.; Wittrick, W.H.; Plank, R.J.: Critical Buckling Loads of Some Prismatic Plate Assemblies. Buckling of Structures. Budiansky, B.(ed) Springer-Verlag (1976) 17-26.

On the Buckling Stability of a Thick Elastic Plate *)

K.N. SAWYERS, R.S. RIVLIN

Lehigh University, Bethlehem, PA., USA

The onset of (flexural) buckling in a very slender, straight elastic column occurs at a certain critical load, the value of which may be calculated on the basis of Euler's classical theory. At the critical point the second variation of the potential energy of the system is positive, and the buckling is deemed stable. In other words, buckling occurs as a smooth bifurcation from the initially straight state to a nearby equilibrium configuration.

That such stable behavior might not necessarily prevail for less-slender elastic members has been demonstrated by investigating the buckling problem for a thick plate of neo-Hookean material [1,2]. With respect to the lowest-order flexural buckling mode, it was found that stability of the underlying critical state ceases to obtain when the aspect ratio (thickness/length) exceeds approximately 0.2. In contrast to the very slender column, the thick plate may also suffer bifurcations of the bulging, or barreling, type. For these, the underlying critical state was shown to be stable for modes of all orders and for all values of the aspect ratio. These results, for both flexure and barreling, are independent of the single elastic constant which completely characterizes a particular neo-Hookean material.

Recent work, extending the above results, has been directed towards determining the dependence of the limits of stable buckling regimes on the elasticity parameters of a plate of general isotropic, incompressible material. The strain-energy function, W, for such a material is taken to depend in an arbitrary manner on the first two principal invariants of strain. The purpose of

*) Abstract of the lecture held in Nümbrecht

this paper is to report on some of the latest results of this investigation.

Certain features that emerge in the formulation of the general problem will be discussed before describing the results of an asymptotic analysis for the case of a thin plate. This analysis shows how the classical buckling problem fits into the framework of the general nonlinear theory. The result of the asymptotic analysis gives an approximation to the value of the aspect ratio, which limits the regime of stable Euler-type buckling, explicitly in terms of a quantity that depends on the first and second derivatives of W with respect to the strain invariants, these derivatives being evaluated in the unstrained state.

---

[1] K.N. Sawyers and R.S. Rivlin, Stability of a thick elastic plate under thrust, J. Elasticity (in the press).

[2] K.N. Sawyers, Stability of a thick neo-Hookean plate, J. Elasticity (in the press).

Large Deflections and Stability of Thin-Walled Beam Structures

H. Obrecht, W. Wunderlich, V. Schrödter

Ruhr-Universität Bochum
Institut für Konstruktiven Ingenieurbau
Bochum, West-Germany

Summary

The buckling behavior of thin-walled beam structures is investigated by means of a theory applicable to the analysis of straight and curved beams undergoing large displacements and rotations. Bifurcation as well as limit point buckling is considered, and attention is focused on cases where loss of stability may be associated with large coupled displacements in bending and in torsion. The influence of the cross-sectional shape, the point of load application and of restrained warping in torsion is explored.

1. Introduction

It is well-known that the global buckling behavior of elastic beams is characterized by two aspects. Firstly, their post-buckling behavior is usually stable. This means that in the post-buckling regime their load-deflection equilibrium curves have non-negative slopes, and that their load-carrying capacity may increase considerably. Secondly, the deflections increase rapidly in the neighborhood of the bifurcation load. Thus, individual elastic beams may be considered safe but, from a practical standpoint, useless at loads above the classical critical value. In these cases the bifurcation load is a good indicator of the actual load-carrying capacity.

In contrast, the buckling behavior of elastic built-up frame structures may be fundamentally different because the influence of higher-order geometric nonlinearities (in particular those due to the prebuckling deformations) often plays a crucial role. As a result, the postbuckling behavior may be unsymmetric with a stable as well as an unstable equilibirum branch. On the stable side the static equilibrium curve rises and may reach a maximum at a load larger than the bifurcation value. Vice versa, on the unstable side stability is lost at a limit point smaller than the bifurcation load. This implies decreasing load-carrying capacity in that postbuckling range and a sensitivity to imperfections. Both, upper and lower, load maxima may be associated with large deflections. In these cases the classical bifurcation load loses much of its significance. Likewise, asymptotic analyses based on the analysis of the structure in the immediate vicinity of the bifurcation point, are not sufficient to completely describe the behavior along the stable branch.

Therefore, a full assessment of the load-carrying behavior of beam structures requires a fully nonlinear analysis which accounts for such important factors as shape of the cross-section, warping, and point of load-application. Such an analysis is presented here on the basis of a beam theory for thin-walled beams with arbitrary cross-sections undergoing large displacements and rotations. Effects of warping constraints and imperfections are included. Elastic structures which lose stability in a global mode with associated large coupled displacements in bending and in torsion are emphasized. Attention is thus focused on geometrically nonlinear effects. Additional factors which may have a significant detrimental influence, such as plastic deformations or interactive local buckling, are specifically excluded from consideration.

2. Basic relationships and beam theory

A comprehensive description of the beam theory used in this investigation was given in [1] . To avoid repetition only several of the basic relationships will be summarized, and a few points of interest will be treated in some more detail.

The theory is applicable to the analysis of general curved beam structures undergoing large, three-dimensional displacements. It fully accounts for the influences of the cross-sectional shape and of restrained warping in torsion. The formulation is developed from a generalized variational principle for finite elastic displacements of Hellinger-Reissner type. It is related to the complementary energy principle due to Fraeijs de Veubeke [2] , and differs from similar principles in that it contains the rotations of principal lines as independent variables. This gives rise to a non-linear moment-equilibrium equation. Concepts relating to the derivation of beam theories from generalized variational principles have been drawn from [3,4] .

A beam theory following similar lines, but excluding the influence of arbitrary cross-sections as well as effects due to non-linear coupling of bending and torsion, has recently been given in [5] .

Basic to the derivation is the use of the polar decomposition theorem as the fundamental kinematical relationship. It connects the deformation gradient tensor $\underline{F}$, a Lagrangian quantity, with the rigid-rotation tensor $\underline{\alpha}$ and the right-stretch tensor $\underline{U}$ as follows

$$\underline{F} = \underline{\alpha}\, \underline{U} = \underline{\alpha}\, (\underline{I} + \underline{h}). \tag{1}$$

In (1) $\underline{U}$ has been expressed in terms of the Lagrangian strain measure $\underline{h}$, often called the "engineering strain" tensor. The latter is related to the familiar Green-Lagrange strain tensor $\underline{\varepsilon}$ via $\underline{\varepsilon} = \underline{h} + \frac{1}{2}\underline{h}^2$. Both tensors have identical principal axes and, for small strains, equal magnitude.

The description of the deformation of general curved beams is greatly simplified by the use of convected coordinates. Deformed base vectors $\underline{g}_a$ and initial, undeformed, base vectors $\underline{G}_a$ are then related by (see Fig. 1)

$$\underline{g}_a = \underline{F}\,\underline{G}_a = F^{b}_{\cdot a}\,\underline{G}_b = U^b_a\,\hat{\underline{G}}_b, \tag{2}$$

where

$$\hat{\underline{G}}_b = \underline{\alpha}\,\underline{G}_b. \tag{2a}$$

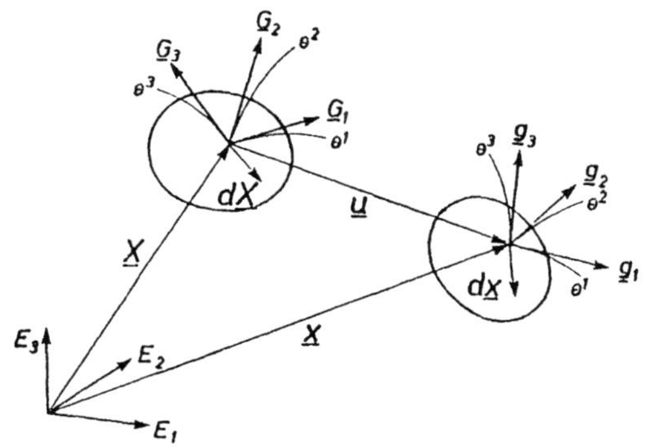

Fig. 1 Base vectors

In (2) $F^{b}_{\cdot a}$ and U^{b}_{a} denote the components of $\underline{F}$ and $\underline{U}$. Both are referred to the initial base vectors. It is seen that the deformed vectors $\underline{g}_a$ are obtained by first performing a rigid rotation of $\underline{G}_b$, resulting in $\underline{\hat{G}}_b$, and then applying the stretch $\underline{U}$.

For the rate of change of the rigidly rotated base vector $\underline{\hat{G}}_b$ about a current, deformed position one finds

$$\dot{\underline{\hat{G}}}_b = \dot{\underline{\alpha}}\, \underline{G}_b = \dot{\underline{\omega}}\, \underline{\hat{G}}_b \, , \tag{3}$$

where

$$\dot{\underline{\omega}} = \dot{\underline{\alpha}}\, \overset{o}{\underline{\alpha}}{}^{T} \, . \tag{3a}$$

There is a slight change in notation here compared to that in [1] . Whereas there rate quantities were not labelled specifically, they are here marked by a dot ($\dot{}$). This is done to adhere to the usual notation when discussing rate-constitutive relationships. As in [1] , current values of quantities are identified by superscripts (^{o}). Thus, in (3a) $\overset{o}{\underline{\alpha}}{}^{T}$ is the transpose of the rigid rotation tensor of the deformed reference configuration, whereas $\dot{\underline{\alpha}}$ is the rate of change of $\underline{\alpha}$ evaluated at $\overset{o}{\underline{\alpha}}$.

The spin tensor $\dot{\underline{\omega}}$ in (3a) is skew-symmetric due to the orthogonality condition $\overset{o}{\underline{\alpha}}\, \overset{o}{\underline{\alpha}}{}^{T} = \underline{I}$. It can be expressed directly in terms of the rates of change of the rotation angles $\dot{\underline{\varphi}}$ about the current vectors $\underline{\hat{G}}_b$ (see Fig. 2 and (4)).

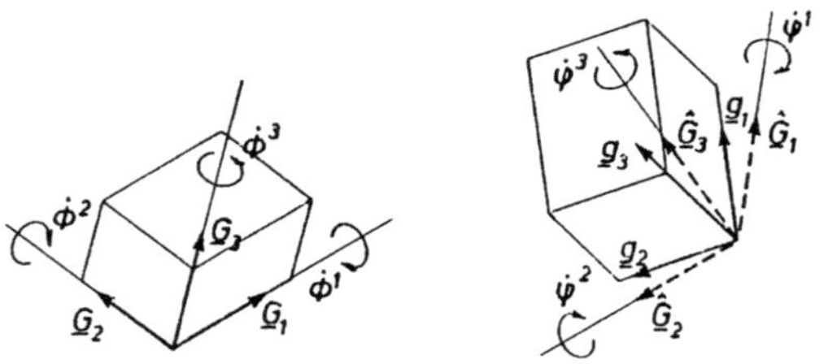

$$\dot{\underline{\omega}} = \begin{bmatrix} 0 & \dot{\varphi}^3 & -\dot{\varphi}^2 \\ -\dot{\varphi}^3 & 0 & \dot{\varphi}^1 \\ \dot{\varphi}^2 & -\dot{\varphi}^1 & 0 \end{bmatrix} \tag{4}$$

Fig. 2 Material element

From Figure 2 it is also seen that if shear strains vanish, and if the vectors $\underline{G}_b$ are orthogonal, then $\underline{g}_b$ will also be orthogonal. Specifically, $\underline{g}_b$ may be taken to coincide with the principal strain triad. This property will be used in the description of the three-dimensional displacement of a beam cross-section.

From a consideration of the strain energy density, written in terms of the symmetric strain tensor $\underline{h}$, one is lead quite naturally to the definition of the "engineering", or "Jaumann", stress tensor $\underline{r}$([2]). It is related to the unsymmetric Lagrangian stress tensor $\underline{t}$, the symmetric second Piola-Kirchhoff stress tensor $\underline{S}$ and the symmetric weighted stress tensor $\underline{\sigma}$, sometimes referred to as "Kirchhoff" stress, in the following way

$$\underline{r} = \underline{t}\ \underline{\alpha} = \underline{S}\ \underline{U} = \underline{F}^{-1}\ \underline{\sigma}\ \underline{\alpha}, \tag{5}$$

where $\underline{\sigma}$ is given in terms of Cauchy stress $\underline{\tau}$ by $\underline{\sigma} = (\rho_o / \rho)\ \underline{\tau}$, and ρ is the current material density. From the first relation in (5) it is seen that for vanishing rotation ($\underline{\alpha} = \underline{I}$) $\underline{r}$ coincides with $\underline{t}$. Hence the term "engineering stress". The second relation in (5) shows that for small strains, i.e. $|\underline{h}| << 1$, the difference between $\underline{r}$ and $\underline{S}$ is minimal for any value of the rotation. Thus, although unsymmetric in general, $\underline{r}$ may be assumed to be symmetric as long as strains remain small. It is therefore not necessary to invoke material isotropy when justifying the use of $\underline{r}$, rather than its symmetric part, in a standard semilinear constitutive relationship.

A central requirement in the formulation of incremental constitutive laws is that the stress rate measures employed be symmetric and objective, i.e. insensitive to material spin. One such measure in frequent use is the "Jaumann rate" of $\underline{\sigma}$: $\overset{*}{\underline{\sigma}} = \underline{\dot{\sigma}} - \underline{\dot{\omega}}\underline{\sigma} + \underline{\sigma}\ \underline{\dot{\omega}}$. Similarly, a "Jaumann rate" of engineering stress $\overset{*}{\underline{r}}$ may be defined by

$$\overset{*}{\underline{r}} = \underline{\dot{r}} + \overset{o}{\underline{r}}\ \underline{\dot{h}}\ . \tag{6}$$

Like $\underline{\dot{r}}$ it is objective. It is, however, also symmetric, which $\underline{\dot{r}}$ is not, as long as (total) strains $\underline{h}$ remain small. It may be shown that the components of $\overset{*}{\underline{r}}$ and $\overset{*}{\underline{\sigma}}$ coincide and that in a uniaxial stress state both reduce to the true stress rate. These properties make both rate measures particularly suitable for the formulation of incremental constitutive relations. A hypo-elastic, semilinear rate-constitutive relationship connecting the conjugate quantities $\overset{*}{\underline{r}}$ and $\underline{\dot{h}}$ may thus be written as follows

$$\overset{*}{r}{}^{ab} = L^{abcd}\ \dot{h}_{cd}, \tag{7}$$

where L^{abcd}, the tensor of instantaneous moduli, is assumed symmetric under ab $\leftrightarrow$ cd interchange. From (7) it follows that

$$\dot{r}^{ab} = \tilde{L}^{abcd}\,\dot{h}_{cd}\,, \tag{8}$$

where
$$\tilde{L}^{abcd} = L^{abcd} - \frac{1}{2}\,(G^{ac}\,\overset{o}{r}{}^{db} + G^{ad}\,\overset{o}{r}{}^{cb}) \tag{8a}$$

depends directly on the current stress state and G^{ac} is the undeformed metric. For uniquely invertible relationship (7) the following rate potentials

$$W(\dot{\underline{h}}) = \frac{1}{2}\,L^{abcd}\,\dot{h}_{ab}\,\dot{h}_{cd} \qquad W^*(\overset{*}{\underline{r}}) = \frac{1}{2}\,M_{abcd}\,\overset{*}{r}{}^{ab}\,\overset{*}{r}{}^{cd} \tag{9}$$

where
$$\overset{*}{r}{}^{ab} = \frac{\partial W}{\partial \dot{h}_{ab}} \qquad \text{and} \qquad \dot{h}_{ab} = \frac{\partial W^*}{\partial \overset{*}{r}{}^{ab}} \tag{9a}$$

as well as the contact transformation

$$W(\dot{\underline{h}}) + W^*(\overset{*}{\underline{r}}) = \overset{*}{\underline{r}}\,\dot{\underline{h}} \tag{10}$$

exist. In (9) the tensor of instantaneous compliances M_{abcd} is the inverse of L^{abcd}.

For stresses which are small compared to the instantaneous values of the moduli of the material, as is the case for elastic materials which are considered here, the second term in (8a) is small compared to the first. Then the complementary energy density in (9) reduces to the simple form

$$W^*(\dot{\underline{r}}) = \frac{1+\nu}{2E}\left[\dot{r}^a_b\,\dot{r}^b_a - \frac{\nu}{1+\nu}\,\dot{r}^c_c\,\dot{r}^d_d\right]. \tag{11}$$

It is noted that in the above discussion only the assumptions of small (compared to unity) strains and small (compared to the tangent moduli) stresses were made. No other assumptions implying symmetry or isotropy, usually invoked at this point, are necessary to justify (11). It is also noted that the same line of argument does not hold in the case of plasticity where the tangent moduli and stresses may have equal magnitude. Then the tensor $\tilde{L}^{abcd}$ in (8a) as well as L^{abcd} may become singular.

The rate-field-equations governing the general nonlinear deformation of an elastic body follow as Euler equations from the requirement that the following multi-field functional J_{CB} (see[1]) be stationary with respect to variations of the displacement rate $\dot{\underline{u}}$, the spin $\dot{\underline{\omega}}$, and the stress rate $\dot{\underline{r}}$, i.e. that

$$\begin{aligned} \delta J_{CB} = \delta\Big\{ \int_{V_o} \Big[& -W^*(\dot{\underline{r}}) + \mathring{\underline{\alpha}}^T \, \underline{\nabla} \, \dot{\underline{u}} \, \dot{\underline{r}} + \mathring{\underline{F}} \, \dot{\underline{r}} \, \mathring{\underline{\alpha}}^T \, \dot{\underline{\omega}}^T + \underline{\nabla} \, \dot{\underline{u}} \, \mathring{\underline{r}} \, \mathring{\underline{\alpha}}^T \, \dot{\underline{\omega}} \\ & + \mathring{\underline{\alpha}}^T \, \mathring{\underline{F}} \, \mathring{\underline{r}} \, \dot{\underline{\omega}} \, \dot{\underline{\omega}} - \rho_o \, \dot{\underline{f}} \, \dot{\underline{u}} \Big] dV_o - \int_{S_p} \dot{\bar{\underline{p}}} \, \dot{\underline{u}} \, dS \\ & - \int_{S_u} \dot{\underline{p}} \, (\dot{\underline{u}} - \dot{\bar{\underline{u}}}) \, dS \Big\} = 0 \, . \end{aligned} \tag{12}$$

In (12) $W^*(\dot{\underline{r}})$ is the complementary energy density defined in (11), $\underline{f}$ is the body force, and $\underline{p}$ the surface traction. The superscript $(\bar{\ })$ indicates that the value of the respective quantity is prescribed, and it is understood that all terms appearing in the integrands denote scalar functions.

From (12) one obtains the following local field equations:

$$\underline{\nabla} \, (\dot{\underline{r}} \, \mathring{\underline{\alpha}}^T + \mathring{\underline{r}} \, \mathring{\underline{\alpha}}^T \, \dot{\underline{\omega}}^T) + \rho_o \, \dot{\underline{f}} \, \dot{\underline{u}} = 0 \tag{13a}$$

$$(\underline{I} + \mathring{\underline{h}}) \, \dot{\underline{r}} + \dot{\underline{\omega}} \, \mathring{\underline{\alpha}} \, \mathring{\underline{F}} \, \mathring{\underline{r}} + \mathring{\underline{\alpha}}^T \, \underline{\nabla} \, \dot{\underline{u}} \, \mathring{\underline{r}} = \text{symmetric} \tag{13b}$$

$$\mathring{\underline{\alpha}}^T \, \underline{\nabla} \cdot \dot{\underline{u}} + \mathring{\underline{\alpha}}^T \, \dot{\underline{\omega}}^T \, \mathring{\underline{F}} - \frac{\partial Q^*}{\partial \dot{\underline{r}}} = 0 \tag{13c}$$

$$\underline{N} \, (\dot{\underline{r}} \, \mathring{\underline{\alpha}}^T + \mathring{\underline{r}} \, \mathring{\underline{\alpha}}^T \, \dot{\underline{\omega}}^T) - \dot{\bar{\underline{p}}} = 0 \quad \text{on } S_p \tag{13d}$$

$$\dot{\underline{u}} - \dot{\bar{\underline{u}}} = 0 \quad \text{on } S_u, \tag{13e}$$

where (13a) is the force equilibrium equation, (13b) the moment equilibrium equation, (13c) a disguised form of the polar decomposition theorem, (13d) the traction, and (13e) the displacement boundary condition. The above relationships (12,13) are given in their so-called "total Lagrangian (T.L.)" form in which all quantities are referred to the initial, undeformed state, and integration is performed over the initial volume and the initial surface areas. It is noted that due to the choice of variables the moment equilibrium equation appears explicitly as a symmetry condition involving the rates of stress and displacement as well as the spin.

By elementary operations, which essentially involve a change of independent variables, related functionals may be derived [5,6]. The essential point here is that J_{CB} in (12) is a generalized functional which depends on both kinematic and stress variables, a fact which is central to the derivation and numerical solution procedure given in [1] . Thus it differs from the complementary functional derived in [2], and its rate form used extensively e.g. in [5,6], where satisfaction of the force equilibrium equation (13a) is required à priori. At this point it may also be mentioned that, although a complementary viewpoint is implicit in the use of a generalized functional such as J_{CB} above we will not make connection with approaches, such as that in [8] , which treat stability problems by direct use of complementary energy principles.

Starting from (12), the derivation of the beam theory is involved but straightforward [1] . It is based on the following assumptions: normal stresses are taken to have a bilinear distribution across the cross-section; stress resultants, including the warping bimoment, are defined as integrals of the components of $\underline{r}$ over the cross section; the cross-section remains plane and is only rotated during deformation (see Figure 3).

The latter assumption follows from the neglect of transverse shear strains. The remaining axial strain is then a principal strain. Consequently, the triads $\underline{A}_K$ and $\underline{a}_K$ (see Fig.3) are orthogonal and coincide with the principal directions. They are thus related by $\underline{a}_K = \overset{\circ}{\underline{\alpha}}{}_K^{\cdot C}\, \underline{A}_C$, where $\overset{\circ}{\alpha}{}_K^{\cdot C}$ are the components of the rotation tensor $\overset{\circ}{\underline{\alpha}}$. The kinematic relationship is then derivable from the expression for the position vector of a point in an arbitrary, deformed cross-section [1]

$$\underline{x} = \overset{c}{\underline{x}} + \eta\, \underline{a}_2 + \zeta \underline{a}_3 - \omega(\eta,\zeta)\Psi\, \underline{a}_1 . \tag{14}$$

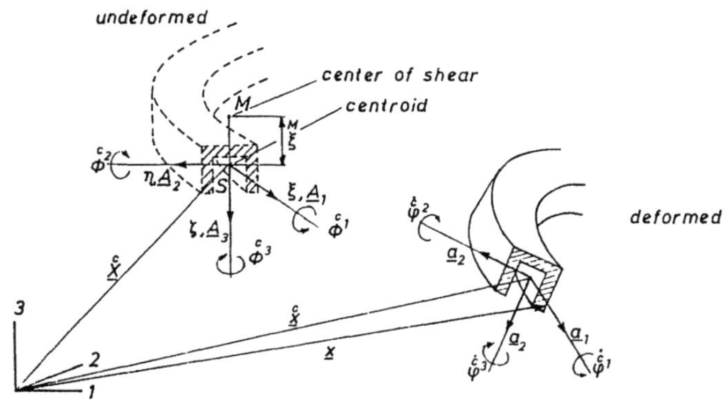

Fig. 3 Initial and deformed beam configurations

In (14) $\overset{C}{\underline{x}}$ is the position vector of the deformed reference axis and the last term accounts for warping of the cross-section.

Insertion of these assumptions into the general three-dimensional principle (12) leads to a related one-dimensional principle, the Euler equations of which govern the large displacement behavior of beams: $\delta J_{CB}(\dot{\underline{u}}, \dot{\underline{\varphi}}, \dot{\underline{r}}) = 0 \quad \Rightarrow \quad \delta J^{*}_{CB}(\dot{\underline{u}}^{C}\ \dot{\underline{\varphi}}^{C}\ \dot{\Psi}, \dot{\underline{V}}\ \dot{\underline{M}}\ \dot{M}_{\omega}) = 0.$ (15)

Like (12) it is a multi-field principle containing kinematic (rates of the displacement and rotation angle vectors, and warping function) as well as static variables (rates of the force resultant and moments vectors, and warping bi-moment).

The numerical solution technique does not employ trial functions but is based on the direct use of the Euler equations of (15). Casting them in the form of a linear set of ordinary differential equations of first order they may be integrated numerically to any degree of accuracy to give the transfer matrix of an element. Simple transformations then lead to the corresponding stiffness matrix which may be used in a standard finite element procedure. For further details the reader is referred to [1,7].

3. Numerical examples

The following examples are chosen to illustrate the application of the beam theory outlined above, and to discuss some of the factors which have a decisive influence on the load-carrying behavior of beam structures.

3.1 Lateral buckling of a simply supported I-beam

In the first example the elastica problem is treated. As shown in Fig. 4, a simply supported beam with a lateral sinusoidal initial imperfection is loaded by an axial force applied at a horizontally movable hinge.

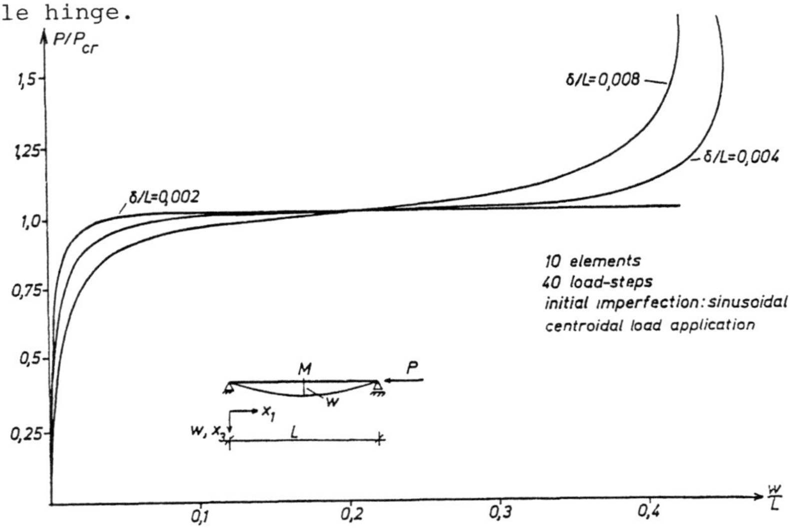

Fig. 4: Load-deflection curves

The load-deflection curves shown in Fig. 4 for various imperfection amplitudes illustrate the well-known behavior whereby in the neighborhood of the critical load the lateral deflections increase dramatically to reach a maximum of about 50 % of the beam length at a load level of approximately 1.5 P_r. At that point the beam reaches its maximum stiffness. For higher loads the stiffness varies much less and the lateral deflections decrease. The latter part of the curve is not shown here. It is interesting to note that this overall behavior is essentially the same for all imperfection levels, and that the loading range in which the deflections increase rapidly is relatively narrow for the imperfections given. Moreover it is seen that the maximum value of the deflection decreases with increasing imperfection amplitude.

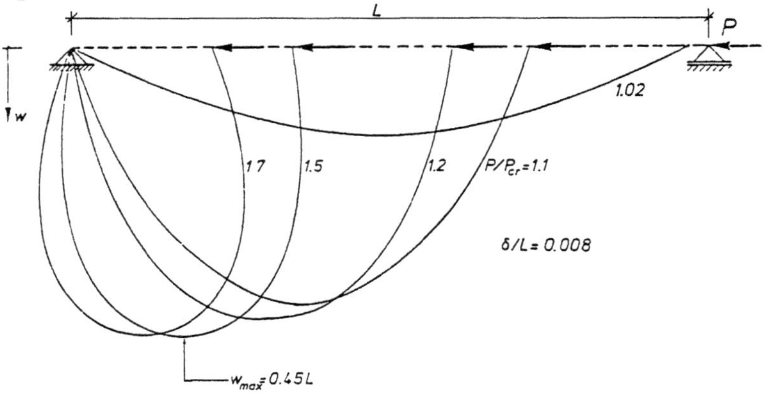

Fig. 5: Deflected shapes at various load levels

Figure 5, where the deformed beam is plotted for various load levels, shows that deflections as well as rotations become quite large in this case. At $P/P_{cr} = 1.7$ the rotation angle at the left hinge exceeds 120 degrees.

3.2 Postbuckling behavior of a simple two-bar frame

In the second example a numerical analysis of the well known Roorda-frame [9], shown in Fig. 6, is given. Although, like in the preceding case, only bending deformations in the plane of the frame are considered, the nonsymmetric postbuckling behavior of this structure presents an intriguing challenge from both a theoretical and numerical standpoint

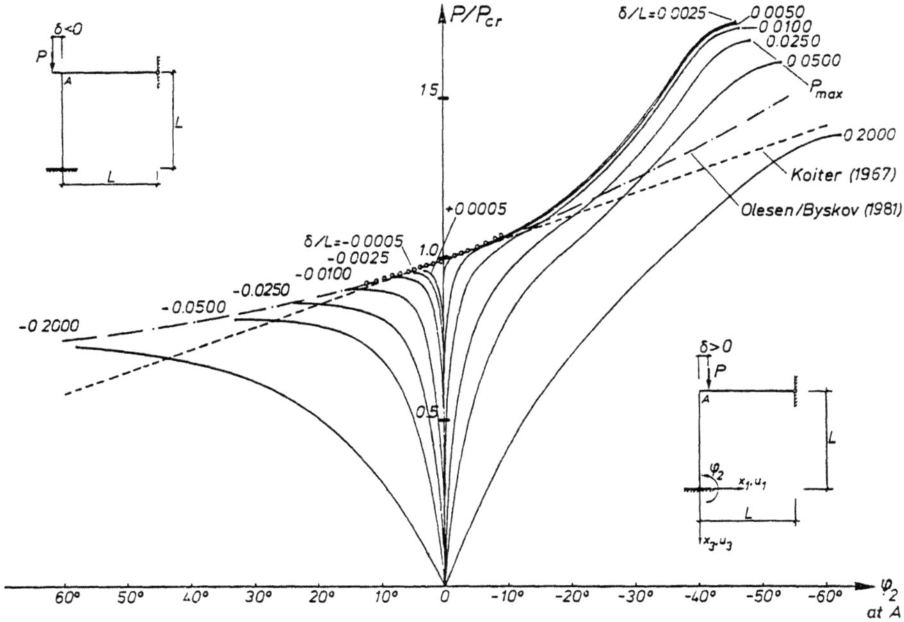

Fig. 6: Postbuckling behavior of two-bar frame

In [10] Koiter gave the first analytical treatment of this problem. Using his general asymptotic theory of initial postbuckling behavior and imperfection sensitivity he was able to give a qualitative as well as quantitative theoretical explanation for the behavior observed by Roorda in his carefully performed tests[9]. Koiter's asymptotic linear relationship for the postbuckling equilibrium path of the perfect frame (shown in Fig. 6 as a dashed straight line through $P/P_{cr} = 1$) as well as his relationship giving the sensitivity to initial negative load eccentricities (shown as a dashed curve on the lower left of Fig. 7) showed excellent agreement with the experimental results. This agreement is convincing support for the validity of his approach to the buckling behavior of structures. The second dashed curve in Fig. 6 represents the asymptotic postbuckling relationship recently given by Olesen and Byskov in [11]. It contains linear as well as quadratic terms in the rotation angle φ at the corner, and in the initial postbuckling regime it is seen to give better agreeement with the experimental results (indicated by circles). In particular, it better reflects the observation that in the postbuckling regime the slopes of the curves increase due to stiffening of the frame.

In a series of papers Kounadis and Simitses [12] have shown that, if the nonlinear interaction between the axial shortening of the vertical column and the compressive, or tensile, strain in the horizontal bar is fully taken into account, no true bifurcation point but only limit points are found, and that these limit values of the load are not, even if normalized by the classical critical load, independent of the slenderness ratio of the column. Another interesting finding was that for positive load eccentricities (to the right of corner A) smaller than a certain limiting value (which also depends on the slenderness ratio) the corner point first rotates clockwise, and later reverses direction so that failure occurs counter-clockwise at a load somewhat below P_{cr}. This result explained why the left imperfection sensitivity curves in Fig. 7 do not lie tangent to the vertical axis but are slightly offset to the right. This offset is thus not entirely attributable to unavoidable imperfections as was first suspected in [9,10].

In the various recent numerical analyses of this problem (see e.g. [13,14]) its stability aspects, and in particular the limit loads of the stable postbuckling branches, were not explored in detail.

Our numerical results are given in Fig. 6 as plots of load vs. corner-rotation for a wide range of positive and negative load eccentricities δ. All curves are terminated at their limit points,

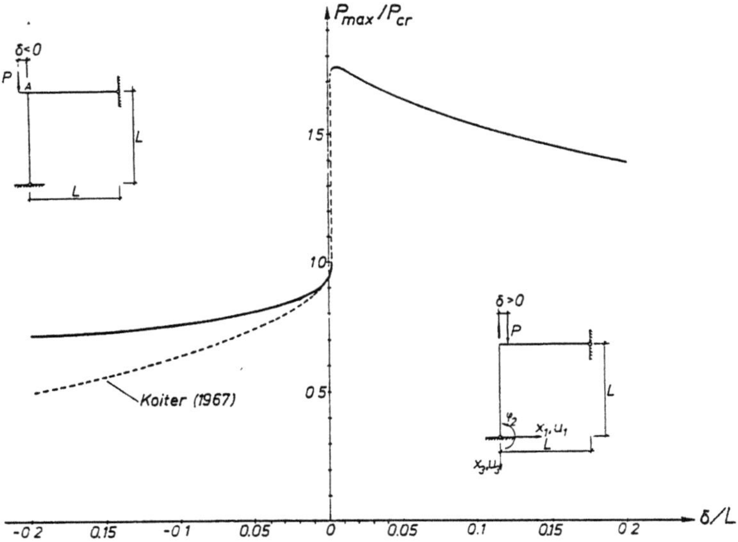

Fig. 7: Imperfection sensitivity of two-bar frame

and the classical critical load P_{cr} is used as a normalizing parameter. In the analysis the slenderness ratio of the bars was taken to be the same as that in Roorda's tests.
The unstable post-buckling behavior and the sensitivity of the limit loads to negative initial excentricities (δ<0) is clearly seen from the left parts of Figures 6 and 7.Depending on the value of δ/L the load maxima may be as low as 72 % of P_{cr}. As usual, the sensitivity is most pronounced for very small imperfections and it is there that the numerical (solid) and asymptotic(dashed) curves in Fig. 7 show their sharpest decrease. The fact that for this structure the asymptotic relationship gives lower values than the numerical results, and hence overestimates imperfection sensitivity, is somewhat exceptional. Usually, e.g. for shells under compression, the asymptotic results yield upper, rather than lower, bounds to the limit loads.

Curves for positive excentricities (δ>0) are plotted on the right parts of Figures 6 and 7. Among these the curve for δ/L = 0.0025 agrees particularly well with the results of the tests and the asymptotic analyses. It is apparent that the upper quadratic asymptotic curve follows the available test results very well, but fails to adequately describe the considerable relative increase in stiffness at higher loads. The points of inflection, and ultimately the limit points, would require the evaluation of higher order terms, but this is clearly beyond the scope of an asymptotic theory.

The curve for $\delta/L = + 0.0005$, just to the left of the vertical axis, is noteworthy because it shows the effect pointed out in [12]. Although the excentricity is positive buckling occurs in a counter-clockwise direction at a limit load smaller than P_{cr}. No effort was made, however, to determine numerically the precise limiting value of δ for which this reversal first takes place.

Another interesting result is that the limit loads for $\delta>0$ are not significantly larger than P_{cr}. Judging from simple cases, e.g. the elastica, one might have expected that very high postbuckling loads are possible. In addition, there is an imperfection sensitivity for the stable postbuckling branches as well, as is seen from the right hand part of Fig. 7. Over the range of positive δ/L values considered, it appears to be even more severe than for negative values of δ. The connection between the upper and lower curves is indicated by the vertical dashed line.

The maximum load that can be carried by this frame is thus 1.75 P_{cr}. However, its deflections became quite large even before the load-carrying capacity has been exhausted. Figure 8, where deflected shapes are plotted for a positive and negative excentricity of equal magnitude, gives a good illustration of this result.

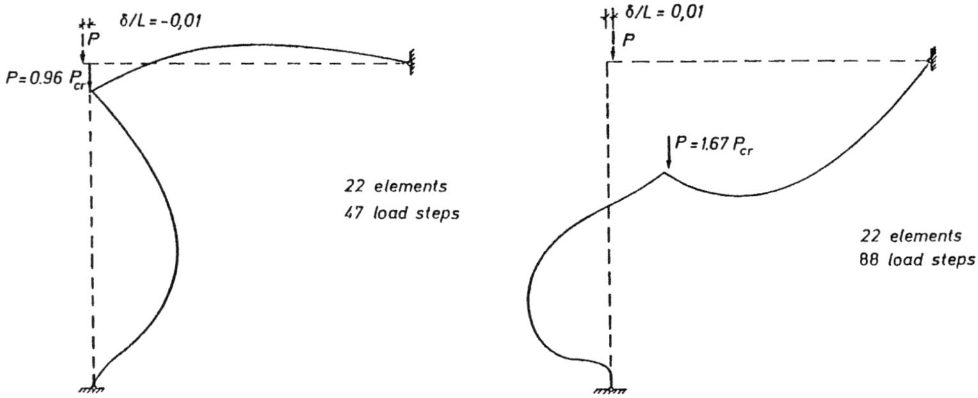

Fig. 8: Deflected shapes at $P \simeq P_{max}$

Figures 8 and 9 provide an explanation for the nonsymmetry of the buckling behavior. For $\delta>0$ the horizontal bar is in tension during most of its deformation history, and thus exerts a stiffening influence. For $\delta<0$, on the other hand, the axial force in the horizontal bar is mainly compressive, and thus a destabilizing factor.

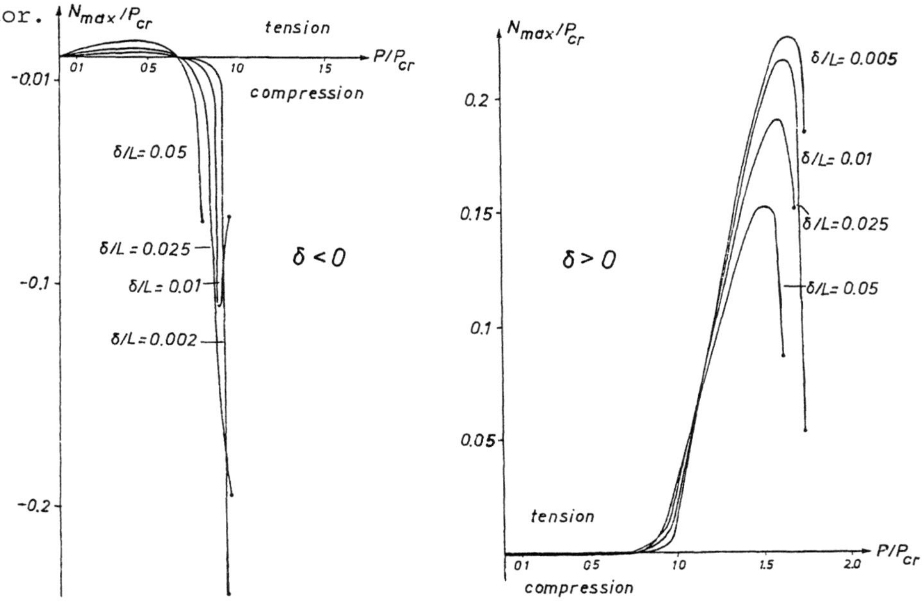

Fig. 9: Maximum axial force in horizontal bar

3.3 Torsional buckling of a cantilever

The third example is a cantilever beam with a circular cross-section, open at the top, and loaded by a transverse force at the tip (see Figs. 10,11). This problem, which was recently treated by Reissner [15] , is interesting because of the non-coincidence of the cross-sectional shear center and centroid locations, and the strong influence of the point of load application on the buckling behavior. The open cross-section considered here is somewhat exceptional because, compared with other thin-walled sections, the distance between the centroid S and the shear center M is extremely large. In Figure 10 normalized bifurcation loads P_{cr} are plotted versus the normalized distance of the loading point from the shear center. The two dashed curves give the critical loads for the case when the warping restraint at the built-in end is taken into account, whereas in the calculation

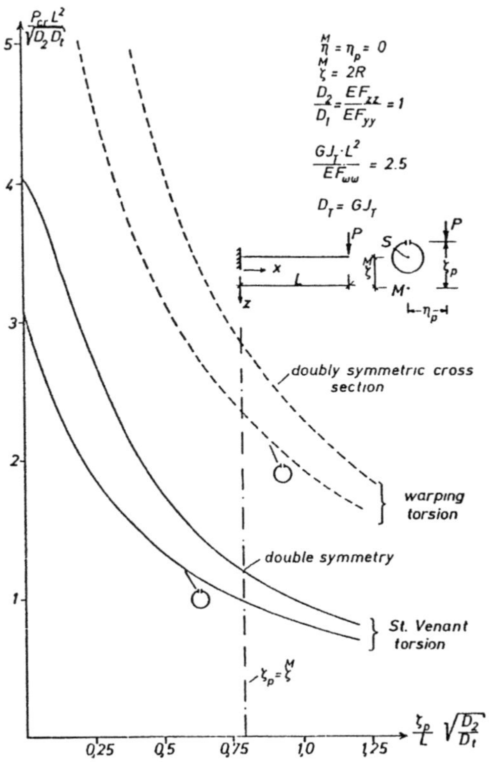

Fig. 10: Bifurcation loads

of the lower solid curves this effect was neglected. These results show very well that a considerable increase in load carrying capacity may be achieved by inhibiting the warping of open cross-sections.

The influence of the point of load application is even stronger. For both sets of curves the critical value decreases substantially with increasing vertical distance of the loading point from the shear center. Non-coincidence of the centroid and the shear center is reflected in the distance between the two individual curves in each solid and dashed set. The respective upper curves hold for doubly symmetric cross-sections whereas the lower ones apply to the open circular section. Results for less extreme cross-sections would lie in between the two pairs of curves. It is thus seen that between the two factors just discussed the degrading effect of the position of the load is most severe, and that for a given relative loading point, there is an additional decrease if an unfavorable cross-section, such as the present one, is chosen.

Figure 11 shows numerical results as plots of the load versus the rotation angle at the tip for various values of initial lateral deflection. The load is taken to act at the centroid and warping is considered in all cases. As is the first example the deflections show a rapid growth as the load approaches the critical value P_{cr}. Here too, stable postbuckling behavior, with zero initial slope for the perfect structure, and a continual increase in stiffness for loads larger than P_{cr} is observed. No maximum

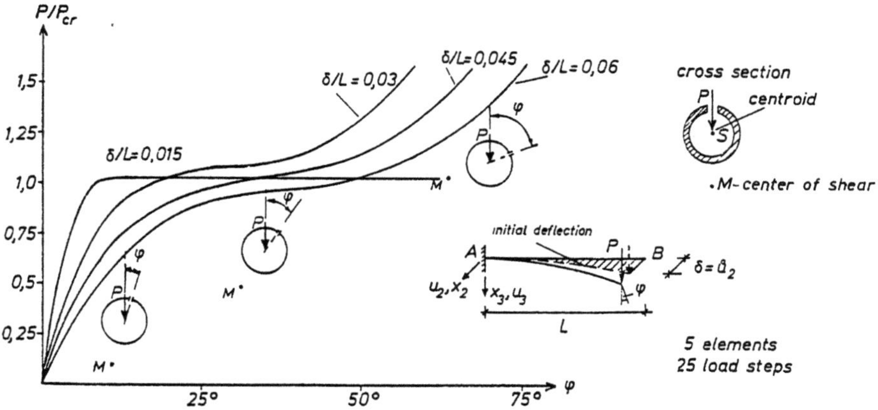

Fig. 11: Torsional buckling of imperfect cantilever

value of the rotation is found, however, in the loading range considered. In addition, in all cases the behavior is decidedly nonlinear from the beginning. This is due to the significant, nonlinearly increasing, torsional moment resulting from the excentricity of P with respect to the shear center M (see Fig. 11). But while in this case the rotation of the cross-section becomes considerable the bending deflections remain quite small, and the load-deflection relation is approximately linear up to loads of 1.5 P_{cr} (see Fig. 11, 12).

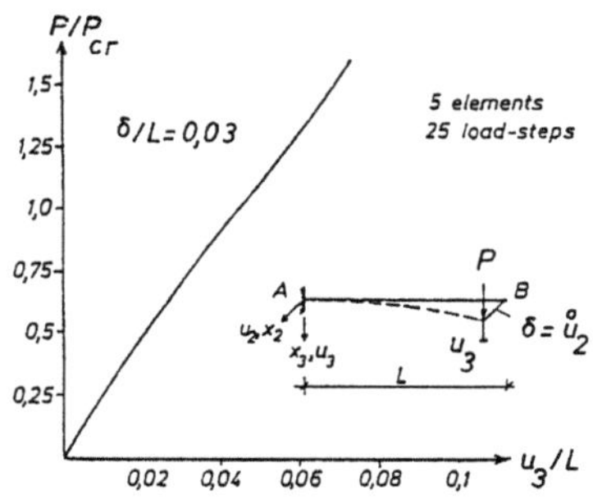

Fig. 12: Load vs. tip-deflection

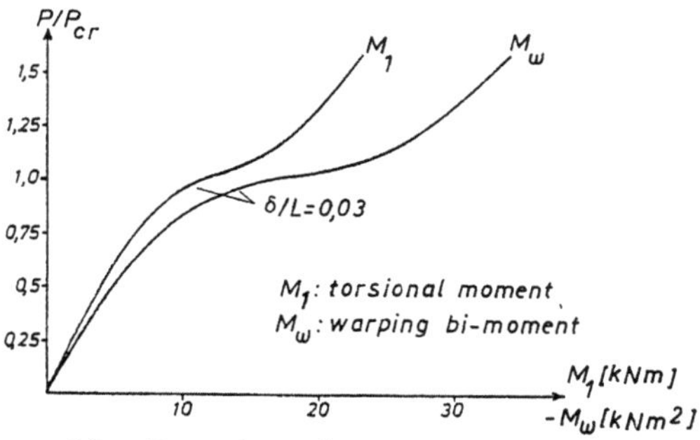

Fig. 13: Torsional moments at built-in end

The strong influence of warping already seen in Fig. 10 is further exemplified by Fig. 13 where the dependence of the torsional moment M_1 and the warping bi-moment M_ω on the applied load is plotted. Although the two moments have different dimensions it is again obvious that warping is by no means negligible.

3.4 Torsional-flexural buckling of a plane frame

In the last example the deflections due to bending and to torsion have approximately equal magnitude. The structure considered is a vertically loaded plane frame with a lateral, plane initial deflection of maximum amplitude δ as shown in Fig. 14. The loads are increased proportionally, and limit loads are reached

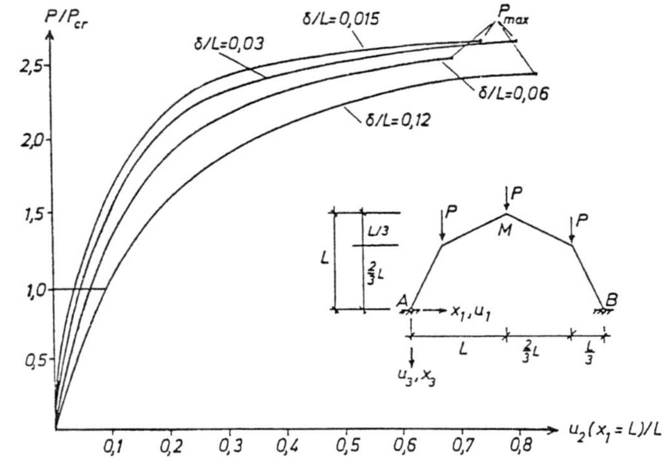

Fig. 14: Load-deflection curves

after the development of considerable lateral and torsional deformations. The deflected shape for δ/L=0.03, plotted in Fig. 15, shows that for the present symmetric loading pattern the deflec-

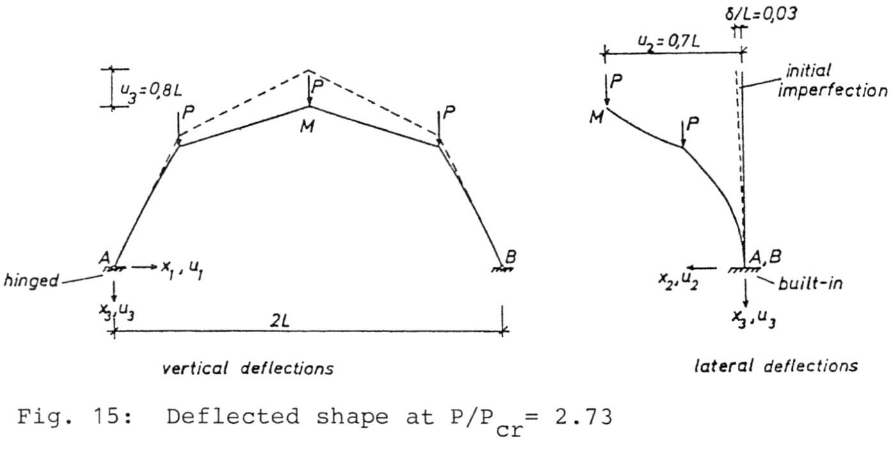

Fig. 15: Deflected shape at $P/P_{cr} = 2.73$

tions remain symmetric throughout. Antisymmetric shapes lead to lower limit loads, but this aspect is not pursued here. Likewise, as explained in the introduction, plastic deformations which would certainly occur in an actual structure at deflections of this magnitude are not accounted for in this investigation. The load vs. lateral deflection curves for various imperfection levels in Fig. 14 show that, in contrast to the previous examples, there is no significant change in load-carrying behavior at or near the classical critical value. Rather, overall deflections remain more or less moderate up to load-levels of 2 P_{cr}, and limit loads may be as high as 2.3 P_{cr}, even if initial imperfection amplitudes are quite large. Figures 16 and 17 show the dependence

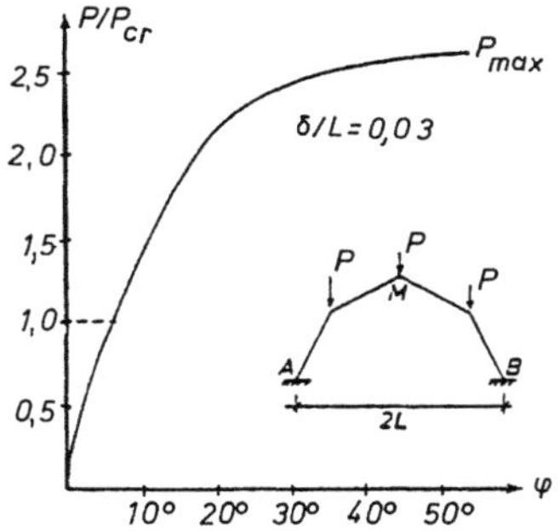

Fig. 16: Lateral tilt angle at M

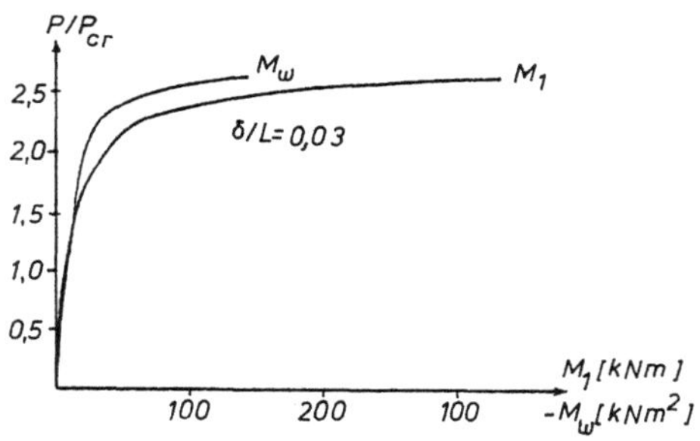

Fig. 17: Torsional moments at M

of the tilt angle at the top of the frame, and of the torsional and warping bi-moments on the applied load intensity. Again, warping plays an important role. Concerning the magnitude of M_1 and M_ω it is noted that they are larger by a factor of 10 than in the previous example.

References

1. Wunderlich, W., Obrecht, H.: "Large spatial deformations of rods using generalized variational principles". In: Nonlinear Finite Element Analysis in Structural Mechanics. Wunderlich,W., Stein, E., Bathe, K.J., (eds.). Springer-Verlag, Berlin (1981) 185-216.

2. Fraeijs de Veubeke, B.: "A new variational principle for finite elastic displacements". Int. J. Engng. Sci. 10 (1972) 745-763.

3. Wunderlich, W.: "On a computer-oriented formulation of stability problems" (in German). In: Konstruktiver Ingenieurbau in Forschung und Praxis - Festschrift Zerna und Institut KIB. Krätzig, W.B., Roik, K., Kotulla, B., Wittek, U. (eds). Werner-Verlag, Düsseldorf (1976) 111-119.

4. Beverungen, G.: "Geometrically nonlinear stress and stability analysis of spatially curved rods"(in German). Mitteilung Nr. 76-13 (1978), Techn. Wiss. Mitteilungen. Institut für Konstruktiven Ingenieurbau, Ruhr-Universität Bochum.

5. Murakawa, H., Reed, K.W., Atluri, S.N.,Rubenstein, R.: "Stability analysis of structures via a new complementary energy method". Comp. Struct. 13 (1981) 11-18.

6. Atluri, S.N.: "On some new general and complementary energy theorems for the rate problems of finite strain, classical elasto-plasticity. J. Struct. Mech. 8 (1980).

7. Wunderlich, W.: "Incremental formulation for geometrically nonlinear problems". In: Formulations and Computational Algorithms in Finite Element Analysis. Bathe, K.J., Oden, J.T., Wunderlich, W. (eds.). MIT Press (1977) 193-240.

8. Koiter, W.T.: "Complementary energy, neutral equilibrium and buckling". Proc.Kon. Ned. Ak. Wet., Series B 79 (1976) 183-200.

9. Roorda, J.: "Stability of structures with small imperfections" Proc. ASCE 91, J. Engng. Mech. Div. EM1 Feb (1965) 87-106.

10. Koiter, W.T.: "Post-buckling analysis of a simple two-bar frame". In: Recent Developments in Applied Mechanics - The Folke Odqvist Volume. Broberg, B. et al. (eds.). Almqvist and Wiksell, Stockholm (1967) 337-354.

11. Olesen, J.F., Byskov, E.: "Accurate determination of asymptotic postbuckling stresses by the finite element method". DCAMM Rep. No. 215 July (1981). The Danish Center for Applied Mathematics and Mechanics, The Technical University of Denmark.

12. Simitses, G.J., Kounadis, A.N.: "Buckling of imperfect rigid-jointed frames". Proc. ASCE 104, J. Engng. Mech. Div. EM3 June (1978) 569-586.

13. Frey, F., Cescotto, S.: "Some new aspects of the incremental total Lagrangian description in nonlinear analysis". In: Finite Elements in Nonlinear Continua. Bergan, P.G. et al. (eds.) Tapir, Trondheim (1978) 323-343.

14. Karamanlidis, D., Honecker, A., Knothe, K.: "Large deflection finite element analysis of pre- und postcritical response of thin elastic frames". In: Nonlinear Finite Element Analysis in Structural Mechanics. Wunderlich, W., Stein, E., Bathe, K.J. (eds.). Springer Verlag, Berlin (1981) 217-235.

15. Reissner, E.: "On a generalization of Kirchhoff's theory of space-curved beams". Lecture presented at the Institut für Konstruktiven Ingenieurbau, Ruhr-Universität Bochum, May (1981), and private communication.

On the Stability and Large Deflection Analysis of Elastic Bars

Y.C. TOKLU

Department of Engineering Sciences
Middle East Technical University, Ankara, Turkey

Summary

A numerical method is presented for the analysis of cantilevers governed by the Euler-Bernoulli differential equation. The method is applied to initially straight inextensional bars of uniform cross section under axial and lateral loadings. The deflected shapes and two important loading combinations, one corresponding to the first bifurcation, the other corresponding to the onset of an inflection point on the bar, are determined.

Introduction

In the theory of bending of perfectly flexible elastic bars, the linearization of the Euler-Bernoulli differential equation results in the small deflection theory which is known to be insufficient for a thorough stability or deflection analysis. The non-linear form of the Euler-Bernoulli equation is to be handled to develop a more general theory called the large deflection theory. There are considerable difficulties in solving this non-linear problem and therefore, in the literature, solutions are available only for a limited number of cases of which cantilevers cover a significant part. In what follows, attention will be focused on the deflection analysis and the stability of initially straight cantilevers satisfying the assumptions of the Euler-Bernoulli equation. Conservative loads will be considered in the analysis.

Consider the cantilever as initially horizontal. The case of a vertical concentrated load at the free end has been solved with integrations leading to solutions in terms of elliptic integrals [1,2]. The same method is extended to include the effect of a horizontal load at the free end in [3,4,5]. A similar procedure is applied for the case of a bending moment and a horizontal compressive load at the free end [6].

The first solution presented for the case of uniformly distributed vertical loads was a power series solution [7]. An analytical solution was later found for a special distributed load [3]. An exact solution is also presented for the slopes of the elastic curve for uniformly distributed vertical loads [8].

For more complicated loadings or for non-uniform cross sections approximate solutions based on various numerical integrations have been presented [9-14].

As far as the stability aspect of the problem is concerned, a limited number of solutions can be found in books on elastic stability (See [15] for instance). Tang [16], in the investigation of the stability of simply supported beams freely lying on a rigid foundation, has considered simultaneous action of horizontal and vertical concentrated loads. The results obtained with the aid of elliptic integrals show the effect of the vertical load on the critical horizontal load.

The purpose of this paper is to present a numerical method which enables the determination of deflected shapes of inextensional elastic bars under different loadings in pre-and post-buckled positions. Another aim is to determine the load combinations which correspond to some critical cases. Although the method can be applied to more general problems, attention will be focused on the initially straight uniform bars.

Formulation

Consider the cantilever loaded as shown in Fig.1. With the convention shown in Fig.2, the Euler-Bernoulli differential equation governing the system can be written as

$$EI \frac{d\theta}{dS} = -M(S) \tag{1}$$

where EI denotes the flexural rigidity.

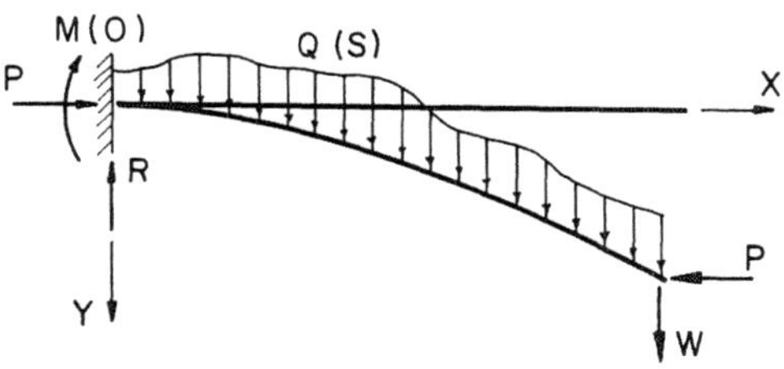

Fig.1. Loading considered on the cantilever.

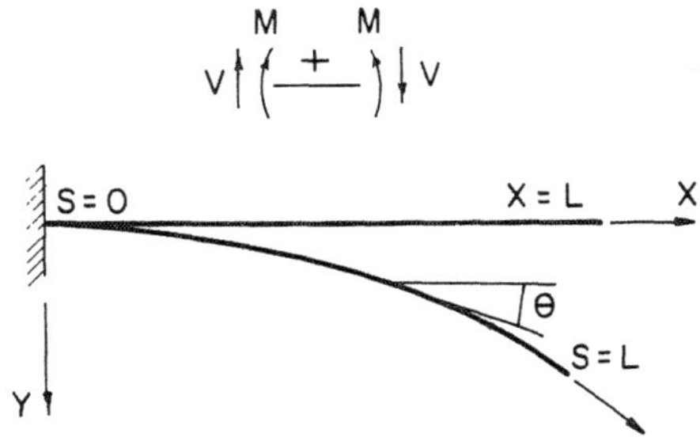

Fig.2. Geometry of the deflected bar and the sign convention for moments and shears.

The method proposed is based on an approximation involving the variation of the angle θ with respect to the arc length S. It is assumed that the bar is made of n segments of equal length, h=L/n, and that θ is linearly related to the arc length over these segments, i.e., for the segment i

$$\theta = \theta_i + a_i t, \qquad 0 \leq t \leq h, \quad i=1,2,...,n \tag{2}$$

with

$$a_i = (\theta_{i+1} - \theta_i) / h, \qquad i=1,2,...,n \tag{3}$$

where θ_i, i=1,2,...,n+1 show the slopes at the nodes, node 1 and node n+1 being located at the points S=0, and S=L, respectively.

For the sake of simplicity, it is further assumed that the distributed load Q can be expressed in a similar way:

$$Q = Q_i + b_i t, \qquad 0 \leq t \leq h, \quad i=1,2,...,n \tag{4}$$

with

$$b_i = (Q_{i+1} - Q_i) / h, \qquad i=1,2,...,n. \tag{5}$$

It is not necessary to assume a linear variation in Q for the elaboration of the method, but under this assumption the integrations can be performed exactly over segments of the bar.

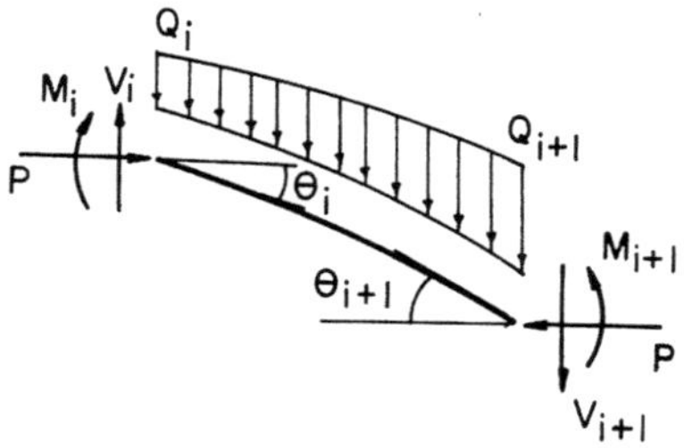

Fig.3. Segment i between nodes i and i+1 under linearly varying distributed load.

Now consider the segment i shown in Fig.3. The integration of $dX=dS\cos\theta$ and $dY=dS\sin\theta$ using (2) and (3), yields

$$X = \begin{cases} X_i + t\cos\theta_i & \text{for } a_i=0 \\ X_i + [\sin(\theta_i+a_i t) - \sin\theta_i]\,/\,a_i, & \text{for } a_i\neq 0 \end{cases} \tag{6}$$

and

$$Y = \begin{cases} Y_i + t\sin\theta_i & \text{for } a_i=0 \\ Y_i + [\cos\theta_i - \cos(\theta_i+a_i t)]\,/\,a_i, & \text{for } a_i\neq 0 \end{cases} \tag{7}$$

where $0 \leq t \leq h$. With the use of (6) and (7) the coordinates of the nodal points can be determined from

$$\Delta X_i = \begin{cases} h\cos\theta_i & \text{for } a_i=0 \\ (\sin\theta_{i+1} - \sin\theta_i)\,/\,a_i, & \text{for } a_i\neq 0 \end{cases} \tag{8}$$

$$X_{i+1} = X_i + \Delta X_i$$

and

$$\Delta Y_i = \begin{cases} h \sin\theta_i & \text{for } a_i=0 \\ (\cos\theta_i - \cos\theta_{i+1}) / a_i & \text{for } a_i \neq 0 \end{cases} \qquad (9)$$

$$Y_{i+1} = Y_i + \Delta Y_i$$

with the boundary conditions $X_1=Y_1=0$.

Again referring to Fig.3, the equilibrium conditions for the segment i can be written. Shear forces can be obtained from

$$V_i = V_{i+1} + h(Q_i+Q_{i+1}) / 2, \qquad i=2,3,\dots,n \qquad (10)$$

with $V_{n+1}=W$ and $V_1=R$ (See Fig.1). The expression for the moments over the segment i, and for $0<t<h$ is

$$M(t) = M_i+V_i[X(t)-X_i]+P[Y(t)-Y_i]-\int_{r=0}^{t}(Q_i+b_ir)[X(t)-X(r)]dr \qquad (11)$$

The integral in (11) can be taken exactly using (6). Further, with some rearrangement, the expression for the moments at the nodal points $i=1,2,\dots,n$ can be found as

$$M_{i+1} = M_i+V_i\Delta X_i+P\Delta Y_i-(Q_i+hb_i/3)h\Delta X_i/2, \quad \text{for } a_i=0 \qquad (12\text{ a})$$

and

$$M_{i+1} = M_i+(V_i+b_i/a_i^2)\Delta X_i+(P+Q_i/a_i)\Delta Y_i$$
$$- [(Q_i+b_ih/2)\sin\theta_{i+1}-b_i\cos\theta_{i+1}/a_i]h/b_i \qquad \text{for } a_i\neq 0 \qquad (12\text{ b})$$

where $M_{n+1}=0$.

Now the Euler-Bernoulli equation (1) can be integrated on the segment i to yield

$$EI\Delta\theta_i = -\int_i M(t)\,dt \qquad (13)$$

where $\Delta\theta_i=\theta_{i+1}-\theta_i$. The integration in (13) can be performed exactly with the substitution of M(t) from (11) to give

$$-\ EI\Delta\theta_i = M_i h+h(V_i\Delta X_i+P\Delta Y_i)/2-h^3(Q_i-b_i\Delta X_i/4)/6$$

for $a_i=0$, (14 a)

and

$$-\ EI\Delta\theta_i = M_i h+h\left[(P+Q_i/a_i)\cos\theta_i-(V_i+b_i/a_i^2)\sin\theta_i\right]/a_i$$
$$+\ h/a_i^2(hb_i/2+Q_i)\cos\theta_{i+1}-2hb_i/a_i^3\sin\theta_{i+1}$$
$$+\ \Delta Y_i/a_i(V_i+3b_i/a_i^2)-\Delta X_i/a_i(P+2Q_i/a_i),$$

for $a_i\neq 0$ (14 b)

The related boundary condition is $\theta_1=0$.

In the solution of these equations, the non-dimensional forms are used and the results are presented in terms of the following non-dimensional quantities:

$x = X(L)/L$, $y=Y(L)/L$, coordinates of the displaced free end,

$p = PL^2/EI$, the horizontal load at the free end,

$w = WL^2/EI$, the vertical load at the free end,

$q = QL^3/EI$, the vertical uniformly distributed load.

Computational Procedure

An iterative procedure can be based upon the equations given above. The steps in this procedure are as follows:

1- Choose a set of values θ_i, $i=2,3,\ldots,n+1$ with $\theta_1=0$.

2- Determine the geometry of the bar using the equations (3), (6), (7) beginning from $X_1=Y_1=0$.

3- Determine the moments at the nodal points using (12) and starting with $M_{n+1}=0$.

4- Calculate the new set of rotation angles from (4) starting with $\theta_1=0$ and using $\theta_{i+1}=\theta_i+\Delta\theta_i$.

5- Compare the new set of angles with the preceeding one. If the change is negligible than the solution is obtained. Otherwise return back to step 2 to perform a new iteration.

The algorithm outlined above is effective in the calculations for non-inflectional deflections. For the inflectional cases two types of modifications can be applied:

1- If the inflection point is predetermined by user, this point is considered as a nodal point. At each iteration a load parameter is calculated such that the moment at this point vanishes.

2- If the loads are kept constant, additional nodal points are considered at the approximate places where moment vanishes.

The choice of the number of segments, n, determines the accuracy of the method. The results presented below are obtained with n=50. Comparisons with exact solutions [15] have shown that the geometry of the bar determined by the proposed method has an error of about 0,02%. The error decreases with increasing n(0.1% for n=20 and 0.006% for n=100).

For deflection analysis, the first choice for the deflection angles is observed to be immaterial before the first bifurcation point corresponding to the first Euler load. After this point, the initial choice determines whether an upward or downward deflected shape will be obtained.

To obtain the first buckling loads two different approaches are possible. The first approach is based on a trial-and-error search for the minimum horizontal load P that will make the bar assume an upward deflected shape for a given downward vertical loading.

For the second approach, consider the point X=A, Y=0 on the X-axis such that the moments of the loads acting on the deflected bar add up to zero. The dimensionless parameter a=A/L gives the location of this point. For a given vertical loading a relation is observed between p and a (See Fig.4). On this figure the loads p which correspond to the first Euler load are marked with small arrows. It is seen that an iterative technique can be used here to obtain a minimum value for p, denoted as p_1, in a range for the values of a. This approach was much more effective than the previous one to obtain the first buckling load.

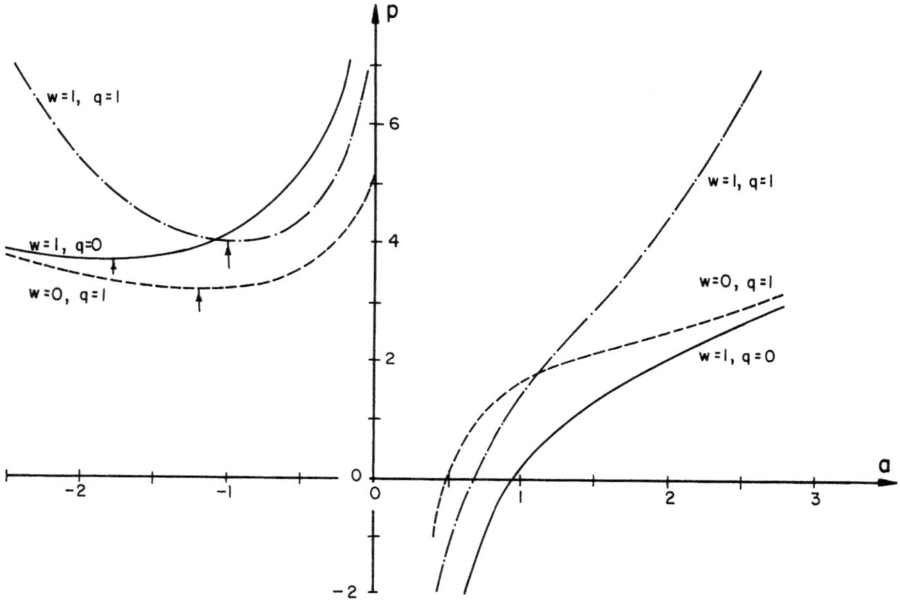

Fig.4. Relation between the horizontal load, $p=PL^2/EI$, and the zero-moment-point, $a=A/L$, on the X axis for three vertical loadings. The arrows indicate the points where p shows a local minimum corresponding to p_1.

Second buckled shape of a cantilever with no vertical load is characterized by an inflection point at the one third of the bar near the fixed end with an axial load of $P=9\pi^2EI/4L^2$. However the onset of an inflection point on such a bar takes place whenever $P=\pi^2EI/L^2$ [15]. Under this condition the bending moment at the fixed end vanishes and that point becomes an inflection point so that the cantilever behaves as a simply supported beam. It can be seen from the results presented in [15] that the value of P increases monotonically as the inflection point moves towards the free end. For laterally loaded bars the present analysis gives comparable results. The compressive horizontal load of minimum magnitude which onsets an inflection point on the bar makes the fixed end a zero-moment-point. As the inflection point moves away from the fixed end the magnitude of P increases monotonically in a range where results can be obtained (See Fig.5). After a region where the iterations would not converge, the horizontal load becomes a tensile force again of increasing magnitude as a'=A'/L approaches 1, where the point S=A' is the inflection point. In this analysis the horizontal compressive load that makes the fixed end an inflection point will be denoted as P_2, or in dimensionless form $p_2=P_2L^2/EI$.

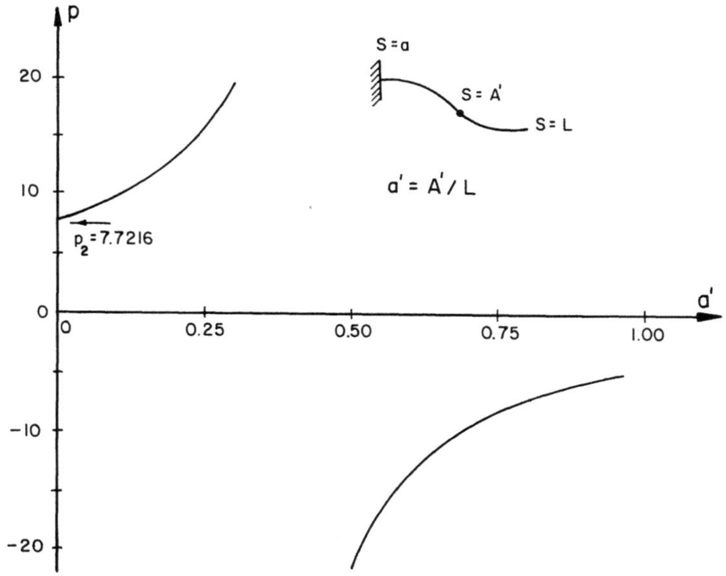

Fig.5. Relation between the horizontal load $p=PL^2/EI$, and the position of the inflection point on the bar a'=A/L, for w=1, q=1

<u>Results</u>

The results obtained using the technique outlined are presented in Figures 6 to 9. In Fig.6, the relation between the horizontal load, p, and the position of the free end for a vertical loading such that w=1, q=1 is shown. Here another advantage of the technique proposed can be discussed. Whenever an equilibrium position of the bar is obtained, the total potential energy can be calculated from

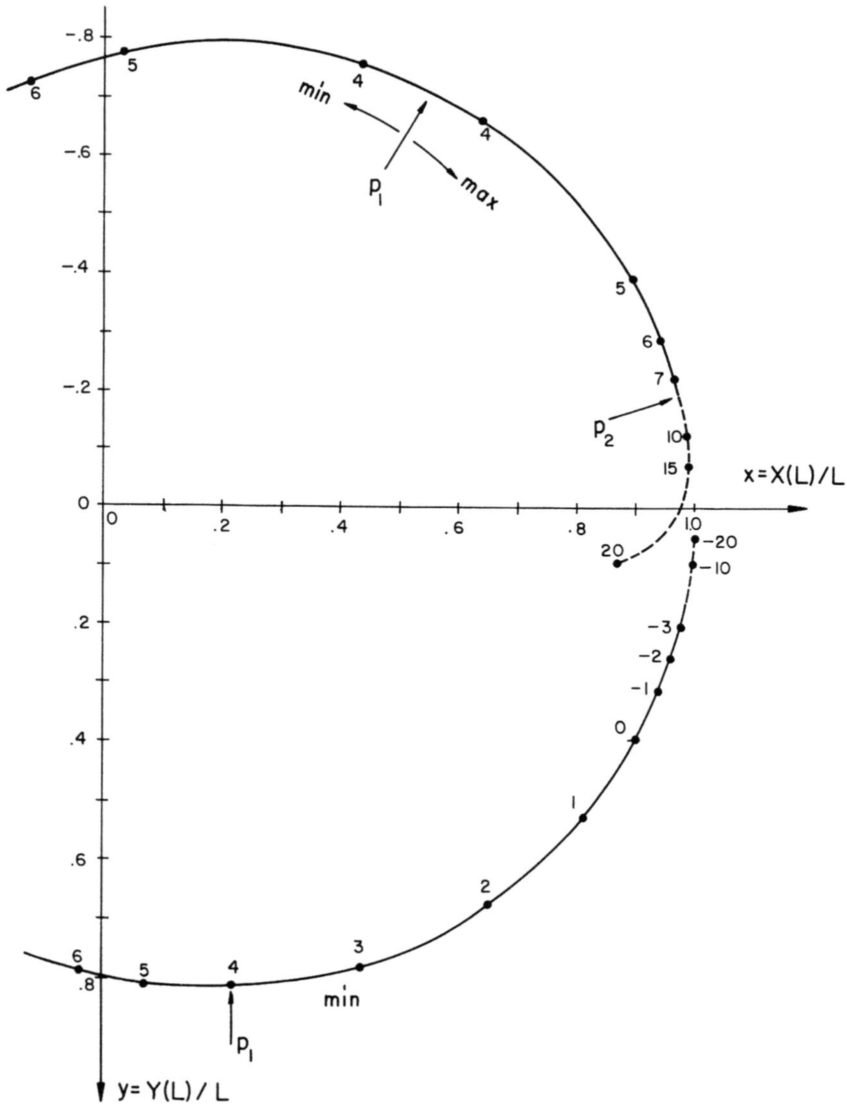

Fig.6. The position of the free end of the cantilever as a function of p for w=1, q=1. The numbers on the curve denote values of p.

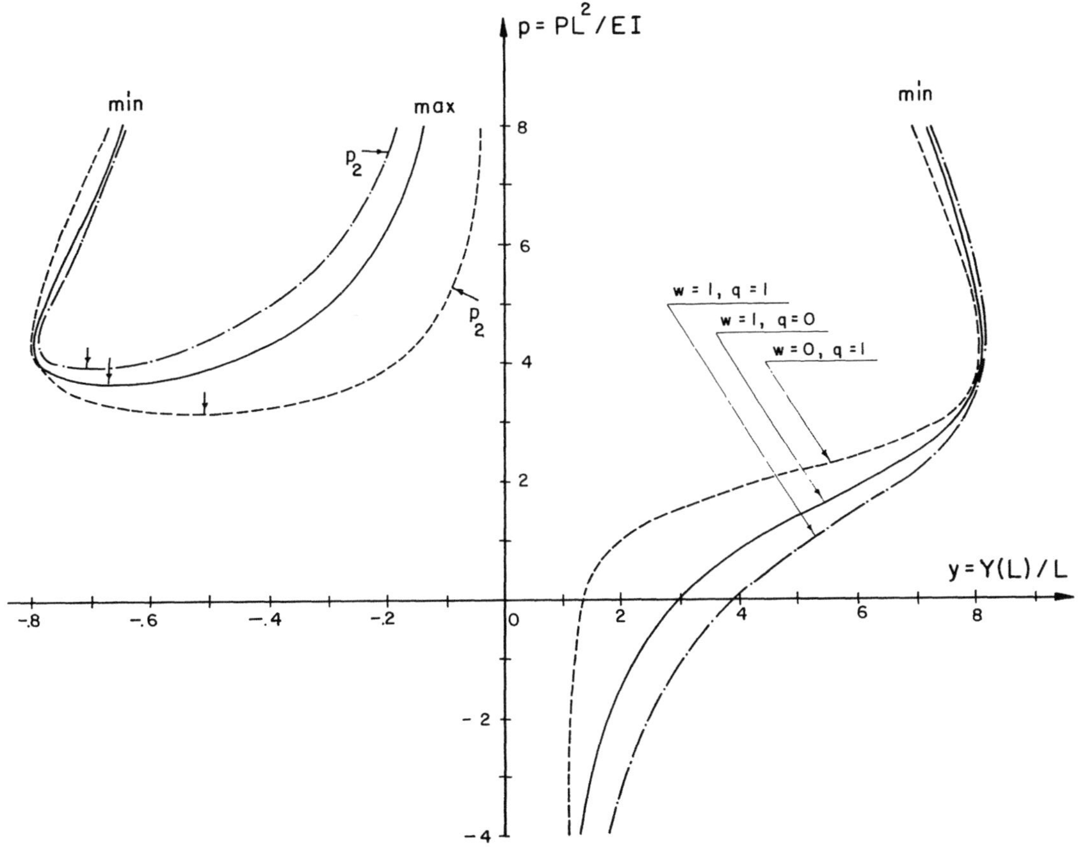

Fig.7. Relation between the vertical end deflection, y, and the horizontal load, p, for three vertical loadings.

$$U = \frac{EI}{2}\int_0^L \left(\frac{d\theta}{dS}\right)^2 dS - \int_0^L Q(S)Y(S)dS - WY(L) - P(L-X(L)) \quad (15)$$

Then by giving small perturbations to the geometry of the bar the change in U can be calculated after which one can determine whether this position corresponds to a minimum or maximum total energy. In Fig.6 minimum and maximum energy regions are marked. The dashed parts correspond to inflectional geometries.

In Fig.7 the relation between the end deflection and the horizontal load is shown for three different types of vertical loadings. With small arrows the loads corresponding to first Euler load, p_1, and those corresponding to the onset of an inflection point on the bar, p_2, are shown. The branches are marked whether they correspond to minimal or maximal energies.

The critical loads p_1 and p_2 are shown in Table I and in Figures 8 and 9.

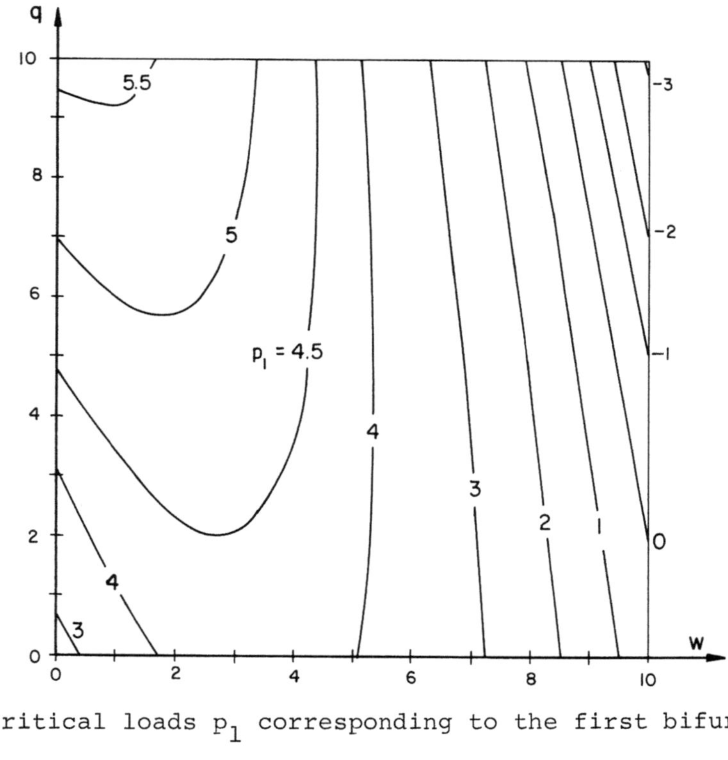

Fig.8. Critical loads p_1 corresponding to the first bifurcation.

Table I. Critical loads $p_1=P_1L^2/EI$ (above) and $p_2=P_2L^2/EI$ (below) for various combinations of uniformly distributed load and concentrated tip load, $q=QL^3/EI$, $w=WL^2/EI$.

w \ q	0	1	2	3	4	5	6	7	8	9	10
0	$(\pi^2/4)$ (π^2)	3.2102 5.4441	3.6284 5.4706	3.9698 5.5139	4.2676 5.5729	4.5352 5.6459	4.7799 5.7311	5.0060 5.8265	5.2162 5.9300	5.4121 6.0394	5.5949 6.1526
1	3.6849 9.8348	3.9588 7.7216	4.2038 6.9805	4.4271 6.6214	4.6330 6.4280	4.8240 6.3246	5.0021 6.2766	5.2923 6.2658	5.3232 6.2807	5.4674 6.3137	5.6008 6.3592
2	4.1064 9.7202	4.2907 8.3171	4.4615 7.5843	4.6203 7.1452	4.7683 6.8640	4.9062 6.6791	5.0345 6.5578	5.1533 6.4804	5.2625 6.4344	5.3622 6.4108	5.4519 6.4030
3	4.2634 9.5274	4.3922 8.4618	4.5118 7.7964	4.6228 7.3479	4.7252 7.0320	4.8193 6.8036	4.9048 6.6360	4.9815 6.5121	5.0490 6.4198	5.1069 6.3506	5.1545 6.2978
4	4.2292 9.2537	4.3142 8.3802	4.3917 7.7743	4.4615 7.3330	4.5235 7.0008	4.5773 6.7448	4.6226 6.5438	4.6587 6.3831	4.6851 6.2521	4.7010 6.1429	4.7056 6.0490
5	4.0298 8.8950	4.0758 8.1410	4.1144 7.5795	4.1454 7.1465	4.1681 6.8036	4.1821 6.5261	4.1867 6.2968	4.1810 6.1035	4.1642 5.9368	4.1352 5.7893	4.0929 5.6552
6	3.6729 8.4455	3.6808 7.7687	3.6808 7.2375	3.6721 6.8088	3.6540 6.4550	3.6258 6.1567	3.5865 5.9000	3.5348 5.6744	3.4697 5.4715	3.3895 5.2846	3.2927 5.1081
7	3.1552 7.8973	3.1231 7.2700	3.0816 6.7564	3.0297 6.3259	2.9664 5.9576	2.8904 5.6360	2.8004 5.3494	2.6948 5.0886	2.5716 4.8460	2.4287 4.6152	2.2635 4.3908
8	2.4630 7.2399	2.3860 6.6414	2.2970 6.1336	2.1948 5.6936	2.0776 5.3048	1.9437 4.9544	1.7909 4.6321	1.6167 4.3297	1.4179 4.0400	1.1907 3.7567	0.9306 3.4740
9	1.5706 6.4583	1.4395 5.8716	1.2923 5.3575	1.1267 4.8982	0.9402 4.4799	0.7298 4.0915	0.4915 3.7238	0.2208 3.3685	-0.3794 3.0186	-0.7860 2.6671	-1.1705 2.3074
10	0.4338 5.5322	0.2321 4.9404	0.0067 4.4061	-0.2462 3.9144	-0.5316 3.4533	-0.8556 3.0123	-1.5226 2.5822	-1.9392 2.1545	-2.3376 1.7207	-2.7050 1.2720	-3.0288 0.7988

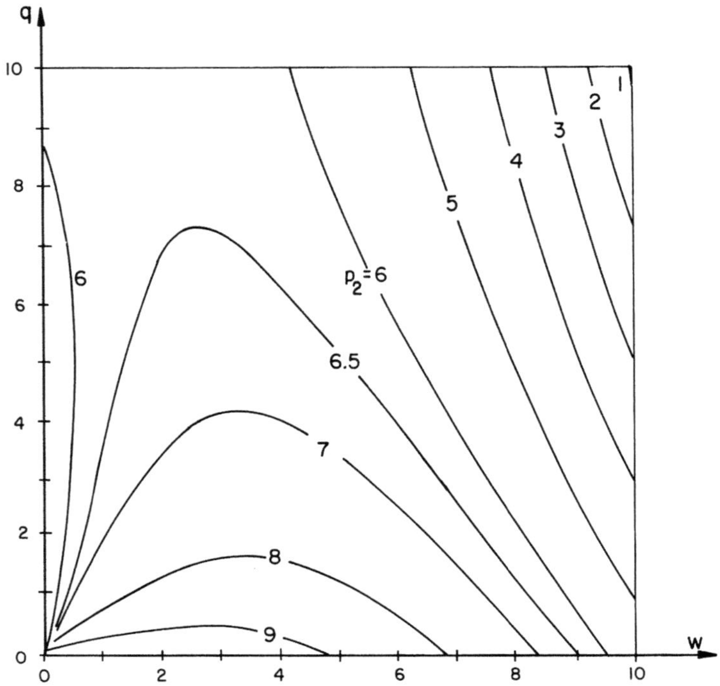

Fig.9. Critical loads p_2 corresponding to the onset of inflection points.

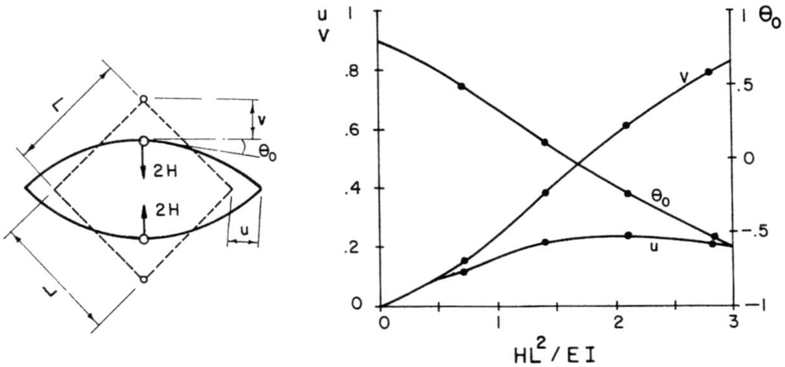

Fig.10. Deformation of a diamond-shaped pinned-fixed frame, curves are obtained from Mattiasson [5]. Points marked are obtained from the present analysis.

Finally a comparison is presented in Fig.10 with a recent result concerning diamond-shaped pinned-fixed frames [5]. Our results corresponding to the case q=0, w=p have been adapted easily for this problem by means of a 45° rotation.

Remarks

The technique proposed is based on simple equations and can be cast into computer programs very easily. Since the method does not require solutions of large systems of equations, small size computers can be used in the calculations. Although the method is applied here to a relatively simple problem, it can easily be extended to cover the cases of non-uniform sections, initial curvatures and more general loadings.

References

1. Barten, H.J.: On the deflection of a cantilever beam. Quart. Appl. Math. 2(1944) 168-171 and 3(1945) 275-276.

2. Bisshopp, K.E.; Drucker, D.C.: Large deflection of cantilever beams. Quart. Appl. Math. 3(1945) 272-275.

3. Mitchell, T.P.: The nonlinear bending of thin rods. Trans. ASME, J. Appl. Mech.26(1959) 40-43.

4. Frisch-Fay, R. Flexible bars. London: Butterworths 1962.

5. Mattiasson, K.: Numerical results from large deflection beam and frame problems analysed by means of elliptic integrals. Int. J. Numer. Meth. Engng. 17(1981)145-153.

6. Hymans, F.: Flat spring with large deflections. Trans. ASME, J. Appl. Mech. 13(1946) 223-230.

7. Rohde, F.V.: Large deflections of a cantilevered beam with uniformly distributed load. Quart. App. Math. 11(1953)337-338.

8. Wang, T.M.: Non-linear bending of beams with uniformly distributed loads. Int. J. Non-lin. Mech. 4(1969) 389-395.

9. Wang, T.M.; Lee, S.L.; Zienkiewicz, O.C.: A numerical analysis of large deflections of beams. Int. J. Mechanical Sci. 3(1961) 219-228.

10. Kemper, J.D.: Large deflections of tapered cantilever beams. Int. J. Mechanical Sci. 10(1968) 469-478.

11. Iyengar, N.G.R.: Note on non-linear bending of tapered cantilever beams. Int. J. Mechanical Sci. 14(1972) 395-397.

12. Verma, M.K.; Krishna Murty, A.V.: Non-linear bending of beams of variable cross-section. Int. J. Mechanical Sci. 15(1973) 183-187.

13. Wang, T.M.; Lee, C.H.: Tapered cantilevers with varying distributed loads. Proc. ASCE, J. Engng. Mech. Div. 99(1973) 919-925.

14. Prathap, G.; Varadan, T.K.: Finite deflection of tapered cantilevers. Proc. ASCE, J. Engng. Mech. Div. 102(1976) 549-552.

15. Timoshenko, S.P.; Gere, J.M. Theory of elastic stability. New York: McGraw-Hill 1961.

16. Tang, M.C.: Zur Stabilitat querbelasteter frei auf starrer Unterlage liegender Stabe und Platten. Der Stahlbau 35(1966) 112-118.

Imperfection Sensitivity of a Nearly Double Bifurcation Point

M. Potier-Ferry

Laboratoire de Mécanique Théorique
Université Pierre et Marie Curie
4, place Jussieu
7532o Paris, France

1. Introduction

Consider the mechanism with two degrees of freedom which is pictured in figure 1. Assume that the system is symmetric with respect to the x-axis and the y-axis. If the straight springs are soft, the buckling mode is symmetric (with respect to the y-axis). When these springs become sufficiently hard, the buckling mode is antisymmetric. In the present lecture, I discuss the transition between these two modes. A second example is given by the buckling of a flexible simply supported beam which rests on an elastic foundation (figure 2). I assume that the non-linear terms come only from the response of the foundation and that they have a hardening effect. The governing equation can be written in the following form:

$$\frac{d^4X}{dx^4} + \frac{d^2X}{dx^2} + kX + X^3 = 0 \qquad \text{if } |x| < \pi/2$$

$$X = \frac{d^2X}{dx^2} = 0 \qquad \text{if } x = \pm\,\pi/2 .$$

If the coefficient k is smaller than 4, the buckling mode is the symmetric function $U_1 = [\cos x]/\sqrt{2\pi}$. When k is between 4 and 36, the buckling mode becomes antisymmetric: $U_2 = [\sin 2x]\,/\sqrt{\pi 2}$.

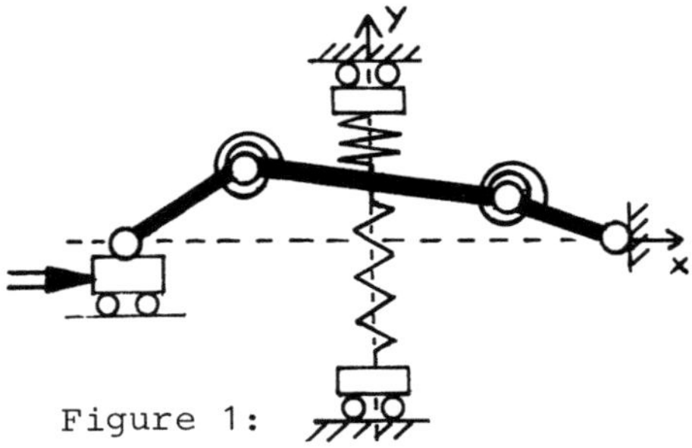

Figure 1:

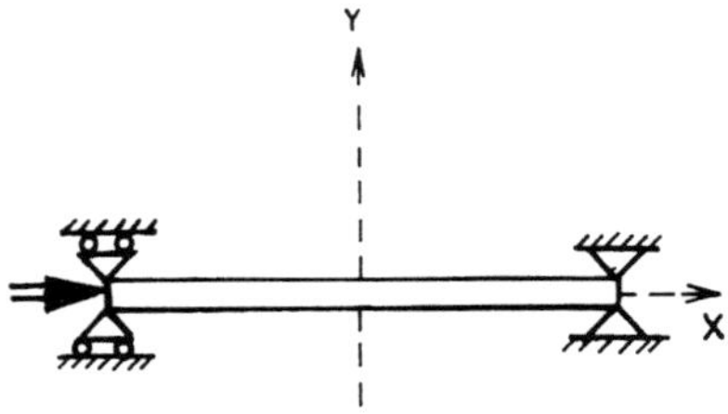

Figure 2:

Third example: a rectangular simply supported plate which is subjected to a uniaxial compression. I described this problem in a previous paper [8] . It is well known that the buckling mode varies in the same manner as in the previous examples, as a function of the aspect ratio of the plate.

These three systems are characterized by existence of a potential energy Φ (.) which is a function of the deflection X. By virtue of the symmetry properties, this functional does not change if one applies to the deflection, first the reflexion R_1 with respect to the x-axis, secondly the reflexion R_2 with respect to the y-axis, and therefore, globally, the reflexion R_3 which combines R_1 and R_2.

$$\Phi(X) = \Phi(R_1 X) = \Phi(R_2 X) = \Phi(R_3 X)$$

$$R_1 X = - X \qquad (R_2 X)(x) = X(-x) \qquad R_3 = R_1 R_2 = - R_2 \tag{1}$$

When the buckling mode is symmetric, one can apply the most classical bifurcation theory at simple eigenvalues and one establishes the existence of two bifurcating branches. If the imperfections are taken into account, the bifurcations do not persist and the postbuckling behaviour is almost symmetric (figure 3). If the buckling mode is antisymmetric, one shows by the same argument that the fundamental path goes to an almost antisymmetric behaviour (figure 4). Two questions arise from this remark. How do these response curves evolve as functions of the aspect ratio? Have not we forgotten other solutions by using this simple argument?

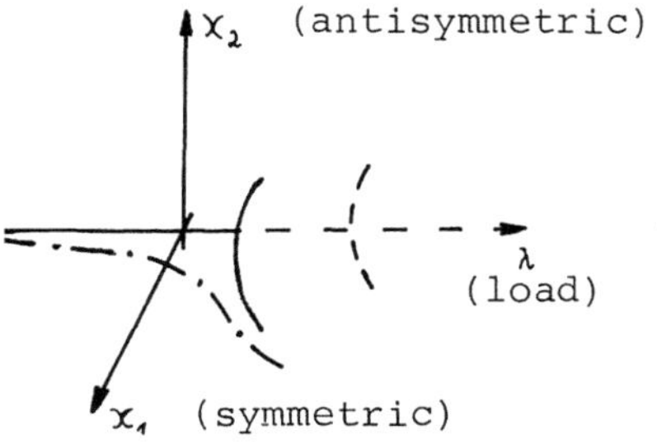

Figure 3:

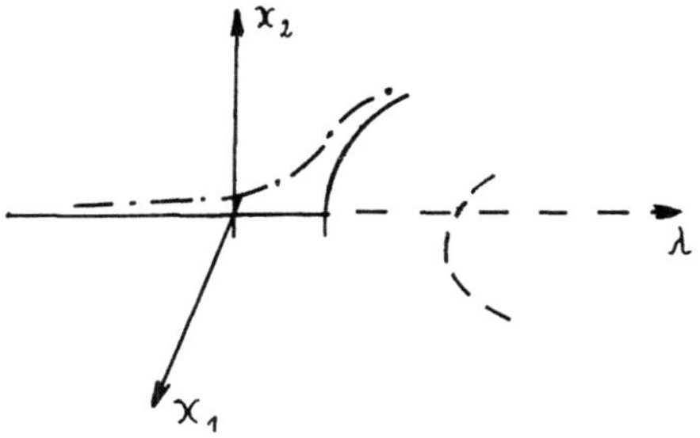

Figure 4:

————	Stable branch of the perfect system
- - - - -	Unstable branch of the perfect system
- · - · -	Stable branch of the imperfect system

2. Perfect systems

I recall here what arrives during the transition between figure 3 and figure 4 for a perfect system. This question was set by Bauer, Keller and Reiss [1] , who suggested that a double eigenvalue can lead to secondary bifurcations. This problem was studied independently by several people [6] , [11] , [10] .

Let the potential energy Φ be a function of the deflection X, of the load increment $\mu = \lambda - \lambda_c$ and of a parameter τ_1 which measures the variation of the aspect ratio. For the beam problem, one defines $\tau_1 = k - 4$ and the critical load at the coincidence $\lambda_c = 5$. The deflection X is the sum of its projections on the buckling modes U_1, U_2 and of a function Y, which is orthogonal to U_1 and U_2:

$$X = x_1 U_1 + x_2 U_2 + Y. \qquad (2)$$

In the terminology of Thompson and Hunt [13] , the real numbers x_1 and x_2 are the active coordinates and the components of Y are the passive coordinates. One eliminates these passive coordinates by taking the local minimal value of the potential energy with respect to Y:

$$\min_{Y} \ \Phi(x_1U_1 + x_2U_2 + Y, \mu, \tau_1). \tag{3}$$

I established that, for a large class of elastic systems, equation (3) has a unique small solution $Y(x_1, x_2, \mu, \tau_1)$ which depends smoothly on its arguments [8] . Next one replaces Y by its minimal value in the potential energy, which now is only a function of x_1, x_2 and the parameters μ and τ_1 :

$$\phi(x_1, x_2, \mu, \tau_1) = \Phi(x_1U_1 + x_2U_2 + Y(x_1, x_2, \mu, \tau_1), \mu, \tau_1). \tag{4}$$

One obtains the two bifurcation equations by taking the extremal value of the reduced energy functions ϕ :

$$\frac{\partial\phi}{\partial x_1} = \frac{\partial\phi}{\partial x_2} = 0. \tag{5}$$

This classical method has been introduced by Koiter in elastic stability theory [5] and it is called Lyapunov-Schmidt reduction in the mathematical literature. But what is the effect of the underlying symmetries? One can establish that, with additional restrictions on the splitting (2), the reduced energy function actually depends on the square of x_1 and x_2 [11]. Furthermore, ϕ has the following form:

$$\phi(x_1, x_2, \mu, \tau_1) = \frac{ax_1^4 + 2bx_1^2x_2^2 + cx_2^4}{4} - \mu\frac{x_1^2 + x_2^2}{2} + \tau_1\frac{\alpha_1x_1^2 + \alpha_2x_2^2}{2}$$
$$+ \text{higher order terms.} \tag{6}$$

The behaviour depends on five coefficients a, b, c, α_1, α_2, but mainly on the coefficients of the quartic terms. One easily finds the shape of the response curves if one neglects the higher order terms. This approximation was justified in [10]. For the beam problem, one has:

$\alpha_1 = 1 \quad \alpha_2 = 1/4 \quad a = 3/2\pi \quad b = 3/4\pi \quad c = 3/32\pi.$

The response curves are pictured in figure 5 when τ_1 is negative. In a first step, the postbuckling behaviour is symmetric, in accordance with an argument of Part 1.

Next there is a snap through and one goes to an antisymmetric solution. There is a branch of unstable unsymmetric solutions which connects the primary branches and one observes a hysteresis phenomenon. When τ_1 becomes positive, the secondary branch and the hysteresis disappear; the postbuckling behaviour is antisymmetric.

For the plate problem, the numerical values of the coefficients are:

$$\alpha_1 = -\alpha_2 = 3\pi^2/8 \qquad a = 1{,}56.10^{-3} \qquad b = 2{,}91.10^{-3}$$
$$c = 0{,}645.10^{-3}.$$

When τ_1 is negative (figure 6) the postbuckling behaviour is symmetric; there is a secondary bifurcation on the antisymmetric branch and a restabilization of this branch. When τ_1 is positive, the behaviour is similar but with an exchange of both primary branches. When the numerical values of a, b, c are different, it can happen that the unsymmetric solutions of figure 5 are stable or that the secondary bifurcation point of figure 6 is on the first primary branch.

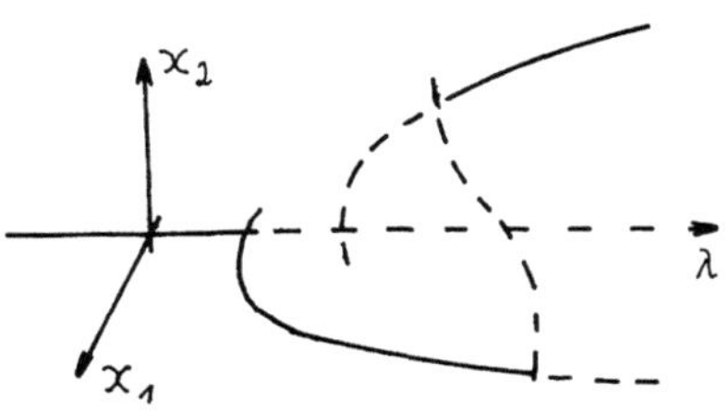

Figure 5: to be completed by symmetrie

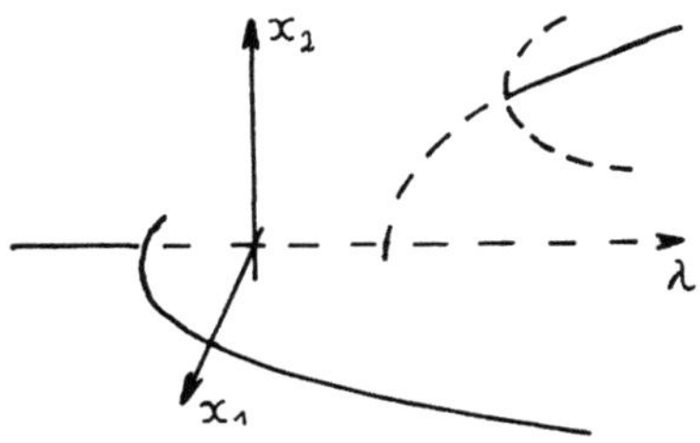

Figure 6:

3. Classification of singular points on a response curve

For a better understanding of what follows, it is necessary to explain briefly the classification of singular points. This classification is not that of the Elementary Catastrophe Theory [12] . It is that given by Golubitsky and Schaefer in a recent paper [3] and it is consistent with the ideas of elastic stability theory [4] [5] [13] [14].

Catastrophe Theory does not assign a peculiar role to the loading parameter λ. It is interested in singularities in terms of the state variable X. On the contrary, Golubitsky and Schaefer studied singularities of a response curve in terms of both variables X and λ. In the present classification, I discuss only conservative systems.

Systems without symmetry

In elastic stability, it is well known that singular points are generally limit points. These are characterized by the property that the second variation of the potential energy is degenerate ($D_x^2 \Phi \,.\, U = 0$) and by three generic conditions:

the eigenvector (or eigenfunction) U is simple, (7)

$$\frac{\partial}{\partial \lambda} D_x \Phi \,.\, U \neq 0, \tag{8}$$

$$D_x^3 \Phi(U, U, U) \neq 0. \tag{9}$$

One says that the limit point is a singular point of codimension one, which means that it is generally observed. More precisely, this means that, if one perturbates the potential energy by letting $\Phi = \Phi\,(X, \mu, \tau_1, \ldots \tau_k)$, the limit point continues to exist when one allows for a small variation in the parameter $\lambda = \lambda(\tau_1, \ldots \tau_k)$. In the same manner one says that a singularity is of codimension two, if, after perturbation, the same singularity continues to exist when one allows for a small variation in the two parameters $\lambda = \lambda_1\,(\tau_2, \ldots \tau_k)$ and $\tau_1 = \tau_1\,(\tau_2, \ldots \tau_k)$. Of course this implies that the perturbations is generic in terms of τ_1.

One obtains the singular points of codimension two by discarding one of the generic conditions (7), (8), (9). If one discards (9), then one finds the singular limit point that is pictured in figure 7. The effect of a one-parameter perturbation is also represented and one sees that a singular limit point delimits the onset of a hysteresis phenomenon.

The characterization must be completed by generic conditions. If one discards the condition (8), one finds the asymmetric (or transcritical) bifurcation point. Anybody knows that after perturbation, it can lead also to a hysteresis phenomenon. Note that, with the same conditions, the singular point can be an isolated solution, which delimits the onset of a closed path. The last singular point of codimension two is a double-limit point, but it is not useful in this study. The famous symmetric bifurcation point is only of codimension three.

Systems with symmetry

This classification is completely changed if one studies a system in the presence of a symmetry. The potential energy is assumed to be invariant under reflexions R and one considers perturbations which do not alter this symmetry:

$$\Phi(RX, \lambda, \tau) = \Phi(X, \lambda, \tau) \qquad R^2 = Id \qquad R \neq Id \tag{10}$$

By the invariance property (10), the second variation of Φ nearly a symmetric state (i.e. $RX = X$) can be splitted and the eigenvector may be symmetric or antisymmetric (i.e. $RU = -U$). In the first case, the solutions in a neighbourhood remain symmetric and the previous classification is valid. But if U is antisymmetric, there exists a persisting branch of symmetric solutions $X_0(\lambda,\tau)$. If one applies the Lyapunov-Schmidt reduction in the following manner:

$$X = X_o(\lambda,\tau) + x\,U(\tau) + Y \quad , \quad \langle Y, U(\tau) \rangle = 0,\ x \in R \quad ,$$

one obtains a reduced energy function $\phi(X,\mu,\tau)$ which is even with respect to x. By virtue of this evenness, the bifurcation equation has the following form:

$$\frac{\partial\phi}{\partial x} = x(ax^2 - b\mu - c\tau_1 + h.o.t.) = 0 \tag{11}$$

and one generally finds a symmetric bifurcation point. In addition, because x is factorized, the symmetric bifurcation point is not altered by a perturbation that maintains the symmetry. It is of codimension one. There are several types of singularities of codimension two. First, if the coefficient a in equation (11) is equal to zero, the bifurcation equation becomes:

$$x(a'x^4 - b\mu - c\tau_1 - c'\tau_1 x^2 + \text{h.o.t.}) + d\tau_2 = 0 \tag{12}$$

where the invariance property (10) is altered by the τ_2-perturbation, but not by the τ_1-perturbation. This singularity is the butterfly of Catastrophe Theory and the response curve is pictured in fugure 8. After a symmetric perturbation ($\tau_1 \neq 0$), hysteresis is possible. Figure 8 shows also that a two-parameter perturbation of this butterfly leads to a singular limit point. If the coefficient b in (11) is zero, the bifurcation equation becomes:

$$x(ax^2 - b'\mu^2 - c_1\tau_1 + \text{h.o.t.}) + d\tau_2 = 0 \tag{13}$$

The response curves of the unperturbed system is given in figure 9, when a and b' are positive. This singularity was called "trifurcation" by Plaut [7] , but I prefer the name of bifurcation point degenerate with respect to the parameter λ . The effect of a symmetric perturbation is illustrated in figures 9 and 10. One sees that a two-parameter perturbation leads to an asymmetric bifurcation point. When ab' is negative, there is a closed path which intersects twice the zero solution. Simple examples of these singularities in elasticity can be found in [3] page 88, [7] . Lastly, there are other types of singularities, especially double limit points, that are not useful in this paper

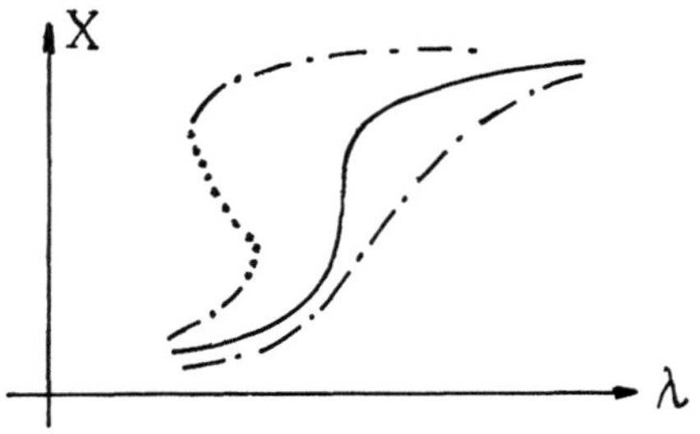

Figure 7:

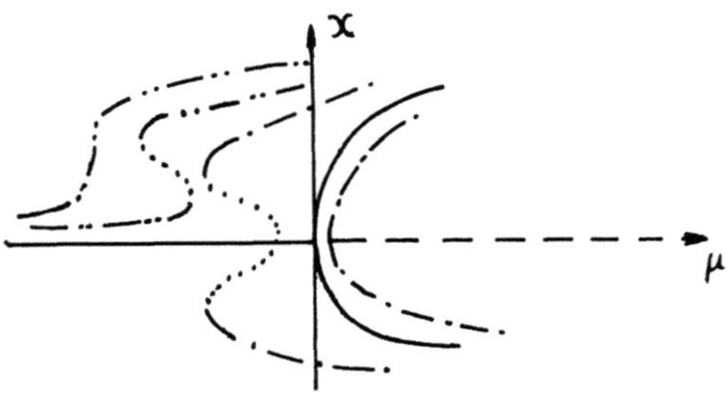

Figure 8:

—— perfect system (stable) —·—· $\tau_1 \neq o$, $\tau_2 = o$, stable
—·— imperfect system (stable) —··—·· $\tau_1 \neq o$, $\tau_2 \neq o$, stable
········ imperfect system (unstable) ········ $\tau_1 \neq o$, unstable

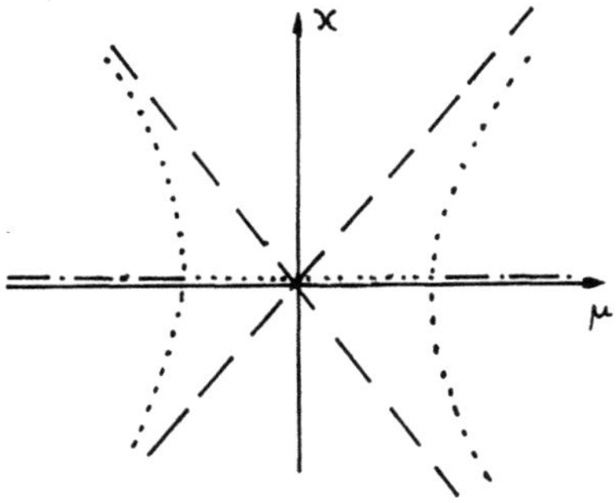

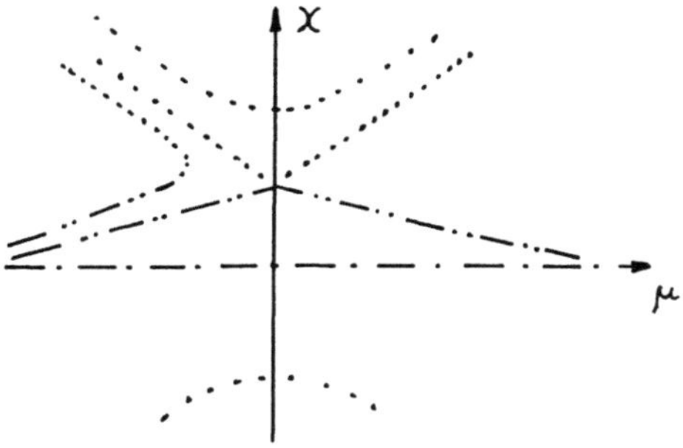

Figure 9: Figure 1o:

—— —— $\tau_1 = \tau_2 = o$ (unstable)

4. Imperfect systems

I come back to the study of the transition between two modes of imperfect bifurcation. To take into account the imperfections of the system, one introduces an additional unknown energy function as also a new parameter τ_2 which measures the size of the imperfections:

$$\Phi(X,\mu,\tau_1,\tau_2) = \Phi_i\,(X,\mu,\tau_1) + \tau_2\,\Phi_{ad}\,(X,\mu,\tau_1,\tau_2)$$

where Φ_i denotes the potential energy of the ideal system. On applying the Lyapunov-Schmidt reduction, one finds the new reduced energy:

$$\phi(x_1,x_2,\mu,\tau_1,\tau_2,r) = \frac{ax_1^4 + 2bx_1^2x_2^2 + cx_2^4}{4} - \mu\frac{x_1^2 + x_2^2}{2} + \tau_1\frac{\alpha_1x_1^2 + \alpha_2x_2^2}{2} \tag{14}$$

$$+ \beta\tau_2\,(x_1\cos r + x_2\sin r) + \text{higher order terms.}$$

A new parameter appears: the angle r which measures the direction of the imperfections with respect to both buckling modes. One can eliminate one of the three parameters τ_1, τ_2, r by the following scaling:

$$\beta\tau_2 = \varepsilon^3,\ \mu = m\varepsilon^2,\ \tau_1 = t\varepsilon^2,\ x_1 = \varepsilon X_1,\ x_2 = \varepsilon X_2$$

Next one computes the reduced energy in terms of the new variables:

$$\phi(X_1,X_2,m,t,r,\varepsilon)\ /\ \varepsilon^4 = \frac{aX_1^4 + 2bX_1^2X_2^2 + cX_2^4}{4} - m\frac{X_1^2 + X_2^2}{2} + t\frac{\alpha_1X_1^2 + \alpha_2X_2^2}{2}$$

$$+ X_1 \cos r + X_2 \sin r + 0\ (\varepsilon). \tag{15}$$

The term of order ε in the right hand side of (15) will be neglected. For the sake of simplicity, I restricted my topics to the discussion of the shape of the fundamental path according to the two remaining parameters: first, the parameter t that is defined in (14) and is proportional to variation of the aspect ratio ; secondly, the direction of the imperfections r.

To begin with, I observe that, when t is large, the behaviour is that of the perfect system with a slight perturbation. For the beam problem, the fundamental path is represented in figure 11 when t is negative and large. This curve is very close to that of figure 5. One finds a symmetric postcritical behaviour, which is temporary. Next, there is a snap-through and one goes to an almost anti-symmetric behaviour. When t is positive and large, there is no snap-through. Therefore, one guesses that a singular limit point can appear during the transition. For the plate problem, this snap-through cannot occur because there is no secondary branching from the first mode (cf. figure 6). When t is negative and large, the path goes to the symmetric branch while for positive t it goes to the antisymmetric one. Hence a bifurcation point could explain this diversion, but, generally, a bifurcation is transcritical and leads to hysteresis. Therefore, a singular limit point seems necessary to stop this hysteresis. In the case $c < b < (ac)^{1/2}$, the postbuckling behaviour is antisymmetric in both cases and no singularity is needed to explain this transition.

In a second step, one chooses the angle r equal to zero, which means that the imperfection is symmetric. The bifurcation equations become simpler and one can separate an equation for symmetric solutions:

$$m = \alpha_1 t + aX_1^2 + 1/X_1 \quad , \qquad X_2 = 0 \ . \tag{16}$$

and an equation for unsymmetric solutions:

$$(b-c)m = (b^2-ac)X_1^2 - c/X_1 - t(c\alpha_1-b\alpha_2), (b-c)X_2^2 = (b-a)X_1^2-1/X_1 - t(\alpha_1-\alpha_2). \tag{17}$$

The symmetric branch has the same shape as in the neighbourhood of a simple eigenvalue. For the beam problem, all the response curves of (16) (17) are plotted in the figures 12, 13, 14, for several typical values of t. In the case of figure 12, there is an unstable closed loop of unsymmetric solutions, that intersects twice the symmetric branch. When t increases, the loop disappears (figure 13) and, at the coalescence, the singular point is a parameter-degenerate point of bifurcation. But the most interesting curves are below the plane $X_1 = 0$. One observes a transition between a symmetric response and an almost antisymmetric response. In the cases of figures 12 and 13, there is a subcritical point of bifurcation and a snap-through. Next, the bifurcation becomes supercritical (figure 14) and the snap-through disappears. A singularity of the butterfly type explains this transition.

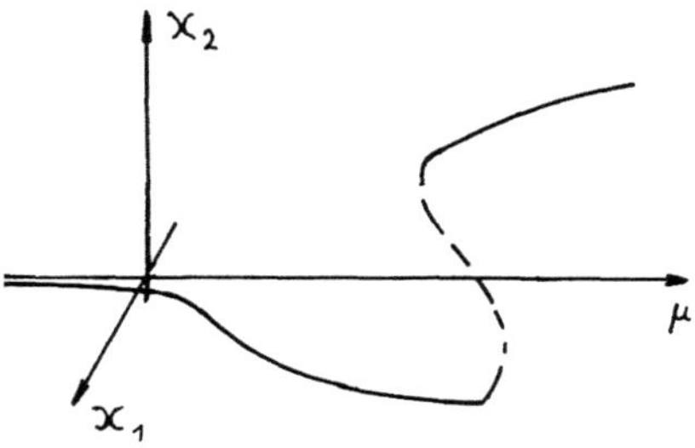

Figure 11:

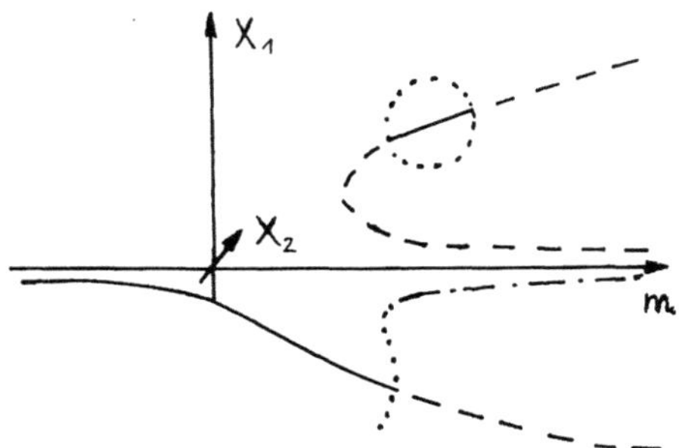

Figure 12: ——— - - -symmetric
—·—·—· ·········unsymmetric

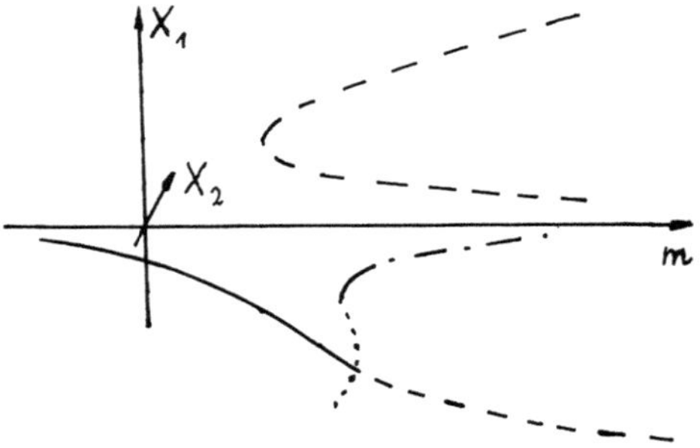

Figure 13:

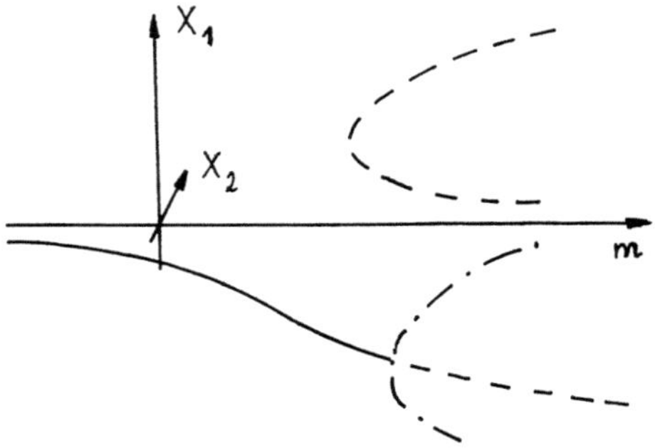

Figure 14:

The complete results for the beam problem are given on figure 15. The fundamental path goes always to an almost antisymmetric solution, but there is a snap-through on the left hand side region. When (t,r) is on the separating curve, there is a singular limit point on the fundamental path. This curve was computed numerically. This calculation is easy, because one can obtain explicitly the equation, by using the characterization of these singular points:

$$D_X\phi = 0 \quad , \quad D_X^2\phi \, . \, U = 0 \quad , \quad D_X^3\phi \, (U,U,U) = 0.$$

The dashed curve in figure 15 indicates the existence of an isolated point and the coalescence of a closed loop. The equation of this dashed curve is deduced from the following equations:

$$D_X\phi = 0 \quad , \quad D_X^2\phi \, . \, U = 0 \quad , \quad \frac{\partial}{\partial m} D_X\phi \, . \, U = 0 \, .$$

For the plate problem, the scheme is that one which was guessed (figure 16). In the left (resp. right) hand-side region, one goes to a symmetric (resp. antisymmetric) response and there is no limit point on the fundamental branch. The two separating curves correspond to the existence of a singular limit point or of a transcritical bifurcation point. When (t,r) is between both curves, there are two limit points on the fundamental path and a snap-through occurs during the loading. Hence, for any direction of the imperfections, a hysteresis phenomenon exists when the aspect ratio t is in some interval. But this interval is very small and it is magnified in figure 16 for the sake of illustration. Note that both separating curves have a point of intersection E, which corresponds to a symmetric bifurcation. This is consistent with the classification of Part 3, because a singularity of codimension three must be isolated in this scheme. Furthermore, the separating curves represent singularities of codimension two. Of course, this remark is not valid when the imperfections are symmetric or antisymmetric. In this case, the second classification of Part 3 must be applied. The points A and C in figures 15 and 16 correspond to a parameter-degenerate point of bifurcation and B and D to a butterfly singularity.

The study of this type of imperfect bifurcation was initiated by Chow, Hale and Mallet-Paret [2] with t = o and fixed r. List [6] investigated the case t ≠ o and the hysteresis phenomenon is implicit in his work. I found this hysteresis phenomenon when t is equal to zero [9], but this property remained rather mysterious. The introduction of the aspect ratio t has allowed me to explain it more satisfactorilly by considering the special cases of a large t and of a symmetric perturbation.

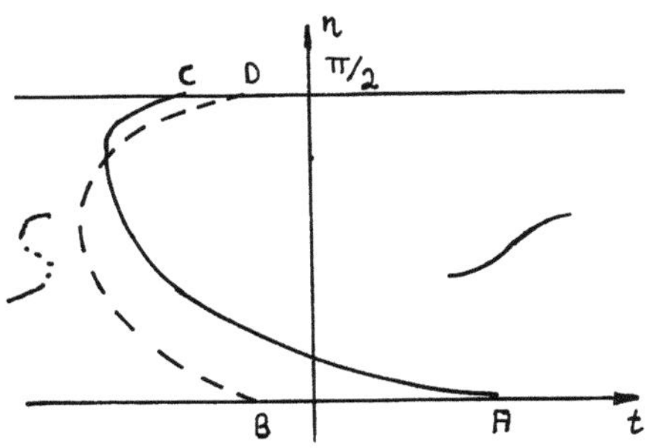

Figure 15:

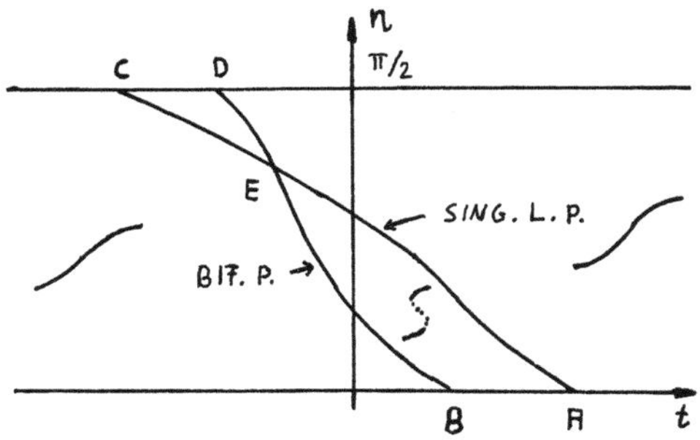

Figure 16:

References

1. Bauer, L.; Keller, H.B.; Reis, E.L.: Multiple eigenvalues lead to secondary bifurcation.S.I.A.M. Review. 17 (1975) 1o1-121.

2. Chow, S.N.; Hale, J.K.; Mallet-Paret, J.: Applications of generic bifurcation II, Arch. Rat. Mech. Anal. 62 (1976) 2o9-235

3. Golubitsky, M.; Schaeffer, D.: A theory of imperfect bifurcation via singularity theory. Comm. Pure Appl. Math. 32 (1979) 21-98.

4. Huseyin, K.: Non linear theory of elastic stability. Leyden: Noordhoff 1975.

5. Koiter, W.T.: On the stability of elastic equilibrium. Dissertation. Delft 1945. English translation, NASA Tech. Transl. F 1o, 833 (1967).

6. List, S.E.: Generic bifurcation with application to the Von Kamen equations. J. Diff. Equ. 3o (1978) 89-118.

7. Plaut, R.H.: Postbuckling analysis of nonconservative elastic systems. J. Struct. Mech. 4 (1976) 395-416.

8. Potier-Ferry, M.: Bifurcation et stabilité pour des systèmes dérivant d'un potentiel. J. Mécanique 17 (1978) 579-6o8.

9. Potier-Ferry, M.: Perturbed bifurcation theory. J. Diff. Equ. 33 (1979) 112-146.

1o. Potier-Ferry, M.: Multiple bifurcation, symmetry and secondary bifurcation. In: Non linear problems of analysis in geometry and mechanics, Research Notes in Math. 46. Boston, London, Melbourne: Pitman 1981, 158-167.

11. Schaeffer, D.; Golubitsky, M.: Boundary conditions and mode jumping in the buckling of a rectangular plate. Comm. Math. Phys. 69 (1979) 2o9-23o.

12. Thom, R.: Stabilité structurelle et morphogénèse. Reading: Benjamin 1972.

13. Thompson, J.M.T.; Hunt, G.W.: A general theory of elastic stability. London: Wiley 1973.

14. Thompson, J.M.T.; Hunt, G.W.: Towards a unified bifurcation theory. Zeit. Angew. Math. Phys. 26. (1975) 581-6o4.

Snap-Through of a Shallow Arch under Pulsating Load

K.-Y. HUANG and R. H. PLAUT

Department of Civil Engineering
Virginia Polytechnic Institute and State University
Blacksburg, Virginia 24061, USA

Summary

A shallow, elastic arch subjected to a distributed vertical load is considered. The load has a static part and a dynamic part which varies harmonically in time. Conditions for dynamic snap-through are investigated. The results are presented as plots of the critical magnitude of the harmonic load versus its frequency. Some features of these curves are related to resonance conditions.

Formulation

A number of authors have analyzed arches subjected to pulsating loads, and a list of references can be found in Ref. [1]. Most of them, however, only consider small motions and examine conditions of resonance. Two exceptions are Refs. [2] and [3], which discuss snap-through of an arch into an inverted position during its motion. Here, we investigate this phenomenon further. The frequency of the load is varied, producing a locus of critical load amplitudes. Viscous damping is included, and several spatial distributions of the harmonic load are considered.

The ends of the arch are pinned at a distance L apart and are constrained against translation. The arch has mass per unit length m, cross-sectional area A, bending stiffness EI, and radius of gyration $r = \sqrt{I/A}$. We let X denote the horizontal axis through the ends, while T denotes time. The unloaded shape of the arch, $Y_o(X)$, and the static load, $Q_s(X)$, are assumed to be sinusoidal. The equilibrium configuration under Q_s is denoted $Y_s(X)$, the harmonic load is $Q_d(X,T)$, and the motion under the total load $Q(X,T) = Q_s(X) + Q_d(X,T)$ is $Y(X,T)$.

We define the nondimensional quantities

$$x=X/L,\ y_o=Y_o/(2r),\ y_s=Y_s/(2r),\ y=Y/(2r),$$
$$q_s=Q_sL^4/(2EIr),\ q_d=Q_dL^4/(2EIr), \tag{1}$$
$$q=QL^4/(2EIr),\ t=(T/L^2)\sqrt{EI/m}.$$

Also, we let $w(x,t)$ denote the nondimensional motion about the equilibrium state, i.e.,

$$w(x,t) = y(x,t) - y_s(x). \tag{2}$$

Then the equation of motion is assumed to be [1]

$$\ddot{w} + 2c\dot{w} + w'''' + 2w'' \int_0^1 [(y_o')^2-(y_s')^2]dx$$
$$-2(w''+y_s'') \int_0^1 [(w')^2 + 2w'y_s']dx + q_d = 0 \tag{3}$$

where dots and primes denote partial derivatives with respect to t and x, respectively, and c is the nondimensional damping coefficient. The boundary conditions are $w=0$ and $w''=0$ at $x=0$ and $x=1$.

We choose the unloaded arch shape $y_o(x)$, static load $q_s(x)$, and equilibrium shape $y_s(x)$ as

$$y_o(x) = \lambda \sin \pi x\ , \quad q_s(x) = \pi^4 p \sin \pi x\ ,$$
$$y_s(x) = b \sin \pi x, \tag{4}$$

where $\lambda>0$, $p>0$, $b>0$, and b is a root of the equation

$$b^3 + (1-\lambda^2)b + p - \lambda = 0. \tag{5}$$

We then represent the motion in the series

$$w(x,t) = \sum_{j=1}^{\infty} u_j(t) \sin j\pi x. \tag{6}$$

If we substitute (6) into (3) and make the result orthogonal to $\sin n\pi x$, we obtain the set of equations

$$\ddot{u}_n + 2c\dot{u}_n + \omega_n^2 u_n + \pi^4(\sqrt{2}b\delta_{1n} + 2n^2 u_n)\sum_{j=1}^{\infty} j^2 u_j^2 + 2\sqrt{2}\pi^4 n^2 b u_1 u_n - 2K_n \cos\Omega t = 0, \quad n=1,2,\ldots \tag{7}$$

where δ_{1n} is the Kronecker delta and

$$\int_0^1 q_d(x,t)\sin n\pi x\, dx = -\sqrt{2}\, K_n \cos\Omega t. \tag{8}$$

Therefore Ω and K_n represent the frequency and modal amplitudes of the harmonic load, respectively. The natural frequencies ω_n are given by [4]

$$\omega_1^2 = \pi^4(1+3b^2-\lambda^2), \quad \omega_k^2 = \pi^4 k^2(k^2+b^2-\lambda^2), \quad k=2,3,\ldots \tag{9}$$

The Budiansky-Roth criterion [5] is used to determine critical loads. For a given load level, we measure the maximum response by the quantity

$$d = \max_{t\geq 0}\left[\int_0^1 w^2(x,t)\, dx\right]^{1/2}. \tag{10}$$

As the load is increased, the critical value is defined when d jumps abruptly to a value which is sufficiently large to indicate inversion of the arch during the motion. In the present study, when the load is above its critical value, two types of behavior are observed: after snapping several times to an inverted position, the arch either settles into a steady-state oscillation about the equilibrium configuration y_s or about the inverted configuration.

Results

For arches which are very shallow (e.g., $\lambda<1$), the response measure d usually does not exhibit a sudden jump. For higher arches, the curves of critical load versus frequency have similar characteristics for different values of λ, and results for $\lambda=5$ are presented here. Also, changes in the damping coefficient do not produce significant changes in the general characteristics of the results, and $c=0.1$ is chosen.

As a first example, we assume the harmonic load is sinusoidal over the arch, so that all the modal amplitudes K_n for the harmonic load are zero except K_1. Then $u_j=0$, $j\geq 2$, is a solution of (7) for $n\geq 2$, and we obtain a single equation in u_1. Critical values of K_1 are plotted versus the load frequency Ω in Fig. 1. The solid curve corresponds to the case of no static load, so that $b=\lambda$. For the dashed curve, the static load has $p=11.83$ in (4), and the arch rise parameter for the equilibrium configuration is $b=4.75$ from (5).

The natural frequency ω_1 is 70.48 for $\lambda=5$, $b=5$, and 65.23 for $\lambda=5$, $b=4.75$. In an asymptotic analysis for small harmonic loads, resonance conditions occur when Ω equals $\omega_1/3$, $\omega_1/2$, ω_1, $2\omega_1$, and $3\omega_1$[1]. In Fig. 1, we observe that the curves have dips near these values of Ω, indicating an effect of these resonances on the critical load for dynamic snap-through. Actually, the dips occur to the left of these resonance frequencies. This may be partially due to the fact that the arch exhibits a softening behavior for these sets of arch rise parameters λ and b, and the frequency-response curves bend toward the left.

The addition of the static load in Fig. 1 tends to shift the curve of critical loads lower and to the left. These features are not surprising. With a static load present, we might expect that the additional harmonic load required for snap-through would be lower. Also, in light of the discussion in the previous paragraph, the shift of the natural frequency to the left, and hence the resonance frequencies, might be expected to shift the results leftward. However, we note that the critical load amplitude K_1 for a given load frequency Ω is not always decreased by the addition of the static load (e.g., for $62 < \Omega < 91$).

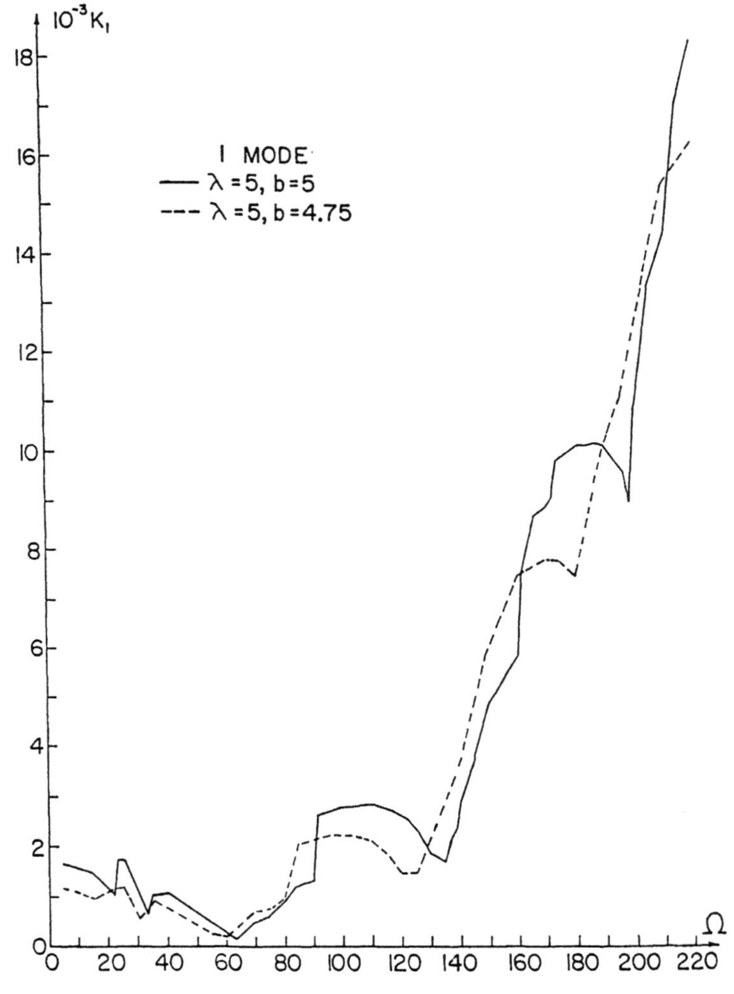

Fig. 1. Critical load versus load frequency: $K_1 \neq 0$

In Fig. 1, we also observe the dramatic increase in critical load as the load frequency becomes high. This is a typical feature.

For the second example, we consider the same combinations of arch rise parameters λ and b, but we include two modes in the spatial distribution of the harmonic load, with $K_2=0.5K_1$. The response of the arch then involves two functions, u_1 and u_2, governed by the equations corresponding to n=1 and n=2 in (7). Critical values of K_1 are plotted versus Ω in Fig. 2.

The values of ω_1 are the same as before, while ω_2 is 39.48 for $\lambda=5$, b=5, and 24.67 for $\lambda=5$, b=4.75. In the asymptotic analysis, resonances occur at $\omega_n/3$, $\omega_n/2$, ω_n, $2\omega_n$, and $3\omega_n$, for n=1 and n=2, and also at the combinations $(\omega_1 \pm \omega_2)/2$, $\omega_1 \pm \omega_2$, $2\omega_1 \pm \omega_2$, $2\omega_2 \pm \omega_1$, and $\omega_1 - 2\omega_2$ [1]. Perhaps the greater number of resonances in this example causes the curves in Fig. 2 to be much more erratic than those in Fig. 1, with many peaks and dips. The general form is similar, however: the lowest critical loads occur when Ω is in a range near the natural frequencies, and very high critical loads occur for high values of Ω. Critical values are much lower in Fig. 2 (for a given load frequency), since the load is asymmetric in this case and symmetric in the previous one.

Finally, results for a case with five modes in the spatial distribution of the harmonic load and in the response are presented in Fig. 3. Here, $\lambda=5$, b=5, $K_2=0.5K_1$, $K_3=0.24K_1$, $K_4=-0.3K_1$, and $K_5=-0.4K_1$. The relevant natural frequencies are $\omega_1=70.48$, $\omega_2=39.48$, $\omega_3=88.83$, $\omega_4=157.91$, and $\omega_5=246.74$. The critical value of K_1 is the lowest in the region of the first two natural frequencies. In fact, the curve in Fig. 3 is similar to the solid curve in Fig. 2, as far as critical values are concerned, except at high frequencies Ω. The high values of ω_4 and ω_5, which are involved in this case, may tend to keep the critical loads from increasing drastically, until Ω reaches much higher values.

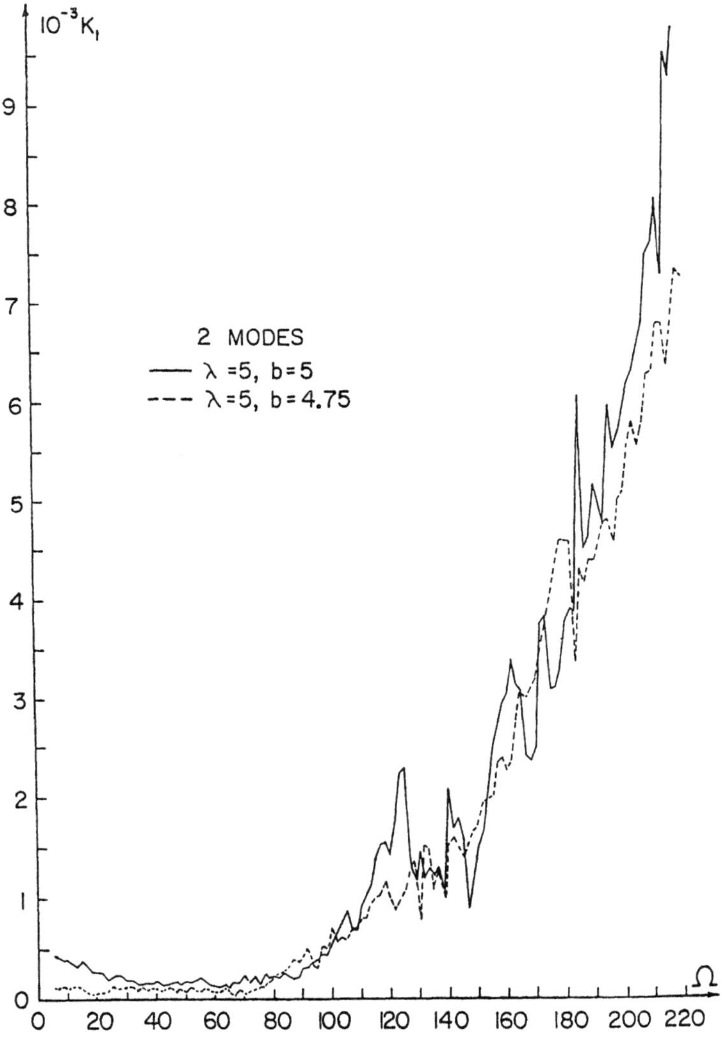

Fig. 2. Critical load versus load frequency: $K_2=0.5K_1$

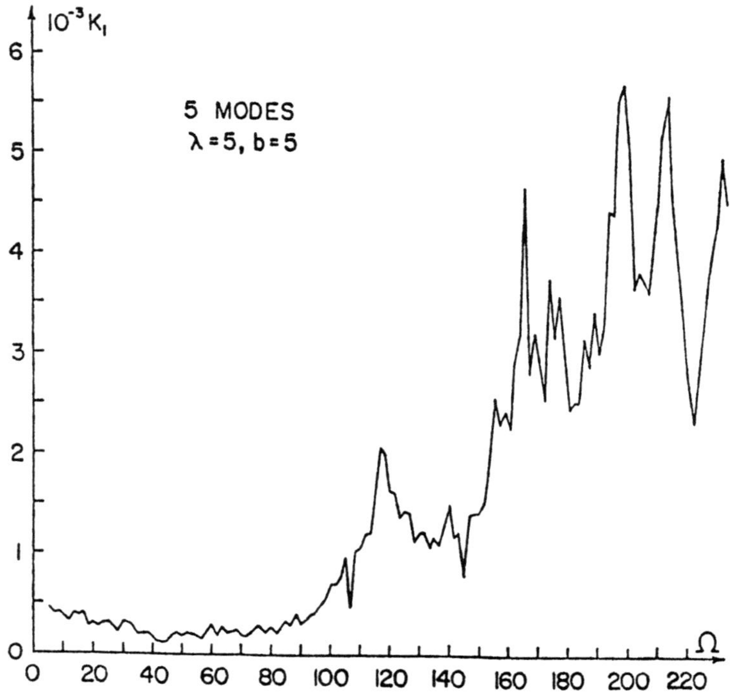

Fig. 3. Critical load versus load frequency: $K_2=0.5K_1$, $K_3=0.24K_1$, $K_4=-0.3K_1$, $K_5=-0.4K_1$

Concluding Remarks

We have considered a shallow, pinned, sinusoidal arch subjected to (i) a sinusoidal static load and (ii) a pulsating load with frequency Ω and spatial distribution proportional to $K_1 \sin\pi x + K_2 \sin 2\pi x + \ldots$. Cases with one, two, and five modal amplitudes of the harmonic load have been analyzed.

The characteristics of the critical load curves depend greatly on the number of modes involved in the response. The magnitudes of the critical loads depend primarily on the initial arch rise, the static load, and whether or not the problem is symmetric. We expect that the general features of the results will be retained for shallow arches with different initial shapes, boundary conditions, static load distributions, and damping.

Acknowledgment

This research was supported by the National Science Foundation under Grant CME79-20781.

References

1. Mook, D. T.; Plaut, R. H.; HaQuang, N.: The response of multidegree-of-freedom systems with quadratic and cubic nonlinearities to harmonic excitation, with application to a shallow arch. SIAM Rev. 24 (1982) to appear.

2. Huang, N. C.: Dynamic buckling of some elastic shallow structures subject to periodic loading with high frequency. Int. J. Solids Strucs. 8 (1972) 315-326.

3. Clemens, H.; Wauer, J.: Durchschlagen vorgekrümmter Stäbe unter harmonisch pulsierender Belastung. Proceedings of the 8th International Conference on Nonlinear Oscillations, Academia, Prague (1978) 191-197.

4. Plaut, R. H.; Johnson, E. R.: The effects of initial thrust and elastic foundation on the vibration frequencies of a shallow arch. J. Sound Vib. 78 (1981) to appear.

5. Budiansky, B.; Roth, R. S.: Axisymmetric dynamic buckling of clamped spherical caps. U.S. NASA Technical Note D-1510 (1962) 597-606.

Effect of Pretwist on the Stability of a Nonuniform Cantilever under Circulatory Force

R. C. KAR and W. HAUGER

Institut für Mechanik, TH Darmstadt, Hochschulstraße 1,
6100 Darmstadt, West Germany

Summary

The non-self-adjoint boundary value problem including the effects of internal and external damping is formulated for the investigation of stability of a pretwisted nonuniform cantilever subjected to a tangential force at the free end. Through the introduction of an adjoint problem a suitable variational principle is used to calculate numerically the approximate values of the critical load. Results are shown through graphs.

Introduction

Nonconservative problems of the theory of elastic stability have attracted the attention of numerous research workers in the past [1 - 4]. Whereas Hui [5], Leipholz [6] and Fischer [7] were concerned with the stability of pretwisted bars under conservative loads, Farshad [8] obtained the general conservative end loads for a pretwisted uniform rod subjected to nonconservative force and couple at the free end by the introduction of an adjoint field.

The present study deals with the investigation of stability of an externally damped pretwisted uniformly tapered viscoelastic cantilever beam of rectangular cross-section under the influence of a follower force at its free end. The non-self-adjoint boundary value problem is formulated and the associated adjoint boundary value problem is introduced. Approximate values of the critical load are determined on the basis of a suitable variational principle for prescribed values of the parameters of the system. In the absence of pretwist and with the assumption that the beam undergoes flexural deformation in x_1x_3 plane only, the present problem reduces to that treated in [9] for tangential loading.

The Equations of Motion

A pretwisted linearly tapered cantilever beam of length L having rectangular cross-section subjected to a tangential force of constant magnitude P at its free end is shown in Figure 1.

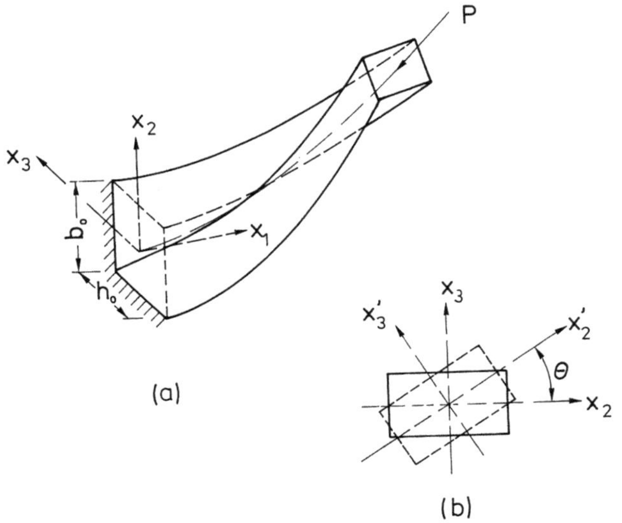

Fig.1.(a) Configuration of the system, (b) Angle of twist

Assuming that the material of the beam is Kelvin-Voigt solid, the differential equations of motion and the boundary conditions for small flexural deformations in the x_1x_2- and x_1x_3-planes, including the effects of internal and external damping, are given in dimensionless form as

$$(1 + \eta^*\lambda)[f(\zeta)\eta'']'' + (1 + \eta^*\lambda)[g(\zeta)\xi'']'' + Q\eta'' + \beta^*\lambda\eta + m(\zeta)\lambda^2\eta = 0, \quad 0 < \zeta < 1, \tag{1}$$

$$(1 + \eta^*\lambda)[q(\zeta)\xi'']'' + (1 + \eta^*\lambda)[g(\zeta)\eta'']'' + Q\xi'' + \beta^*\lambda\xi + m(\zeta)\lambda^2\xi = 0, \quad 0 < \zeta < 1, \tag{2}$$

$$\eta(0) = \xi(0) = \eta'(0) = \xi'(0) = 0, \tag{3}$$

$$f(1)\eta''(1) + g(1)\xi''(1) = 0,$$
$$[f(\zeta)\eta'']'(1) + [g(\zeta)\xi'']'(1) = 0, \tag{4a}$$

$$q(1)\xi''(1) + g(1)\eta''(1) = 0,$$
$$[q(\zeta)\xi'']'(1) + [g(\zeta)\eta'']'(1) = 0, \tag{4b}$$

where $\bar{\eta} = w/L$, $\bar{\xi} = v/L$, $\zeta = x_1/L$, $\tau = ct$ (with $c^2 = EI^o_{2'2'}/\rho A_o L^4$), $\eta^* = E^*c/E$, $\beta^* = \beta_1 cL^4/EI^o_{2'2'}$, $r = I^o_{3'3'}/I^o_{2'2'}$, $Q = PL^2/EI^o_{2'2'}$,

$$f(\zeta) = s_1(\zeta)\cos^2\theta_L\zeta + rs_2(\zeta)\sin^2\theta_L\zeta\ ,$$
$$g(\zeta) = \tfrac{1}{2}[rs_2(\zeta) - s_1(\zeta)]\sin 2\theta_L\zeta\ ,$$
$$q(\zeta) = rs_2(\zeta)\cos^2\theta_L\zeta + s_1(\zeta)\sin^2\theta_L\zeta\ ,$$
$$\bar{\eta}(\zeta,\tau) = \eta(\zeta)e^{\lambda\tau}\ ,$$
$$\bar{\xi}(\zeta,\tau) = \xi(\zeta)e^{\lambda\tau}\ .$$

Besides $v(x_1,t)$ and $w(x_1,t)$ are the transverse deflections of the beam at point x_1 and at time t, ρ is the density, E is the Young's modulus, E^* is the viscous modulus, β_1 is the external damping coefficient, A_o is the cross-sectional area at the root of the cantilever, $I^o_{2'2'}$ and $I^o_{3'3'}$ are the principal moments of inertia of the root cross-section, $m(\zeta)$ and $s_i(\zeta)$ $(i = 1,2)$ are the mass- and stiffness-distributions respectively, θ_L is the pretwist angle at the free end, λ, the eigenvalue parameter, is in general a complex number, and the notation $(\ldots)' = d(\ldots)/d\zeta$.

The Adjoint Problem and Variational Principle

Taking $\phi(\zeta)$ and $\psi(\zeta)$ to be the dependent variables of the adjoint problem, it can be easily shown that the boundary value problem which is adjoint to the original non-self-adjoint boundary value problem in equations (1) - (4) is the following:

$$(1 + \eta^*\lambda)[f(\zeta)\phi'']'' + (1 + \eta^*\lambda)[g(\zeta)\psi'']'' + Q\phi'' + \beta^*\lambda\phi + \lambda^2 m(\zeta)\phi = 0,\quad 0 < \zeta < 1, \tag{5}$$

$$(1 + \eta^*\lambda)[q(\zeta)\psi'']'' + (1 + \eta^*\lambda)[g(\zeta)\phi'']'' + Q\psi'' + \beta^*\lambda\psi + \lambda^2 m(\zeta)\psi = 0,\quad 0 < \zeta < 1, \tag{6}$$

$$\phi(0) = \psi(0) = \phi'(0) = \psi'(0) = 0, \tag{7}$$

$$(1 + \eta^*\lambda)[f(1)\phi''(1) + g(1)\psi''(1)] + Q\phi(1) = 0\ ,$$
$$(1 + \eta^*\lambda)\{[f(\zeta)\phi'']'(1) + [g(\zeta)\psi'']'(1)\} + Q\phi'(1) = 0, \tag{8a}$$

$$(1 + \eta^*\lambda)[q(1)\psi''(1) + g(1)\phi''(1)] + Q\psi(1) = 0,$$
$$(1 + \eta^*\lambda)\{[q(\zeta)\psi'']'(1) + [g(\zeta)\phi'']'(1)\} + Q\psi'(1) = 0. \quad (8b)$$

The variational principle which generates the original and the adjoint problems is

$$\delta\Big[\int_0^1 \{(1 + \eta^*\lambda)[f(\zeta)\eta''\phi'' + g(\zeta)\xi''\phi'' + g(\zeta)\psi''\eta'' + q(\zeta)\xi''\psi'']$$
$$- Q[\eta'\phi' + \xi'\psi'] + [\beta^*\lambda + \lambda^2 m(\zeta)][\eta\phi + \xi\psi]\}d\zeta$$
$$+ Q\{\eta'(1)\phi(1) + \xi'(1)\psi(1)\}\Big] = 0 . \quad (9)$$

The Approximate Solution

For the original boundary value problem given in equations (1) - (4) and its adjoint problem in equations (5) - (8), it is assumed that the functions $\eta(\zeta)$, $\xi(\zeta)$, $\phi(\zeta)$ and $\psi(\zeta)$ can be approximated by

$$\eta(\zeta) = \sum_{r=1}^{N} a_r\eta_r(\zeta) , \quad \phi(\zeta) = \sum_{r=1}^{N} c_r\phi_r(\zeta) ,$$
$$\xi(\zeta) = \sum_{r=1}^{N} b_r\xi_r(\zeta) , \quad \psi(\zeta) = \sum_{r=1}^{N} d_r\psi_r(\zeta) , \quad (10)$$

where N is a positive integer, a_r, b_r, c_r, d_r are arbitrary constants,and $\eta_r(\zeta)$, $\xi_r(\zeta)$, $\phi_r(\zeta)$, $\psi_r(\zeta)$ are coordinate functions. These coordinate functions are chosen to be the set of polynomials, viz.,

$$\eta_r(\zeta) = \xi_r(\zeta) = \zeta^{r+1}(\alpha_1 - 2\beta_1\zeta + \gamma_1\zeta^2) ,$$
$$\phi_r(\zeta) = \zeta^{r+1}(\alpha_2 - 2\beta_2\zeta + \gamma_2\zeta^2) , \quad (11)$$
$$\psi_r(\zeta) = \zeta^{r+1}(\alpha_3 - 2\beta_3\zeta + \gamma_3\zeta^2) , \quad r \geq 1 ,$$

where α_i, β_i and $\gamma_i (i = 1,2,3)$ are defined in such a way that the coordinate functions satisfy not only the geometric boundary conditions in equations (3) and (7), but also those natural boundary conditions which are obtained from equations (4) and (8) by deleting the terms containing λ and g.

Substitution of equation (10) into the variational principle in equation (9) and performance of the indicated variation yield the set of homogenous algebraic equations

$$\sum_{r=1}^{N} (M_{rn}a_r + N_{rn}b_r) = 0 \quad ,$$

$$\sum_{r=1}^{N} (P_{rn}a_r + Q_{rn}b_r) = 0 \quad , \quad n = 1,2,\ldots, N \ ,$$

which, to facilitate numerical calculations, is expressed as a quadratic eigenvalue problem

$$\sum_{i=1}^{2N} (A_{ij}\lambda^2 + B_{ij}\lambda + C_{ij})x_i = o \ , \ j = 1,2,\ldots, 2N \quad , \qquad (12)$$

where

$$x_i = \begin{cases} a_i & \text{if} \quad 1 \leq i \leq N \quad , \\ b_{i-N} & \text{if} \quad N+1 \leq i \leq 2N, \end{cases}$$

and M_{rn}, N_{rn}, P_{rn}, Q_{rn}, A_{ij}, B_{ij} and C_{ij} are suitably defined functions of the system parameters.

The Numerical Results

For the tapered beam shown in Figure 1, the breadth b and the thickness h were assumed to vary as $b = b_o(1 - \alpha\zeta)$ and $h = h_o(1 - \beta\zeta)$, where α and β are the breadth and thickness taper parameters respectively. The critical load was determined numerically from equation (12) by using double precision arithmetic on IBM 370/168 computer at TH Darmstadt for a beam with r = 9 for several values of the breadth taper parameter α, the thickness taper parameter β, the internal damping parameter η^*, the external damping parameter β^* and the pretwist angle Θ_L.

The variation of critical flutter load Q_f with pretwist angle Θ_L for five values of α is shown in Figure 2. For the values of α depicted, the value of Q_f increases monotonically to reach a maximum value as the value of Θ_L increases from zero and thereafter decreases with further increase in the value of Θ_L. At any pretwist angle the value of Q_f decreases due to an increase in the value of α.

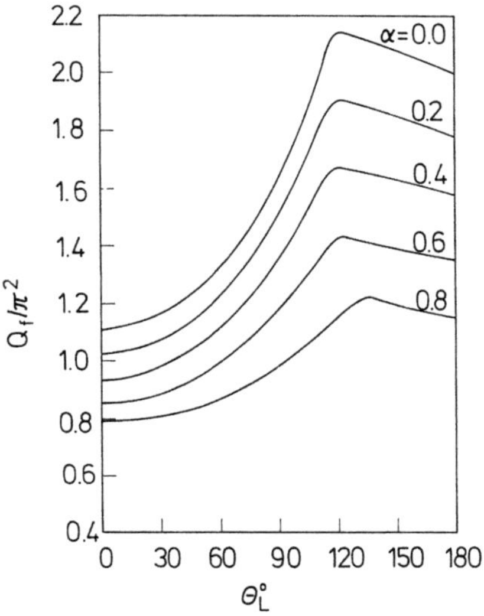

Fig.2. Variation of critical flutter load with angle of pretwist and breadth taper for $\beta = 0$, $\eta^* = 0.01$ and $\beta^* = 10^{-6}$

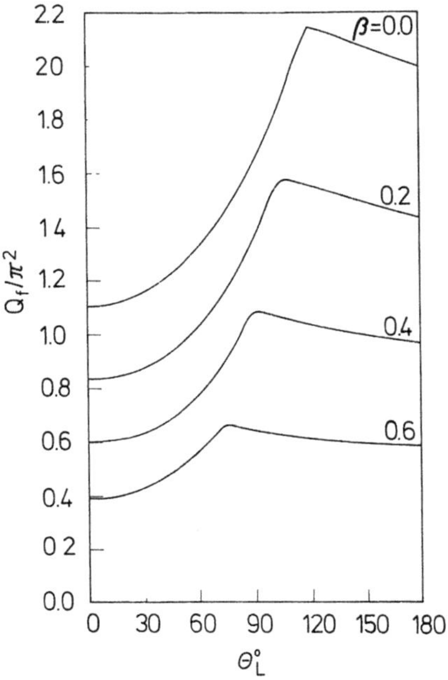

Fig.3. Variation of critical flutter load with angle of pretwist and thickness taper for $\alpha = 0$, $\eta^* = 0.01$ and $\beta^* = 10^{-6}$

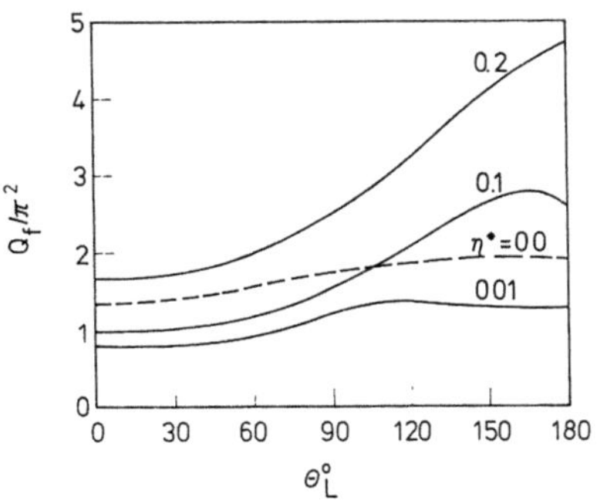

Fig.4. Effects of angle of pretwist and internal damping on the critical flutter load for $\alpha = \beta = 0.2$ and $\beta^* = 10^{-6}$

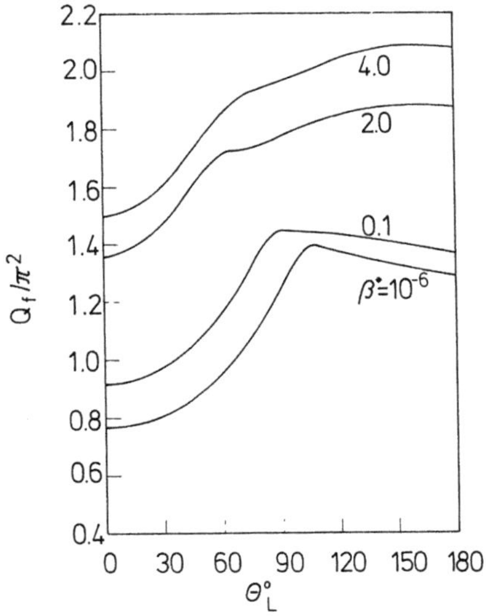

Fig.5. Effects of angle of pretwist and external damping on the critical flutter load for $\alpha = \beta = 0.2$ and $\eta^* = 0.01$

Figure 3 shows the variation of critical flutter load Q_f with pretwist angle Θ_L for several values of β. Though the curves are quite similar to those in Figure 2, it may be observed that the maximum value of the critical flutter load is attained at lower values of Θ_L and the reduction in the values of Q_f is more pronounced with increase in β than with α.

From the plot of the variation of Q_f versus Θ_L for four values of η^* with $\alpha = \beta = 0.2$ and $\beta^* = 10^{-6}$ in Figure 4 the familiar phenomena of destabilization and stabilization are revealed for $\eta^* = 0.01$ and 0.2 respectively. On the other hand $\eta^* = 0.1$ has a stabilizing or destabilizing effect depending upon the value of Θ_L.

The effects of variation of β^* and Θ_L on Q_f for $\alpha = \beta = 0.2$ and $\eta^* = 0.01$ are shown in Figure 5. All curves indicate that Q_f increases to reach a maximum as Θ_L increases from zero and thereafter it decreases at a slower rate, the rate of decrease being lower for higher values of β^*. At any pretwist angle an increase in β^* raises the value of Q_f when the values of other parameters are unchanged.

Lastly, the pretwist angle Θ_L appears to have always a stabilizing effect as depicted in Figures 2 - 5.

Acknowledgment

The support of this research by the Alexander von Humboldt Stiftung through the award of a post-doctoral research fellowship to the first author is gratefully acknowledged.

References

1. Bolotin, V.V.: Nonconservative problems of the theory of elastic stability (editor G. Herrmann). New York: MacMillan 1963.

2. Herrmann, G.: Stability of equilibrium of elastic systems subjected to nonconservative forces. Appl. Mech. Rev. 20 (1967) 103-108.

3. Sundararajan, C.: The vibration and stability of elastic systems subjected to follower forces. The Shock and Vib. Digest 7 (1975) 89-105.

4. Kar, R.C.: Stability of nonuniform cantilever beams subjected to nonconservative Forces. Ph.D. Dissertation, Dept. of Mech. Engg., I.I.T. Kharagpur, 1978.

5. Hui, E.: Knickung verwundener Stäbe unter Druck. Oesterr. Ing.-Arch. 9 (1955) 288-319.

6. Leipholz, H.: Knickung verwundener Stäbe unter Druck einer konservativen, kontinuierlich und gleichmäßig verteilten Belastung. Ing.-Arch. 29 (1960) 262-279.

7. Fischer, W.: Verwundene und versetzte Knickstäbe. Dr.-Ing. Dissertation, Fakultät für Maschinenbau und Verfahrenstechnik der Universität Karlsruhe (TH), 1968.

8. Farshad, M.: On general conservative end loading of pretwisted rods. Int. J. Solids Struct. 9 (1973) 1361-1371.

9. Kar, R.C.: Stability of a nonuniform cantilever subjected to dissipative and nonconservative forces. Computers and Struct. 11 (1980) 175-180.

On the Basic Properties of Stability Regions and Boundaries of Rotating Flexible Shafts

K. HUSEYIN

Department of Systems Design
University of Waterloo
Waterloo, Ontario, Canada

Introduction

The transverse vibrations and stability of gyroscopic systems and, in particular, rotating flexible shafts have been studied by several authors. Extensive lists of references on the subject can be found, for example, in the books [1-3]. In two recent papers attention was focussed on the divergence and flutter boundaries of discrete systems under the influence of additional conservative forces [4], and the effect of combined external and internal damping on these boundaries was examined [5]. Several theorems concerning the basic stability properties of such systems as well as lower/upper bounds were established.

In this paper, rotating flexible shafts subjected to axial conservative as well as external and internal dissipation forces are considered from a continuum point of view, and certain results established earlier for discrete and limited-degree-of-freedom systems are generalized to continuous systems. Symmetric as well as asymmetric shafts with different principal flexural rigidities are analyzed. Certain necessary and sufficient conditions for stability are developed for both continuous and discrete systems. In the case of uniform discretized shaft, some explicit conditions of stability in terms of system matrices are given which can be used directly in the analysis and design of specific problems.

Rotating Flexible Shafts

Consider an elastic shaft rotating with a constant angular velocity ξ and subjected to a constant axial force of magnitude η. Let x denote the axial coordinate, $y_j(x,t)$ the transverse displacements and EI_j the flexural rigidities in two principal directions $(j = 1,2)$, ℓ the length and μ the mass per unit length of the shaft. The differential equations of the motion of the undamped shaft in a reference frame rotating with the angular velocity of the shaft are given [3] by

$$\mu \ddot{y}_1 - 2 \xi \mu \dot{y}_2 + E I_1 y_1'''' + \eta y_1'' - \xi^2 \mu y_1 = 0,$$
$$\mu \ddot{y}_2 + 2 \xi \mu \dot{y}_1 + E I_2 y_2'''' + \eta y_2'' - \xi^2 \mu y_2 = 0 \quad (1)$$

where primes and dots on the y_j $(j = 1,2)$ denote partial differentiation with respect to x and time t respectively. It is assumed that the boundary conditions at each end are either in the form

$$y_j = 0, \quad y_j' = 0 \quad \text{(clamped)}$$

or (2)

$$y_j = 0, \quad y_j'' = 0 \quad \text{(simply supported)}$$

The effect of dissipative forces can be incorporated into these equations by considering velocity dependent forces in rotating as well as stationary coordinates. The internal damping is assumed to be given as $\mu\, d_i\, (\dot{y}_1, \dot{y}_2)\, d\, x$, and the external damping as $\mu\, d_e [(\dot{y}_1 - \xi y_2), \dot{y}_2 + \xi\, y_1]\, d\, x$ where $d\, x$ is an infinitesimal element along the shaft, and d_i and d_e are the coefficients of internal and external damping respectively. The differential equations of the notion are then expressed [3] as

$$\mu \ddot{y}_1 + (d_i + d_e)\, \mu \dot{y}_1 - 2 \xi \mu \dot{y}_2 + E I_1 y_1'''' + \eta y_1'' - \xi^2 \mu y_1 - d_e \mu \xi y_2 = 0$$
$$\mu \ddot{y}_2 + (d_i + d_e)\, \mu \dot{y}_2 + 2 \xi \mu \dot{y}_1 + E I_2 y_2'''' + \eta y_2'' - \xi^2 \mu y_2 + d_e \mu \xi y_1 = 0 \quad (3)$$

with the boundary conditions (2).

Uniform Shaft

Consider first a uniform shaft with $E I_1 = E I_2 \equiv E I$; using the complex displacement

$$v = y_1 + i\, y_2 \tag{4}$$

the equations (2) and (3) can be written in the form

$$\mu \ddot{v} + 2 i \xi \mu \dot{v} + E I v'''' + \eta v'' - \xi^2 \mu v = 0 \tag{5}$$

and

$$\mu \ddot{v} + (d_i + d_e) \mu \dot{v} + 2 i \xi \mu \dot{v} + E I v'''' + \eta v'' - \xi^2 \mu v + i d_e \mu \xi v = 0 \tag{6}$$

respectively.

Introducing now

$$v(x,t) = z(x)\, e^{i\omega t} \tag{7}$$

into (6) yields

$$-\omega^2 \mu z + i \omega (d_i + d_e) \mu z - 2 \omega \xi \mu z + E I z'''' + \eta z'' - \xi^2 \mu z + i d_e \mu \xi z = 0 \tag{8}$$

and the corresponding equation for undamped systems can readily be obtained by setting $d_i = d_e = 0$ in (8). Multiplying (8) by the complex conjugate $\bar{z}$, using the boundary conditions (e.g. simple B.C.) and integrating over the length ℓ results in the scalar equation

$$-\omega^2 + i \omega (d_i + d_e) - 2 \omega \xi + (a - \eta e - \xi^2) + i d_e \xi = 0 \tag{9}$$

where

$$a = E I \int_0^{\ell} \bar{z}'' z'' \, dx$$
$$e = \int_0^{\ell} \bar{z}' z' \, dx \tag{10}$$

and

$$m = \int_0^{\ell} \mu\, \bar{z}\, z\, d\,x = 1.$$

Clearly, z is assumed to be normalized with respect to μ, and a and e are real and positive.

For the undamped system, (9) yields

$$\omega = -\xi \pm (a - \eta\, e)^{\frac{1}{2}} \tag{11}$$

which indicates that if $\eta = 0$ the flutter instability is ruled out and divergence occurs at $\xi^2 = a$. In the presence of the axial force the flutter boundary is characterized by

$$a - \eta\, e = 0 \tag{12}$$

and divergence boundary by

$$a - \eta\, e - \xi^2 = 0. \tag{13}$$

The system is stable if and only if

$$a - \eta\, e > 0, \quad a - \eta\, e - \xi^2 \neq 0 \tag{14}$$

for all eigenfunctions z.

A dissipative system, on the other hand, is stable if and only if

$$k\, d^2 + s\, d\, g - s^2 > 0, \quad d > 0 \tag{15}$$

where k, d, s, and g are associated with the general eigenvalue equation [3, p. 159]

$$\lambda^2 + \lambda\,(d + i\, g) + k + i\, s = 0. \tag{16}$$

Applying this to (9), it is recognized that

$$k = a - \eta\, e - \xi^2, \quad g = 2\,\xi, \quad s = d_e\,\xi, \quad \text{and} \quad d = d_i + d_e > 0. \tag{17}$$

Introducing (17) into (15) results in

$$a - \eta e - \left(\frac{d_i}{d_i + d_e}\right)^2 \xi^2 > 0. \tag{18}$$

It follows that the dissipative system (9) is stable if and only if (18) is satisfied for all eigenfunctions z. In general, flutter instability occurs when the real part of one of the eigenvalues $\lambda = i\,\omega$ crosses the imaginary axis. At the onset of flutter, one can then assume that ω in (9) is real, leading to

$$-\omega^2 - 2\,\omega\,\xi + a - \eta\, e - \xi^2 = 0 \tag{19}$$

and

$$\omega\,(d_i + d_e) + d_e\,\xi = 0. \tag{20}$$

Introducing (20) into (19) yields

$$a - \eta e - \left(\frac{d_i}{d_i + d_e}\right)^2 \xi^2 = 0 \tag{21}$$

which characterizes the flutter boundary. If the flutter boundary is also the stability boundary (where initial loss of stability occurs), then (18) holds for all the remaining eigenfunctions. If we now suppose that the eigenvunctions in (21) and (18) are replaced by the eigenfunctions of the nondissipative system, then

$$a - \eta e \geqq \left(\frac{d_i}{d_i + d_e}\right)^2 \xi^2 \tag{22}$$

is satisfied, and by virtue of (14), it is concluded that the nondissipative system is stable when the dissipative system is at the onset of flutter instability provided $d_i \neq 0$, $d_e \neq 0$. It is also inferred from (19) and (20) that the initial divergence boundary does not exist or rather it consists of the critical point on η axis provided $d_e \neq 0$. If $d_e = 0$, the initial divergence boundary coincides with that of the undamped system given by (13). This confirms the validity of Kelvin-Tait-Chetaev (KTC) theorem for continuous systems as well as finite dimensional systems. On the other hand, if $d_i = 0$, the flutter boundary (21) coincides with that of the nondissipative system. Hence, the following result can be formulated:

Theorem 1. The initial flutter (divergence) boundary of a nondissipative uniform rotating shaft provides an upper (lower) bound for the flutter boundary of the damped system as d_i, d_e are varied.

This result represents a generalization of a similar result established earlier [3,5] for discrete systems and is illustrated in Fig. 1.

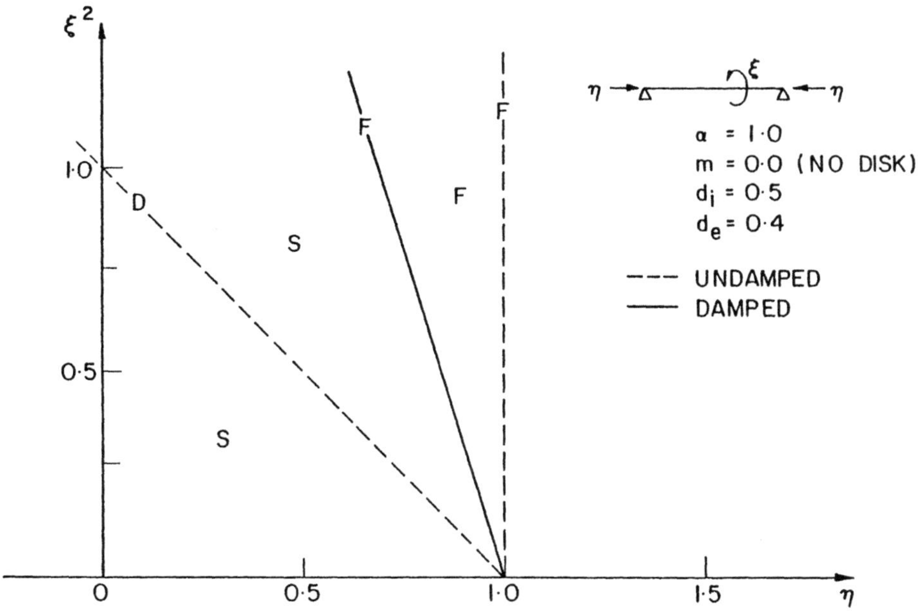

Fig. 1

As-ymmetric Shaft

$$y_j = z_j (x) e^{i\omega t} \tag{23}$$

into (3) leads to

$$-\omega^2 \mu z_1 + i \omega (d_i + d_e) \mu z_1 - 2 i \omega \xi \mu z_2 + E I_1 z_1'''' + \eta z_1''$$

$$- \xi^2 \mu z_1 - d_e \mu \xi z_2 = 0 \qquad (24)$$

$$-\omega^2 \mu z_2 + i \omega (d_i + d_e) \mu z_2 + 2 i \omega \xi \mu z_1 + E I_2 z_2'''' + \eta z_2''$$

$$- \xi^2 \mu z_2 + d_e \mu \xi z_1 = 0 \qquad (25)$$

These two equations can be combined into one as

$$-\omega^2 \mu w + i \omega (d_i + d_e) \mu w - 2 \omega \xi \mu w + \frac{1}{2} E (I_1 + I_2) w''''$$

$$+ \frac{1}{2} E (I_1 - I_2) \hat{w}'''' + \eta w'' - \xi^2 \mu w + i d_e \mu \xi w = 0 \qquad (26)$$

where

$$w = z_1 + i z_2 \quad \text{and} \quad \hat{w} = z_1 - i z_2.$$

Multiplying (26) by the complex conjugate $\bar{w}$ (note that $\bar{w}$ is, in general, not the same as $\hat{w}$), integrating over the length of the shaft and using the boundary conditions yield

$$-\omega^2 + i \omega (d_i + d_e) - 2 \omega \xi + (a_1 + a_2 - \eta e - \xi^2)$$

$$+ i a_3 + i d_e \xi = 0 \qquad (27)$$

where w is assumed to be normalized with respect to μ so that

$$m = \int_0^{\ell} \mu \bar{w} w \, d x = 1$$

and a_1, a_2, a_3 and e are real constants given by

$$a_1 = \frac{1}{2} E (I_1 + I_2) \int_0^{\ell} \bar{w}'' w'' \, d x > 0$$

$$a_2 = \frac{1}{2} E (I_1 - I_2) \int_0^{\ell} (\bar{z}_1'' z_1'' - \bar{z}_2'' z_2'') \, d x \qquad (28)$$

$$a_3 = -\frac{1}{2} E (I_1 - I_2) \int_0^{\ell} (\bar{z}_1'' z''_2 + \bar{z}_2'' z_1'') \, d x$$

$$e = \int_0^{\ell} \bar{w}' w' \, d x$$

The nondissipative system described by

$$-\omega^2 - 2 \omega \xi + (a_1 + a_2 - \eta e - \xi^2) + i a_3 = 0 \tag{29}$$

is stable if and only if

$$a_1 + a_2 - \eta e > 0, \quad a_1 + a_2 - \eta e \neq \xi^2 \quad \text{and} \quad a_3 = 0 \tag{30}$$

for all eigenfunctions. It is worth noting that the first, second and third condition in (30) ensures stability against flutter, critical divergence boundary and divergence respectively. In other words, $a_3 = 0$, for example, excludes the unstable divergence region. The flutter boundary is described by

$$a_1 + a_2 - \eta e = 0, \; a_3 = 0. \tag{31}$$

Applying (15) to (27) and recognizing that $d = d_i + d_e > 0$, $k = a_1 + a_2 - \eta e - \xi^2$, $g = 2 \xi$ and $s = d_e \xi + a_3$, leads to the necessary and sufficient stability condition

$$a_1 + a_2 - \eta e - \frac{(d_i \xi - a_3)^2}{(d_i + d_e)^2} > 0 \tag{32}$$

which must hold for all eigenfunctions. At the onset of flutter instability ω is real, and (27) yields

$$-\omega^2 - 2 \omega \xi + (a_1 + a_2 - \eta e - \xi^2) = 0 \tag{33}$$

and

$$(d_i + d_e) \omega + d_e \xi + a_3 = 0 \tag{34}$$

Substituting for ω from (34) to (33) results in

$$a_1 + a_2 - \eta e - \frac{(d_i \xi - a_3)^2}{(d_i + d_e)^2} = 0 \tag{35}$$

which characterizes the flutter boundary. If (35) represents the initial flutter boundary, the remaining eigenfunctions satisfy (31). It is noted that (35) reduces to (21) in the special case when $I_1 = I_2$, since both a_2 and a_3 vanish under this condition. Suppose again that the eigenfunctions in (35) and (32) are replaced by those of the nondissipative system (assuming small dissipation); we then infer from (30) that at the initial flutter boundary (35), the nondissipative system is stable against flutter. In other words, (35) cannot occur in a flutter unstable region of the nondissipative system; or, when the nondissipative system is at the onset of flutter instability characterized by (31), the dissipative system is already unstable. Hence the following theorem can be formulated:

Theorem 2. The flutter boundary of a nondissipative asymmetric rotating shaft is an upper bound for the flutter boundary of the damped system.

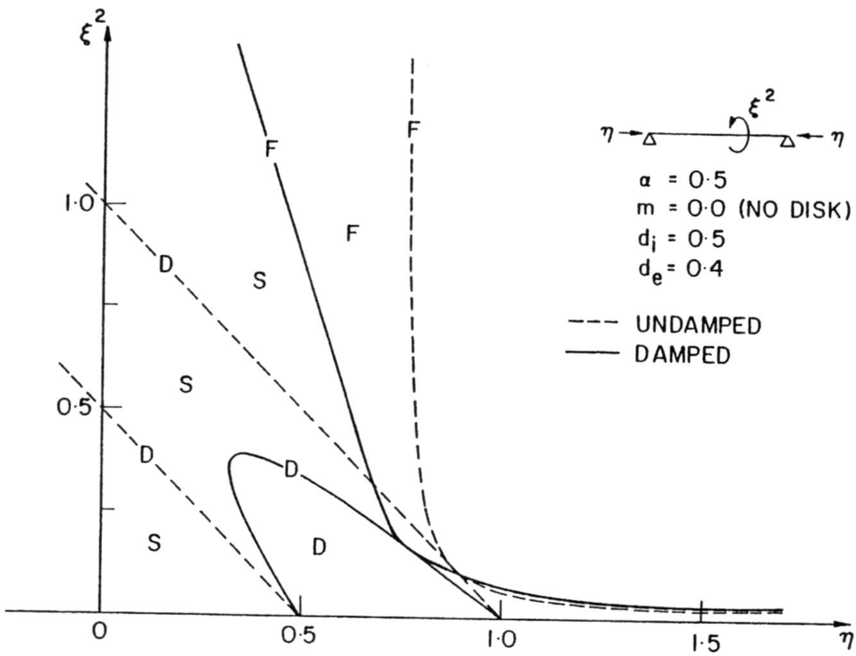

Fig. 2

This result is a generalization of a similar result established earlier for discrete systems with special structure [3,5].

The divergence boundary of the dissipative system is characterized by

$$a_1 + a_2 - \eta\, e - \xi^2 = 0$$
$$a_3 + d_e\, \xi = 0 \tag{36}$$

which indicates that it occurs in the divergence region of the nondissipative system, and <u>it cannot enter the stable region</u> of the latter. These results are illustrated in Fig. 2.

5. DISCRETE SYSTEMS

Consider now the set of assumed solutions for (24) and (25)

$$z_1 = a_k\, \varphi_k\,(x) \qquad (k = 1,2,\ldots,n)$$
$$z_2 = b_k\, \varphi_k\,(x) \tag{37}$$

where the functions $\varphi_k\,(x)$ satisfy the boundary conditions, and summation convention is adopted.

Introducing (37) into (24) and (25), and applying Galerkin's procedure leads [3] to the matrix equation

$$\left\{-\omega^2 \begin{bmatrix} M & 0 \\ 0 & M \end{bmatrix} + i\,\omega\,(d_i + d_e) \begin{bmatrix} M & 0 \\ 0 & M \end{bmatrix} + 2\,i\,\omega\,\xi \begin{bmatrix} 0 & -M \\ M & 0 \end{bmatrix}\right.$$
$$+ \begin{bmatrix} U - \eta\,E - \xi^2\,M & 0 \\ 0 & V - \eta\,E - \xi^2\,M \end{bmatrix}$$
$$\left. + \,d_e\,\xi \begin{bmatrix} 0 & -M \\ M & 0 \end{bmatrix}\right\} \begin{bmatrix} a \\ b \end{bmatrix} = 0 \tag{38}$$

where the matrices M, U, V, and E are given by

$$M = \mu \int_0^\ell \varphi_j \; \varphi_k \, d\, x \; , \qquad U = E I_1 \int_0^\ell \varphi_j'' \; \varphi_k'' \, d\, x$$

$$E = \int_0^\ell \varphi_j' \; \varphi_k' \, d\, x \; , \qquad V = E I_2 \int_0^\ell \varphi_j'' \; \varphi_k'' \, d\, x$$

and a and b represent a_k and b_k respectively.

The results obtained in the preceding sections obviously apply to the discrete system (38) as well, and in this section some stability conditions in terms of the matrices involved are developed.

For the uniform shaft $(U = V)$, (38) can readily be transformed into the scalar equation

$$-\omega^2 \, M \, v + i \, \omega \, (d_i + d_e) \, M \, v - 2 \, \omega \, \xi \, M \, v + (U - \eta \, E - \xi^2 M) \, v + i \, d_e \, \xi \, M \, v = 0 \tag{39}$$

where $v = a + i\, b$.

Premultiplying by $\bar{v}'$ (complex conjugate transpose) yields

$$-\omega^2 \, m + i \, \omega \, (d_i + d_e) \, m - 2 \, \omega \, \xi \, m + (a - \eta \, e - \xi^2 \, m) + i \, d_e \, \xi \, m = 0 \tag{40}$$

where

$$m = \langle \bar{v}, M v \rangle, \quad a = \langle \bar{v}, U v \rangle \quad \text{and} \quad e = \langle \bar{v}, E v \rangle .$$

Clearly (40) is analogous to (9), and following the same reasoning as before one obtains the necessary and sufficient condition for stability as

$$a - \eta \, e - \left(\frac{d_i}{d_i + d_e} \right)^2 \xi^2 \, m > 0 \tag{41}$$

which must hold for all eigenvectors v. *This condition in turn leads to the following result: The dissipative rotating uniform shaft is stable if and only if the matrix*

$$U - \eta E - \left(\frac{d_i}{d_i + d_e}\right)^2 \xi^2 M \tag{42}$$

is positive definite.

The stability boundary (the initial flutter boundary) is reached when (42) becomes singular for the first time as η and ξ^2 are increased from zero. The nondissipative uniform shaft on the other hand, is stable if and only if

$$U - \eta E > 0 \qquad \text{(positive definite)}$$

and (43)

$$|U - \eta E - \xi^2 M| \neq 0$$

which follow immediately from the analogue of (14) for discrete systems.

These explicit results can prove very useful in the analysis and design of specific problems. Similar simple conditions for asymmetric systems, however, are not available, and the direct application of the Routh-Hurwitz criterion to (38), which is equivalent to the general condition (15), may be more fruitful.

The flutter and divergence boundaries shown on the figures were obtained through the application of Galertian procedure and introduction of suitable nondimensional quantities [3].

REFERENCES

[1] BOLOTIN, V.V.: Non-conservative Problems of the Theory of Elastic Stability, McMillan, New York, 1963.

[2] ZIEGLER, H.: Principles of Structural Stability, Blaisdell Publ. Co., 1968.

[3] HUSEYIN, K.: Vibrations and Stability of Multiple-Parameter Systems, Sijthoff & Noordhoff, 1978.

[4] HUSEYIN, K. and PLAUT, R.H.: Divergence and Flutter Boundary of Systems Under Combined Conservative and Gyroscopic Forces, Proc. IUTAM Symp. Dyna. Rotors, Lyngby, Denmark (ed. Niordson), Springer Verlag, pp. 182-205, 1975.

[5] HUSEYIN, K.: Effect of Damping on the Flutter Boundary of Rotating Systems, Int. Conf. Vib. & Rot. Machinery, Cambridge, 1976, I. Mech. Eng. Publication, pp. 133-137, 1976.

Postbuckling Analysis of Divergence and Flutter Instability Structures under Tangential Loads

A. N. Kounadis

National Technical University of Athens
Athens, Greece

1. INTRODUCTION

Problems on Stability of nonconservatively loaded systems have been extensively investigated last years. In particular special attention has been devoted to follower type nonconservative forces. In a series of papers [1-3] published recently, using simple models subjected to follower compressive forces, it was shown that their loss of stability may be of flutter or divergence type. Moreover, it was shown that the type of instability depends on the degree of the joint stiffness and in general on the amount of elastic ends restraint of the compressed bar. For instance, for the three-bar frame shown in Fig.1, it has been proven [1] that flutter instability occurs, when the magnitude of the joint stiffness μ is less than 1.53; otherwise, namely for $\mu > 1.53$, the frame loses its stability through divergence.

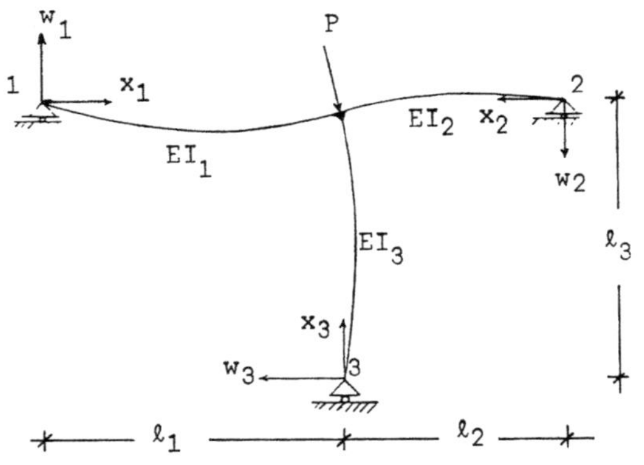

Fig.1. Three-bar frame under a tangential load

For the elastically restrained two-bar frame shown in Fig.2, using a static analysis the following buckling equation can be established

$$\sin k + \frac{k\cos k}{\bar{C}_R} + \frac{k}{3\mu} = 0 \qquad (1)$$

where $k^2 = P\ell_1^2/EI_1$ is the nondimensionalized axial force; $\bar{C}_R = C_R\ell_1/EI_1$ is the nondimensionalized rotational stiffness of the elastic support and $\mu = I_2\ell_1/I_1\ell_2$ is the joint stiffness of the frame.

Discussing this equation it is deduced that for $\mu=1$ flutter instability [2] takes place, if the nondimensionalized rotational constant

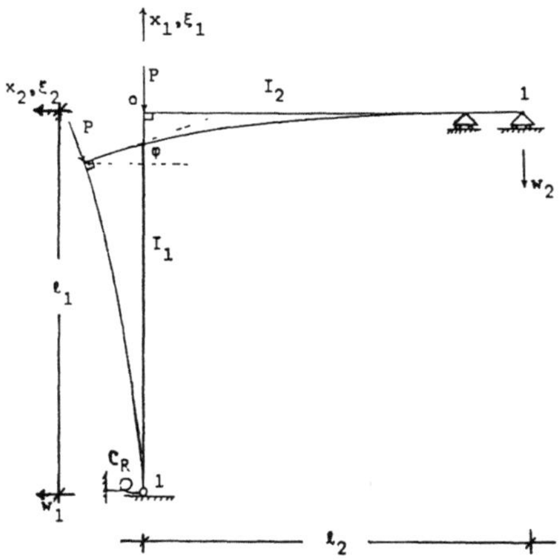

Fig.2 Elastically Restrained Nonconservative Frame

$\bar{C}_R$ is higher than 4.625, whereas for $\bar{C}_R<4.625$ the frame loses its stability through divergence. It is worth observing that, when the support of the frame becomes an immovable hinge ($\bar{C}_R=0$), the frame loses its stability always through divergence [3]. This also occurs for the three-bar frame shown in Fig.1. In this cases the critical load is constant, equal to $\pi^2/4=2.467$ [3,4] regardless of the flexural rigidity and length of the uncompressed horizontal bar.

For the elastically restrained frame shown in Fig.2,

on the basis of a variational approach a lower bound theorem [2] is presented. According to this theorem the load-carrying capacity of the frame under a follower compressive force is higher than that under a conservative (constant-directional) load provided that this frame loses its stability through divergence (i.e. for $\bar{C}_R<4.625$). However, this does not hold for other nonconservatively loaded structures of divergence type. More specifically, for the elastically restrained beam under a follower compressive force shown in Fig.3, on the basis

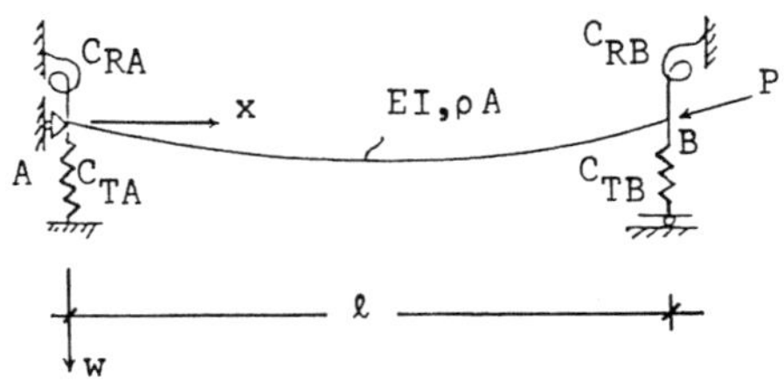

Fig.3. Elastically restrained beam under a tangential force

of the aforementioned lower-bound theorem, it was established that: its load-carrying capacity is higher than that of the corresponding conservative system subjected to a constant-directional compressive force, if the product of the nonconservative component of the load times its respective vertical displacement is positive, provided that the beam-column under the follower load is a divergence type system; otherwise, namely if this product is negative the load-carrying capacity of the beam under a follower load is lower than the corresponding capacity of the same system under a conservative load. It is worth noticing that the sign of this product depends on the values of the constants of elastic end restraints.

Another result of paramount importance observed in all the above systems is the following: When the buckling mechanism changes from flutter to divergence and vice-versa an appreciable jump in their load carrying capacity takes place.

This phenomenon is illustrated in Fig.4, where the curve of the critical load β_{cr} versus the joint stiffness μ of the three-bar frame shown in Fig.1 is plotted. As previously stated, for $\mu<1.53$ this frame loses its stability

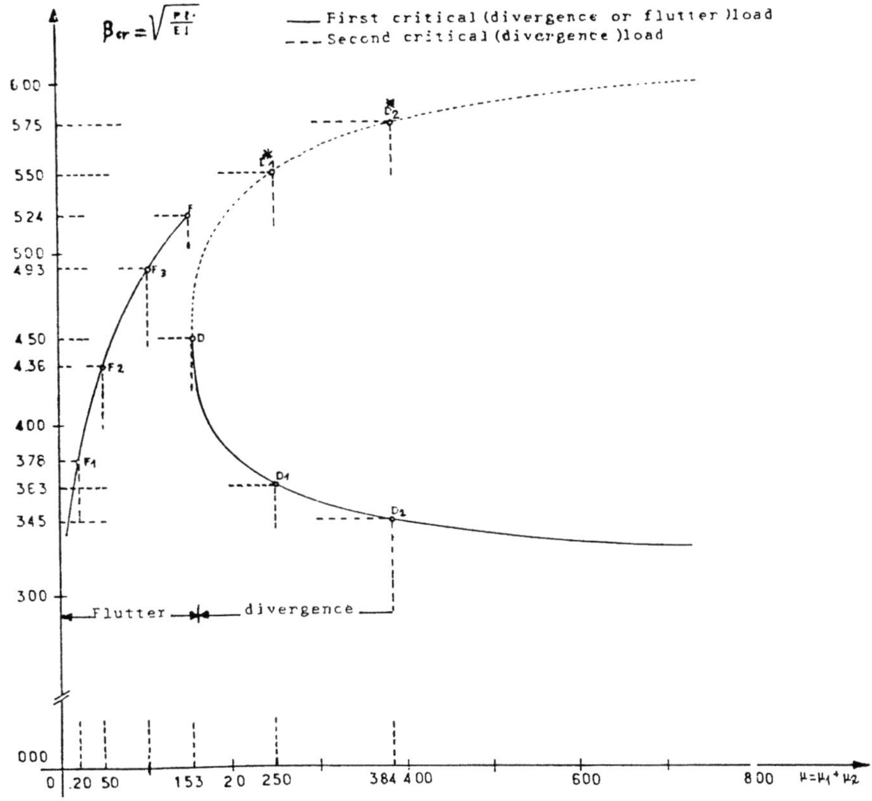

Fig.4. Critical load, β_{cr}, versus $\mu=\mu_1+\mu_2$

through flutter. Note also that the flutter load increases with increasing μ obtaining its maximum value at the point F (β_{cr}=5.24) corresponding to μ=1.53. For $\mu=1.53^+$ a sudden decrease of the load-carrying capacity from the point F to the point D takes place. It is also worth observing that as μ increases the first critical load corresponding to the curve $D_1D_2D_3$... decreases, whereas the second critical load corresponding to the curve $D_1^*D_2^*D_3^*$,... increases. At the point of intersection D, there is a commun vertical tangent for both these curves. This jump can be easily explained by plotting the eigencurve (load versus eigenfrequency) shown in Fig.5. It is clear that

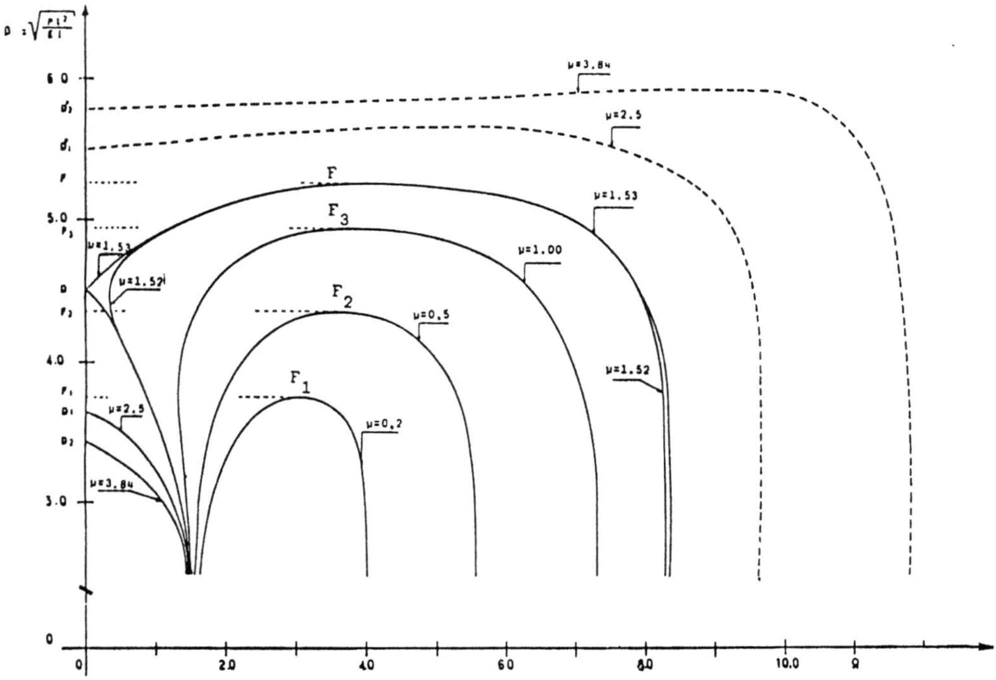

Fig.5 Eigencurves (Load β versus eigenfrequency Ω).
——— First critical (divergence or flutter) load,
----- second critical (divergence) load.

μ=o.2,o.5,1,1.52, where flutter instability occurs, the first and second eigenmodes coincide forming one continous curve. To the maximum points F_1',F_2',F_3' of these curves correspond their critical loads. For μ=1.53 the respective eigencurve degenerates into two branches corresponding to the first and second eigencurves which intersect the vertical axis at the commun point D, where the first divergence load appears as μ varies from (zero) o to 1.53. Thus, a sudden jump in the load-carrying capacity from the value of F to the value D occurs. For μ>1.53

the first and second eigencurves intersect the vertical load axis at distinct points.
A different type of discontinuity in the load-carrying capacity of pure flutter type systems with attached masses is reported in ref [5,6].

The aforementioned findings are obtained on the basis of a linear static (divergence instability) analysis. Reviewing the present state-of-the art, it is observed that the nonlinear response of nonconservatively loaded systems under follower forces has not as yet been thoroughly discussed. For instance some relevant questions which need clarification on the basis of a nonlinear analysis are the following:

1. May the static methods of analysis be employed to investigate the stability of pure flutter type structural systems, if nonlinear terms are taken into account?
2. Does the aforementioned jump phenomenon in the load-carrying capacity (occuring when the buckling mechanism of the above systems changes from flutter to divergence and vice-versa) exist, if a nonlinear stability analysis is employed?
3. Are the magnitudes of critical displacements corresponding to the region, wherein the jump phenomenon is observed consistent with the limitations of the linear stability analysis?
4. Do the types of critical points of equilibrium of conservative systems also appear in the case of nonconservatively loaded systems of divergence type?

To shed some light on these questions the two-bar frame shown in Fig. 6, is analysed on the basis of the linear static, linear dynamic and nonlinear static stability analyses and the obtained results are compared and combined with those corresponding to other simple structures.

On the basis of the linear static stability analysis it is found that, when the magnitude of the joint stiffness $\mu = I_2 \ell_1 / I_1 \ell_2$

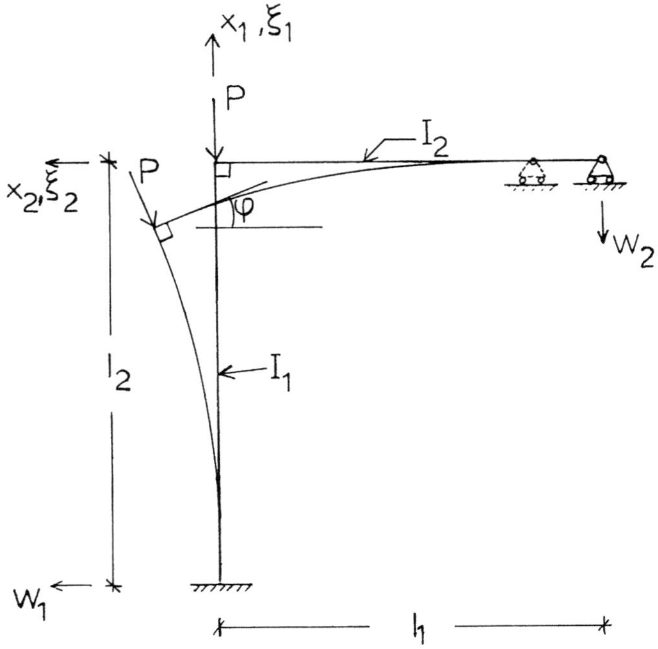

Fig.6. Two-bar frame under a tangential load.

is less or equal to 1.53 the frame loses its stability through flutter, whereas for $\mu > 1.53$ the frame is a divergence type system. These results have also been found for the three-bar frame as stated previously.

2. LINEAR DYNAMIC ANALYSIS

Assuming a free vibratory motion for the last frame the following differential equations can be written

$$W_1'''' + \beta^2 W_1'' - k_1^4 W_1 = 0$$
$$W_2'''' - k_2^4 W_2 = 0 \qquad (2)$$

where

$$\beta^2=P\ell_1^2/EI_1, k_1^4=m_1\omega^2\ell_1^4/EI_1, k_2^4=m_2\omega^2\ell_2^4/EI_2 \quad \text{and}$$

ω= the circular frequency of the steady-state motion.

Integrating these equations and using the support conditions leads to

$$\begin{aligned} W_1(x_1) &= A_1(\sin\bar{\lambda}x_1 - \frac{\bar{\lambda}}{\lambda}\sinh\lambda x_1) + A_2(\cos\bar{\lambda}x_1 - \cosh\lambda x_1) \\ W_2(x_2) &= \bar{A}_1\sin k_2x_2 + \bar{A}_3\sinh k_2x_2 \end{aligned} \tag{3}$$

where

$$\begin{aligned} &\bar{\lambda}=\left[\frac{1}{2}\beta^2+\left(\frac{1}{4}\beta^4+k_1^4\right)^{1/2}\right]^{1/2} \qquad \lambda=\left[-\frac{1}{2}\beta^2+\left(\frac{1}{4}\beta^4+k_1^4\right)^{1/2}\right]^{1/2} \\ &\beta^2=\bar{\lambda}^2-\lambda^2, k_1^4=\bar{\lambda}^2\lambda^2, k_2^4=k_1^4m\rho^2/\mu, \rho=\ell_2/\ell_1, m=m_2\ell_2/m_1\ell_1 \end{aligned} \tag{4}$$

Application of the boundary conditions

$$\begin{aligned} W_1'''(1)+k_1^4mW_1(1)&=0 \\ W_2(1) &=0 \\ W_1'(1)-W_2'(1)&=0 \\ W_1''+\mu W_2''(1) &=0 \end{aligned} \tag{5}$$

by means of equations (3), yields the following frequency equation [7]

$$|\alpha_{ij}|=0 \qquad i,j=1,2,3,4 \tag{6}$$

where

$\alpha_{11}=k_1^4m(\sin\bar{\lambda}-\frac{\bar{\lambda}}{\lambda}\sinh\lambda)-\bar{\lambda}(\bar{\lambda}^2\cos\bar{\lambda}+\lambda^2\cosh\lambda), \alpha_{12}=k_1^4m(\cos\bar{\lambda}-\cosh\lambda)+(\bar{\lambda}^3\sin\bar{\lambda}-\lambda^3\sinh\lambda), \alpha_{13}=\alpha_{14}=0$

$\alpha_{21}=\alpha_{22}=0, \alpha_{23}=\sin k_2, \alpha_{24}=\sinh k_2$

$\alpha_{31}=\bar{\lambda}(\cos\bar{\lambda}-\cosh\lambda), \alpha_{32}=-\bar{\lambda}\sin\bar{\lambda}-\lambda\sinh\lambda, \alpha_{33}=-k_2\cos k_2, \alpha_{34}=-k_2\cosh k_2$

$$\alpha_{41}=\bar{\lambda}(\bar{\lambda}\sin\bar{\lambda}+\lambda\sinh\lambda),\alpha_{42}=\bar{\lambda}^2\cos\bar{\lambda}+\lambda^2\cosh\lambda,\alpha_{43}=\mu k_2^2\sin k_2 \tag{7}$$

$$\alpha_{44}=-\mu k_2^2\sinh k_2$$

The frequency equation can be written by means of eqs (4) in the implicit form

$$F\left[\Omega,\beta^2;\mu;\rho;m\right]=0 \tag{8}$$

where the nondimensionalized frequency Ω is given by

$$k_1=\Omega=\lambda\bar{\lambda} \tag{9}$$

It is known that the limit of stability depends on the nature of the nondimensionalized frequency Ω which for given values of the parameters μ,ρ and m is a function of the nondimensionalized load β^2, i.e.$\Omega=\Omega(\beta^2)$. The frequency equation has always a nontrivial solution regardless of the values of these parameters. Thus, the frame can be analysed either it is a flutter or a divergence type system. Solving numerically the frequency eq. (8), it is found that the eigencurves Ω versus β^2 intersect for $\mu>1.53$ the vertical axis β^2 at points which depend exclusively on μ; i.e. they are independent of the values of the parameters ρ and m. Therefore, for $\mu>1.53$ the instability is of the divergence type, because Ω=o. This result is in complete agreement with that of the (linear) static stability analysis which yields the same divergence critical loads as those obtained by the linear dynamic analysis. For $\mu<1.53$ the frame loses its stability through flutter and its critical load is obtained through the coincidence of the first and second eigenmodes occuring at $\Omega\neq 0$. In this case the static analysis leads to the trivial solution $\beta^2=0$.

3. NONLINEAR STATIC STABILITY ANALYSIS

On the basis of linear curvature-displacement relationship and nonlinear axial strain displacement relationship, the following differential equations governing the equilibrium of the frame can be written

$$
\begin{aligned}
&\xi_1' + \frac{1}{2}W_1'^2 = -\frac{k^2}{\lambda_1^2} \\
&\xi_2' + \frac{1}{2}W_2'^2 = 0 \\
&W'''' + k^2 W_1'' = 0 \\
&W_2'''' = 0
\end{aligned}
\qquad (10)
$$

where ξ_i and W_i denote the axial and lateral displacement of the ith member of the frame, whereas k^2 represents the non-dimensionalized force of the compressed member. Eqs (1o) are associated with the following boundary conditions:

a. kinematic boundary conditions

$$
\begin{aligned}
&W_1(0) = \xi_1(0) = W_1'(0) = W_2(0) = 0 \\
&W_1(1) = \rho\xi_2(1), \rho W_2(1) = -\xi_1(1), W_1'(1) = W_2'(1)
\end{aligned}
\qquad (11)
$$

b. natural boundary conditions

$$
\begin{aligned}
&\xi_2'(0) + \frac{1}{2}W_2'^2(0) = 0 \\
&W_2''(0) = 0 \\
&\beta^2 \sin W_1'(1) - k^2 W_1'(1) - W_1'''(1) = 0 \\
&\rho\beta^2 \cos W_1'(1) - \rho k^2 + \mu W_2'''(1) = 0 \\
&W_1''(1) + \mu W_2''(1) = 0
\end{aligned}
\qquad (12)
$$

Integration of eqs (1o) by means of the support conditions (11) and the second of eqs (12) leads to

$$W_1(x_1)=A_1(\sin kx_1-kx_1)+A_2(\cos kx_1-1)$$

$$W_2(x_2)=\bar{A}_1x_2^3+\bar{A}_2x_2$$

$$\xi_1(x_1) = -\frac{k^2}{\lambda_1^2}x_1-\frac{k^2}{4}\left[A_1^2(3x_1+\frac{\sin 2kx_1}{2k}-\frac{4\sin kx_1}{k})+A_2^2(x_1-\frac{\sin 2kx_1}{2k})+\right.$$

$$\left.+\frac{A_1A_2}{k}(\cos 2kx_1-4\cos kx_1+3)\right]$$

$$\xi_2(x_2)=\frac{1}{\rho}W_1(1)+\frac{1}{2}\left[\frac{9}{5}\bar{A}_1^2(1-x_2^5)+2\bar{A}_1\bar{A}_2(1-x_2^3)+\bar{A}_2^2(1-x_2)\right] \tag{13}$$

where the integration constants $\bar{A}_1,\bar{A}_2$ can be determined as functions of A_1,A_2 and k by means of the last of eqs (11) and the last of eqs (12). Subsequently, inserting the expressions $\bar{A}_1,\bar{A}_2$ into the second and fourth of eqs (12), the following nonlinear system in k,A_1,A_2 is obtained

$$\frac{k^2}{\lambda_1^2}+\frac{k^2}{4}\left[A_1^2(3+\frac{\sin 2k}{2k}-\frac{4\sin k}{k})+A_2^2(1-\frac{\sin 2k}{2k})\right.$$

$$\left.+\frac{A_1A_2}{k}(\cos 2k-4\cos k+3)\right]-\rho\left[A_1\left[k(\cos k-1)-\frac{k^2}{3\mu}\sin k\right]\right.$$

$$\left.-A_2\left[k\sin k+\frac{k^2\cos k}{3\mu}\right]\right]=0$$

$$\beta^2\sin\left[A_1k(\cos k-1)-A_2k\sin k\right]+k^3A_1=0$$

$$\rho\beta^2\cos\left[A_1k(\cos k-1)-A_2k\sin k\right]-k^2\rho+A_1k^2\sin k+A_2k^2\cos k=0 \tag{14}$$

The complete response of the frame, at any level of the applied load β^2, and for all values of the parameter μ,λ_1,ρ is known, if the solution of the nonlinear equation, eqs (14), is known.

In case the magnitude of the joint rotation $w_1'(1)$ up to the critical state is very small, so that we can assume $\sin w_1'(1) \simeq w_1'(1)$ and $\cos w_1'(1) \simeq 1$, the equilibrium of the frame can be established by means of the first of eqs (14), where the constants A_1, A_2 are given by

$$A_1 = \frac{\rho\beta^2(k^2-\beta^2)\sin k}{k^2\left[\beta^2+(k^2-\beta^2)\cos k\right]}$$

$$A_2 = \frac{\rho(k^2-\beta^2)\left[k^2+\beta^2(\cos k-1)\right]}{k^2\left[\beta^2+(k^2-\beta^2)\cos k\right]} \tag{15}$$

From this equation it is clear that the critical load β^2 is a function of the parameters: the joint stiffness μ, the slenderness ratio λ_1 of the compressed member and the length ratio ρ. However, it is worth observing that for $\lambda \to \infty$, the critical load β^2 is dependent only on μ. Thus, the limit of stability of frames with very slender compressed members is independent of the length ratio ρ, whereas it depends on the joint stiffness μ. Consequently, for $\lambda_1 \neq \infty$, it is reasonable to expect differences between the nonlinear critical loads depending on μ, λ_1, ρ and the linear (either static or dynamic) critical loads depending only on μ.

4. NUMERICAL SOLUTION OF NONLINEAR EQ (14)

The numerical solution of the system of nonlinear eqs (14) for given values of the parameters μ, λ_1, ρ at each level of the load β^2 is accomplished by employing the Newton-Raphson numerical scheme. The equilibrium paths are established as plots of the load β^2 versus the horizontal displacement or rotation of the joint. In all cases the frame loses its stability through a limit point.

From Fig.7, one can see the nonlinear equilibrium path as a plot of the load β^2 versus the joint horizontal deflection $w_1(1)$ for $\mu=16, \lambda_1=80$ and $\rho=0.25$. It is clear that the frame loses its stability

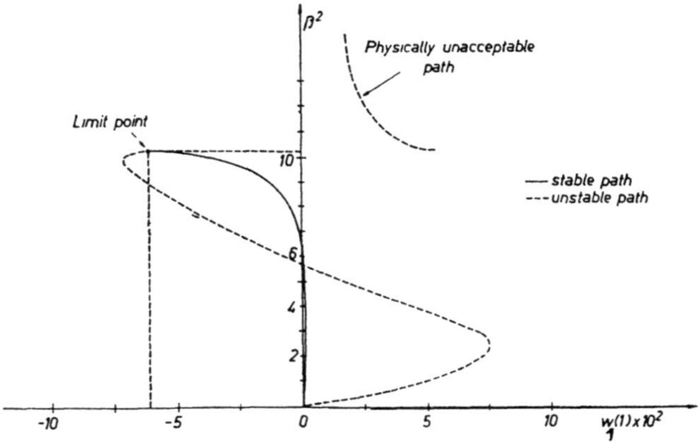

Fig.7. Nonlinear equilibrium path of the load β^2 versus the horizontal deflection $w_1(1)$ of the joint for $\rho=0.25, \mu=16$ and $\lambda_1=80$.

through a limit point. The critical load β^2_{cr} is equal to 10.228 and the corresponding horizontal deflection $w_1(1)$ and rotation $w_1'(1)$ of the joint are -0.0607 and 0.0148, respectively. Therefore, one can make the approximation $\sin w_1'(1) \simeq w_1'(1)$. Note that if the support of the frame becomes hinged the frame loses its stability through an unstable bifurcational [8,9] point.

From Fig.8 it can be seen the dependence of the joint rotation $w_1'(1)$ upon the load β^2 for $\lambda_1=80, \rho=0.25$ and $\mu=16,8,4,2,1.5,1.4,$ $1.3,1.2,1.0$. It is worth observing that for $\mu=16$ and $\mu=8$ the critical joint rotation $w'_{1cr}(1)$ is positive, whereas for $\mu<4$ it becomes negative. The decrease of the parameter μ below 4 implies a considerable increase in the critical (negative) joint rotation $w'_{1cr}(1)$ corresponding to the limit point. For $\mu<1.40$ the corresponding critical joint rotations become extremely large, because the respective limit points tend to infinity. Moreover, as μ approaches some critical value, μ_{cr}, between 1.4 and 1.3, the equilibrium paths degenerate into several curves which cannot be physically accepted. Consequently, for $\mu<\mu_{cr}$ the nonlinear static stability analysis cannot be employed

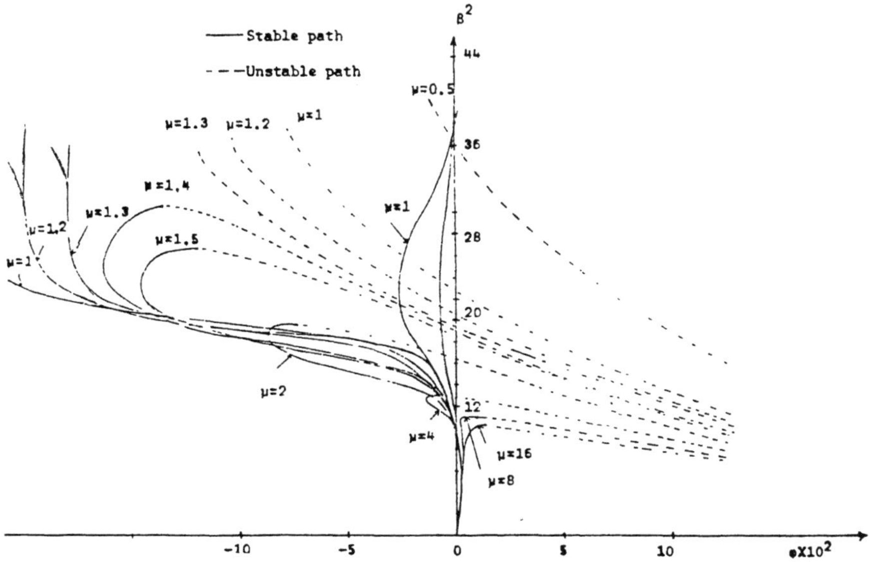

Fig.8. Nonlinear equilibrium paths β^2 vs φ for λ_1=80,ρ=0.25, and μ=0.5,1,1.2,1.3,1.4,1.5,2,4,8, and 16.

and the corresponding region of flutter type instability can be discussed only by using a dynamic analysis. However, two important conclusions are deduced by using the nonlinear static stability analysis: a) The divergence type instability occuring for $\mu>\mu_{cr}$ is associated with a limit point and b) At the region of the jump phenomenon in the linear critical load, the magnitudes of the corresponding displacements are extremely large and therefore the linear (static or dynamic) stability analysis cannot be employed.

5. COMPARISON OF LINEAR AND NONLINEAR SOLUTIONS

As indicated previously, for $\lambda_1\to\infty$ the nonlinear critical load β^2_{cr} depends only on μ as occurs in the case of the linear (static or dynamic) critical load. Excluding the region of the jump phenomenon in the critical load (where both the nonlinear and linear analyses cannot be employed) the nonlinear and linear static analysis as well as the linear dynamic analysis give identical results for the entire region of values of μ (where divergence instability occurs). On the other hand, for $\lambda \neq \infty$ the nonlinear critical load can be substantially higher or lower than the corresponding linear critical load; the latter is the same either the dynamic or static analysis is used.

The difference between the nonlinear and linear critical load is clearly shown in Fig.9 for a frame with $\lambda_1 = 80$ and $\rho = 0.25$. This difference becomes substantial for $1.53<\mu<6$, whereas for $\mu > 6$ it is practically negligible. Moreover, at $\mu = 1.53$ the aforementioned difference is approximately equal to the previous found jump in the linear critical load. From this figure, one

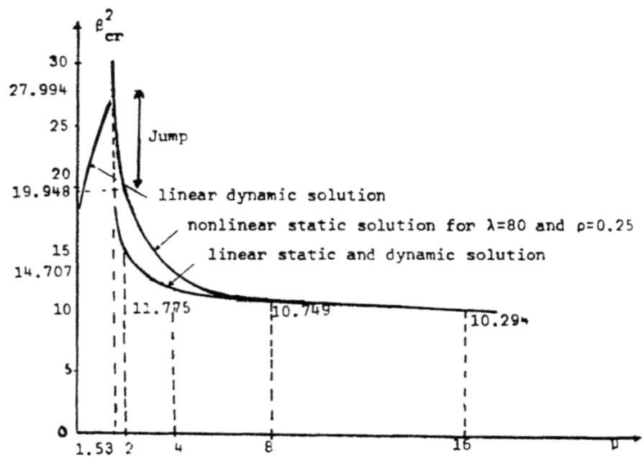

Fig.9. Relationship between load β^2 and joint stiffness μ on the basis of linear static, linear dynamic, and nonlinear static analysis.

can also see that there is not any discontinuity in the nonlinear critical load regardless of the value of the parameter μ. However, as previously stated, the validity of both the nonlinear and linear analysis is questionable as μ approaches its critical value which for $\lambda_1 \to \infty$ is equal to 1.53; otherwise i. e. for $\lambda_1 \neq \infty$, the nonlinear critical load is associated with a varying μ_{cr} which depends on λ_1 and ρ. Hence for $\lambda_1 \to \infty$ and μ taking values at the vicinity of its critical value 1.53, both the nonlinear and the linear **analysis** cannot be employed due to the large magnitudes of the critical state displacements which are not consistent with the corresponding limitations of these theories. It should be also noted that by using the nonlinear static stability analysis, it is established that all the divergence critical states are unstable (limit point instability). For the region $\mu < \mu_{cr}$, where flutter instability takes place, the static (either linear or nonlinear) analysis fail to be employed; thus the flutter region can be discussed by using only the dynamic method of analysis. Moreover, in order to obtain accurate results at the vicinity of μ_{cr} a nonlinear dynamic analysis should be employed.

Finally, from Fig.9, one can also see that the linear static and linear dynamic critical loads coincide for the entire region of divergence instability ($\mu > 1.53$).

6. CONCLUSIONS

In view of the above development, one can summarize the following important conclusions:

1. The static methods (either nonlinear or linear) cannot be employed for investigating the stability of flutter type structural systems which can be analysed only by using dynamic methods of analysis.

2. The jump phenomenon in the critical load does not exist in case of application of the nonlinear static analysis; consequently this jump is a result of the application of linear methods of analysis.

3. At the point of discontinuity in the linear critical load the curves of the first and second divergence critical loads meet each other forming one curve with a common tangent parallel to the load axis.

4. From the nonlinear static stability analysis it is deduced that the critical displacements corresponding to the region where the aforementioned jump phenomenon occurs are extremely large. Thus, the results obtained on the basis of this analysis as those obtained on the basis of the linear analysis are not consistent with the corresponding limitations of these analysis.

5. In the aforementioned region of the jump phenomenon, for a slenderness ratio $\lambda_1 \neq \infty$, there is a considerable difference between the nonlinear and linear critical load; the latter is not on the safe side.

6. According to the linear (static or dynamic) analysis the critical load depends only on the joint stiffness μ and the above jump occurs at $\mu_{cr} = 1.53$. These results are also valid, when the nonlinear static stability analysis is employed provided that $\lambda_1 \to \infty$; otherwise, viz. for $\lambda_1 \neq \infty$ the nonlinear critical load is associated with a varying μ_{cr} which depends on both the parameter λ_1 and ρ.

7. The critical states corresponding to the entire region of divergence instability are unstable. Consequently this frame is imperfection sensitive.

8. The types of critical points (bifurcation and limit point) are also observed in nonconservatively loaded systems of divergence instability.

9. The linear static and linear dynamic critical loads are identical for the entire region of divergence instability representing the actual load-carrying capacity of the system. Thus, the case that for $\mu > 1.53$ (region of divergence) flutter may procede instability by divergence which is possible, at least, theoretically, has not been observed in all problems considered.

7. REFERENCES

1. Kounadis, A. N. and Economou, A. P. "The Effects of the Joint Stiffness and of the Constraints on the Type of Instability of a Frame under Follower Force" Acta Mechanica, Vol. 36, 198o, pp. 157-168.

2. Kounadis, A. N. "Divergence and Flutter Instability of Elastically Restrained Structures under Follower Forces" Int. J. Engng. Sci., Vol. 19, 1981, pp. 553-562.

3. Kounadis, A. N. "On the Static Stability Analysis of Elastically Restrained Structures under Follower Forces" AIAA J. Vol. 18, 198o, pp. 473-476.

4. Panayotounakos, D. E. and Kounadis, A. N. "Elastic Stability of a Simple Frame Subjected to a Circulatory Load" J. of Sound and Vibration, Vol. 64, No. 2, 1979, pp. 179-186.

5. Kounadis, A. N. and Katsikadelis, J. T. "On the Discontinuity of the Flutter Load for Various Types of Cantilevers" Int. J. Solids Structures, Vol. 16, 198o, pp. 375-383.

6. Kounadis, A. N. "Stability of Elastically Restrained Timoshenko Cantilevers with Attached Masses Subjected to a Follower Force" Trans. ASME, J. of Appl. Mech., Vol. 44, 1977, pp. 731-736.

7. Kounadis, A. N. and Avraam, T. P. "Linear and Nonlinear Analysis of a Nonconservative Frame of Divergence Instability" AIAA J., Vol. 1981, pp. 761-765.

8. Kounadis, A. N., Giri, J. and Simitses, G. J. "Nonlinear Stability Analysis of an Eccentrically Loaded Two-Bar Frame" Trans. ASME, J. of Appl. Mech., Vol. 44, 1977, pp. 7o1-7o6.

9. Kounadis, A. N. " The Effects of Some Parameters on the Non-linear Divergence Buckling of a Nonconservative Simple Frame" J. de Mechanique Appliquée, Vol. 3, No. 2, 1979, pp. 173-185.

A Consistent Energy Approach to Defining Stability of Plastic Deformation Processes

H. PETRYK

Institute of Fundamental Technological Research
Polish Academy of Sciences, Warsaw

Summary

A single energy-type definition of stability is used to study different aspects of plastic instability. The definition concerns stability of a quasi-static deformation process but applies also to an equilibrium state. The connections with the bifurcation theory as well as with energy criteria of stability of equilibrium are demonstrated. Hill's extremum principles are re-derived as necessary conditions for stability of a deformation process.

1. Introduction

Instability phenomena in plastic solids may manifest themselves in various ways. One of them is connected with loss of stability of an equilibrium configuration under constant loading. On the other hand, incipient buckling or necking in the plastic range in most cases occur under varying loads and without loss of stability of equilibrium. Such phenomena may be regarded as a symptom of instability of a deformation p r o c e s s [1]. Also the localization of plastic deformation into a thin band can be considered as an instability of plastic flow [2].

Several theoretical approaches to plastic instability problems exist in the literature. However, while for studying stability of equilibrium usually some version of the energy approach is used [3,4,5], the instability of a deformation process is examined in the other ways, e.g. by using Hill's bifurcation theory [3,6], initial imperfection method or some intuitive approach /see [7,8]/. None of these approaches seems to be sufficiently universal to comprise different plastic instability aspects mentioned above.

In the previous paper [9] the autor proposed a definition of stability which applies to quasi-static deformation processes as well as to an equilibrium state. The present paper is a continuation of that work. The definition may be regarded as extension of the energy criterion of stability of equilibrium to processes of deformation under varying loading. The approach is based on the general theory of stability of motion. The motion is understood here as a quasi-static deformation process corresponding to some time-dependent boundary conditions. The term "stability" is used in the following sense: Let the theoretical deformation process be perturbed by an arbitrary disturbance from some class. The theoretical process is called stable in some time interval if the distance between this process and the perturbed one in this time interval is vanishingly small when the disturbance itself is vanishingly small.That definition has to be made precise by defining the class of disturbances considered and the measures of disturbance and of the distance between two processes.

The equations governing the behaviour of the body in a deformation process are never satisfied exactly in reality. This concerns not only the field equations but of course also the initial and boundary conditions as well as constitutive relations. Usually the stability with respect to disturbances of initial conditions only is considered. However, there is a priori no guarantee that a process stable for initial disturbances will also be stable against disturbances of another nature. In the present paper persistent disturbances of general character are taken into account. As the main attribute of all disturbances we assume the amount of energy supplied by a disturbance to the s y s t e m consisting of both the deformed body and the loading device. That energy may be supplied at once or gradually during the process of deformation. Since energy may also be extracted from the system, the maximum value of the energy supplied by a disturbance is taken as a disturbance measure. The disturbances are thought to be represented by a set of additional body forces and surface tractions of arbitrary magnitude, in general varying in time and acting during an arbitrary period of time. By adding those perturbing forces, any kinematically admissible deformation process compatible with initial and boundary kinematical constraints may theoretically be realized. For stable unperturbed processes this always requires some additional energy to be supplied.

The whole approach is formally valid for any type of constitutive relations which do not contain a natural time. There is an open question, how wide is the class of solids or problems for which this approach leads to physically acceptable results. In this paper we are concerned with the conventional rigid-plastic or elastic-plastic solids obeying the normality flow rule. The loads acting on the body are assumed to be derivable from a potential, in general a time-dependent one. Under these assumptions the consequences of the stability definition are examined.

2. Definition of stability [9]

Let us consider a continuous plastic body subjected to an isothermal quasi-static proces of deformation in some time interval. We assume that the constitutive relations for the material do not contain a natural time. Denote by V and S the body volume and surface, respectively, in some reference configuration. Introduce Cartesian spatial coordinates (x_j) and the material coordinates (X_K), $j,K = 1,2,3$, which in the reference configuration coincide. The deformation process $\underset{\sim}{\chi}$ is defined by

$$x_j = \chi_j(X_K, t), \qquad X_K \in V, \tag{1}$$

where t denotes a scalar parameter rather than a natural time, but for simplicity it is referred to as time.

The body surface is split onto two parts, S_u and S_T. Over S_u the displacements history is prescribed by the condition

$$\chi_j(X_K, t) = \bar{x}_j(X_K, t), \qquad X_K \in S_u, \tag{2}$$

where $\bar{x}_j$ are given functions. On S_T the nominal tractions T_j are given, which may depend not only on time but also on the place of a surface point in space. The tractions T_j are assumed to be derivable from a given potential ω,

$$T_j = -\frac{\partial \omega(X_K, x_i, t)}{\partial x_j}, \qquad X_K \in S_T. \tag{3}$$

In the particular case when T_j do not depend on the displacements, $\omega = -\bar{T}_j x_j + c$, where $\bar{T}_j$ and c are functions of X_K and t.

We will use the notion of the loading device, defined as a mechanical system by which the tractions (3) and the kinematical constraints (2) are applied to the body surface. That system may be elastic as well as "rigid" or "infinitely soft" in extreme cases. The loading device transmits energy from external sources to the deformed body, its current potential energy being

$$W^e = \int_{S_T} \omega \, dS \ . \tag{4}$$

The initial configuration of the body at time $t = t_1$ is assumed to be known equilibrium configuration, compatible with the boundary conditions (2) and (3). In the following we will consider only such processes (1) which start from this initial configuration, satisfy at every moment the kinematical boundary conditions (2) and describe sufficiently regular deformations of the body. Such processes will be called kinematically admissible processes.

Let the functions $\bar{x}_j$ and ω be specified in some time interval $[t_1,t_2]$ /or $[t_1,\infty)$/. We assume that there exists a kinematically admissible process $\underset{\sim}{x}^o$ which corresponds to a theoretical quasi-static solution to the problem. That process, the stability of which is to be studied, is called fundamental process, and all corresponding functions or quantities are denoted by the superscript "o". The stresses in the fundamental process, calculated from the constitutive relations, satisfy at every moment the equilibrium conditions /for simplicity we omit body forces/

$$\begin{aligned} s^o_{Kj,K} &= 0 \ , & X_K &\in V \ , \\ s^o_{Kj}\,\nu_K &= T^o_j \ , & X_K &\in S \ , \\ T^o_j &= -\left.\frac{\partial\omega}{\partial x_j}\right|_{x_j=x^o_j} , & X_K &\in S_T , \end{aligned} \tag{5}$$

where s_{Kj} is the /unsymmetric/ nominal stress tensor, the comma denotes partial derivative with respect to the rectangular material coordinates X_K, ν_K is the unit normal to the surface S at the reference configuration, and the summation convention holds.

The fundamental process is thought to be perturbed by some set of additional body forces /per unit reference volume/ $F_j^d(X_K,t)$ and nominal surface tractions $T_j^d(X_K,t)$. The perturbed forces F_j^d and T_j^d are of arbitrary magnitude and may act in the whole time interval $[t_1,t_2]$ /or $[t_1,\infty)$/. The perturbed process $\underset{\sim}{\chi}^*$ is, by definition, any kinematically admissible process such that the stresses, related to the deformation history by the constitutive relations, satisfy at every moment the conditions /equations of perturbed motion/

$$\begin{aligned} s^*_{Kj,K} + F^d_j &= 0 , & X_K &\in V , \\ s^*_{Kj,K}\, \nu_K &= T^*_j + T^d_j , & X_K &\in S , \\ T^*_j &= - \left.\frac{\partial \omega}{\partial x_j}\right|_{x_j = x^*_j} , & X_K &\in S_T, \end{aligned} \tag{6}$$

where the superscript "*" always denotes quantities associated with the perturbed process $\underset{\sim}{\chi}^*$. The process $\underset{\sim}{\chi}^*$ is considered as a quasi-static one, possible inertial forces being included into perturbing forces. The perturbed process can also be considered as a dynamic motion, what will be discussed later.

The disturbance measure ϱ is defined as

$$\varrho = \varrho(\underset{\sim}{\chi}^*, \underset{\sim}{\chi}^o, t) = \sup_{\tau\in[t_1,t]} E^d(\underset{\sim}{\chi}^*, \underset{\sim}{\chi}^o, \tau) \quad , \tag{7}$$

where $E^d(\underset{\sim}{\chi}^*, \underset{\sim}{\chi}^o, \tau)$ is the energy supplied by a disturbance during the perturbed process $\underset{\sim}{\chi}^*$ in the time interval $[t_1,\tau]$. This energy is assumed in the form

$$E^d = W^d + (W^*_{ext} - W^o_{ext}) \equiv W^d + \Delta W_{ext} \quad , \tag{8}$$

where

$$W^d = W^d(\underset{\sim}{\chi}^*, \tau) = \int_{t_1}^{\tau} \left\{ \int_V F^d_j\, v^*_j\, dV + \int_S T^d_j\, v^*_j\, dS \right\} dt \tag{9}$$

is the work done directly by the perturbing forces in the interval $[t_1,\tau]$ ($v_j = \partial\chi_j/\partial t$ denotes velocity), and ΔW_{ext} is the supplement of the energy

$$W_{ext} = W_{ext}(\underset{\sim}{\chi},\tau) = \int_{t_1}^{\tau} \{ \int_S T_j \, v_j \, dS + \frac{dW^e}{dt} \} \, dt \qquad (10)$$

taken in the interval $[t_1,\tau]$ from external sources by the l o a d i n g d e v i c e in the process $\underset{\sim}{\chi}^*$ with respect to that in the fundamental process $\underset{\sim}{\chi}^o$. Here and in the following the prefix Δ denotes the difference of any corresponding quantities in the perturbed and fundamental processes at the same moment of time. For given processes $\underset{\sim}{\chi}^*$ and $\underset{\sim}{\chi}^o$, E^d is a continuous function of time, so we may write $\varrho(t) = \max_{\tau\in[t_1,t]} E^d(\tau)$. $\varrho(t)$ is a non-negative, continuous and non-decreasing function.

From the energy balance we have

$$E^d = E^* - E^o \equiv \Delta E, \qquad (11)$$

where

$$E = E(\underset{\sim}{\chi},\tau) = \int_{t_1}^{\tau} \int_V s_{Kj} \, v_{j,K} \, dV \, dt + \left[W^e \right]_{t_1}^{\tau} \qquad (12)$$

is the total energy which must be supplied to the system: {body + loading device} in the time interval $[t_1,\tau]$ in order to realize the process $\underset{\sim}{\chi}$ in a quasi-static manner. (11) presents an alternative definition of the energy supplied by a disturbance if dynamic effects are neglected. If the same perturbed process $\underset{\sim}{\chi}^*$ were considered as a dynamic motion, then on the right hand side of (11) the kinetic energy of the system at time τ would appear as an additional term. Since the kinetic energy is positive definite, the disturbance measure would be increased, the distance between the processes being unaffected. This means that including inertial forces into perturbing forces we did not overlook a possibility of dynamic instability.

The distance d between the perturbed and fundamental processes is assumed in the form

$$d = d(\underset{\sim}{\chi}^*, \underset{\sim}{\chi}^o, t) = \int_V (x_j^* - x_j^o)(x_j^* - x_j^o) \, dV. \qquad (13)$$

Definition. The process $\underset{\sim}{\chi}^o$ is stable in the time interval $[t_1, t_2]$ if and only if for every number $\varepsilon > 0$ there is another number $\delta > 0$ such that for every perturbed process $\underset{\sim}{\chi}^*$ and all $t \in [t_1, t_2]$

$$\rho\,(\underset{\sim}{\chi}^*, \underset{\sim}{\chi}^o, t) < \delta \quad \text{implies} \quad d\,(\underset{\sim}{\chi}^*, \underset{\sim}{\chi}^o, t) < \varepsilon .$$

The process $\underset{\sim}{\chi}^o$ is called unstable if it is not stable.

This definition with the accompanying definitions of $\underset{\sim}{\chi}^*$, ρ and d constitutes a stability postulate.

3. Necessary conditions for stability of a deformation process

In this section some implications of the stability postulate are examined. In the following we will consider the stability in the small, defined as the stability of the process $\underset{\sim}{\chi}^o$ in every sufficiently small time interval $[\tau, \tau+\delta\tau] \subset [t_1, t_2]$. Evidently, stability in the small is necessary for stability.

Our aim is to examine the particular case when the fundamental and perturbed processes coincide up to time τ and then diverge abruptly, in the sense that the velocity fields in these processes at time τ are different /as velocity at time τ we understand $\lim\limits_{\delta t \to 0} v_j \big|_{\tau+\delta t}$ /. We assume that processes are sufficiently regular in the interval $[\tau, \tau+\delta\tau]$; in particular, the velocities in both processes are continuous and continuously differentiable with respect to X_K.

The energy functional $E(\underset{\sim}{\chi}, t)$ becomes, for a fixed process $\underset{\sim}{\chi}$, a real function of time. Expanding this function into Taylor series we obtain

$$E\,(\tau+\delta t) = E(\tau) + E_1\,\delta t + E_2\,(\delta t)^2 + \mathcal{O}\,(\delta t)^2 ,$$

where

$$E_1 = \frac{dE}{dt} = \int_V s_{ij}\, v_{j,i}\, dV + \int_{S_T} \Big(\frac{\partial\omega}{\partial x_j}\, v_j + \frac{\partial\omega}{\partial t} \Big)\, dS$$

$$= \int_V \sigma_{ij}\, \epsilon_{ij}\, dV - \int_{S_T} T_j\, v_j\, dS + \int_{S_T} \frac{\partial\omega}{\partial t}\, dS,$$

$$E_2 = \frac{1}{2}\frac{d^2E}{dt^2} = \frac{1}{2}\int_V \{\dot{s}_{ij}\, v_{j,i} + s_{ij}\,\dot{v}_{j,i}\}\, dV$$

$$+ \frac{1}{2}\int_{S_T}\{\frac{\partial\omega}{\partial x_j}\,\dot{v}_j + \frac{\partial^2\omega}{\partial x_i\partial x_j}\, v_i\, v_j + 2\frac{\partial^2\omega}{\partial x_j\partial t}\, v_j + \frac{\partial^2\omega}{\partial t^2}\}\, dS$$

$$= \frac{1}{2}\int_V \dot{s}_{ij}\, v_{j,i}\, dV - \int_{S_T}\{\frac{\partial T_j}{\partial t} + \frac{1}{2}\frac{\partial T_j}{\partial x_i}\, v_i\}\, v_j\, dS$$

$$+ \frac{1}{2}\Big[\int_V \sigma_{ij}\dot{v}_{j,i}\, dV - \int_{S_T} T_j\,\dot{v}_j\, dS\Big] + \frac{1}{2}\int_{S_T}\frac{\partial^2\omega}{\partial t^2}\, dS.$$

All quantities in the above expressions are calculated at time τ. The body configuration at time τ is taken as the reference one, the same indexes referring to both the space and material coordinates. σ_{ij} is the Cauchy stress tensor /equal to s_{ij} in the reference configuration/, $\epsilon_{ij} = \frac{1}{2}(v_{i,j} + v_{j,i})$ is the strain rate tensor at time τ, and the dot denotes the material derivative. It is convenient to introduce the velocity functionals

$$J_1(\underset{\sim}{v}) = \int_V \sigma_{ij}\,\epsilon_{ij}\, dV - \int_{S_T} T_j\, v_j\, dS\ , \tag{14}$$

$$J_2(\underset{\sim}{v}) = \frac{1}{2}\int_V \dot{s}_{ij} v_{j,i}\, dV - \int_{S_T}\{\frac{\partial T_j}{\partial t} + \frac{1}{2}\frac{\partial T_j}{\partial x_i}\, v_i\}\, v_j\, dS\ , \tag{15}$$

appearing in the expressions for E_1 and E_2, respectively.

First we consider the case of the rigid-plastic material, for which the stress σ_{ij} depends on the actual strain rate. Since $\partial\omega/\partial t$ at time τ is the same in both the fundamental and perturbed processes, we have

$$\Delta E\Big|_{\tau+\delta t} = \Delta J_1\Big|_{\tau}\ \delta t$$

to first order. If ΔJ_1 is negative at time τ for some velocity field $\underset{\sim}{v}^*$, then ΔE for the corresponding process $\underset{\sim}{\chi}^*$ is negative in a sufficiently small time interval $(\tau, \tau+\delta\tau)$. Thus the disturbance measure $\rho = 0$ at time $\tau+\delta\tau$, and the distance d at time $\tau+\delta\tau$ is a small but finite positive number since the body configuration at time $\tau+\delta\tau$ is different in both processes. This implies instability. Hence the necessary condition for stability of the process $\underset{\sim}{\tau}^o$ in the interval $[t_1,t_2]$ is

$$J_1(\underset{\sim}{v}^*) \geqslant J_1(\underset{\sim}{v}^o) = \int_{S_u} T_j^o v_j \, dS \tag{16}$$

for all $\tau \in [t_1,t_2)$ and for every admissible velocity field $\underset{\sim}{v}^*$ taking the values prescribed over S_u. But this is just the extremum principle due to Hill [10], satisfied automatically if the yield surface is convex and the normality flow rule is adopted. For velocity fields which are additionally compatible with the stress field σ_{ij}^o we have $\Delta J_1 = 0$, and for such fields ΔE need be calculated to second order. We will not consider this problem here.

In the case of elastic-plastic solids we have $\sigma_{ij}^* = \sigma_{ij}^o$, and ΔJ_1 vanishes identically by Green's theorem and the condition $\Delta v_j = 0$ over S_u. Similarily, the term in square brackets in the expression for E_2 is the same for both processes, since $\Delta\dot{v}_j = 0$ over S_u. Thus

$$\Delta E\Big|_{\tau+\delta t} = \Delta J_2\Big|_{\tau}(\delta t)^2$$

to second order, since the remaining terms cancel. In a way analogical to presented above it is easy to show that $\Delta J_2 < 0$ implies instability. Hence the necessary condition for stability of the process $\underset{\sim}{\tau}^o$ in the interval $[t_1,t_2]$ is

$$J_2(\underset{\sim}{v}^*) \geqslant J_2(\underset{\sim}{v}^o) = \frac{1}{2}\left\{\int_{S_u} \dot{T}_j^o v_j \, dS - \int_{S_T} \frac{\partial T_j}{\partial t} v_j^o \, dS\right\} \tag{17}$$

for all $\tau \in [t_1,t_2)$ and for every admissible velocity field $\underset{\sim}{v}^*$ taking the values prescribed over S_u. The inequality (17) is not satisfied automatically, and presents a true stability criterion.

In the examined case of loading, $J_2(\underset{\sim}{v})$ is just the velocity functional considered by Hill [11,6] in connection with the first-order rate problem. The condition (17) presents /without the equality sign/ the extremum principle due to Hill [11] which holds if the sufficient condition for uniqueness is satisfied. Moreover, for many problems of interest it can be shown that the first moment τ^{cr} beyond which (17) ceases to hold coincides with the moment of primary bifurcation. The connections between the present approach and the bifurcation theory are apparent. However, the interpretation of the critical moment τ^{cr} is different. While bifurcation can take place not earlier than τ^{cr}, the process becomes unstable in the present case not later than τ^{cr}. Note that the functional $J_2(\underset{\sim}{v})$ plays here an analogical role as the potential energy functional in examining stability of elastic conservative systems /$J_2(\underset{\sim}{v})$ is stationary at $\underset{\sim}{v} = \underset{\sim}{v}^o$ [11]/.

4. Stability of an equilibrium configuration

The definition of stable deformation process can be directly used to examine stability of a fixed# equilibrium configuration. For this purpose we consider the stability of a particular fundamental process in which the velocities vanish identically, the functions $\bar{x}_j$ and ω being independent of time. In that case we have $W^*_{ext} \equiv W^o_{ext} \equiv 0$ and $E^d \equiv W^d \equiv E^*$ at any moment. The disturbance measure reduces to the maximum value of the work W^d done by perturbing forces. The stability of equilibrium may be examined in infinite time interval $[t_1, \infty)$.

From (17) we obtain that the necessary condition for stability of a fixed equilibrium configuration of elastic-plastic body is

$$\int_V \dot{s}_{ij}\, v^*_{j,i}\, dV - \int_{S_T} \frac{\partial T_j}{\partial x_i}\, v^*_i\, v^*_j\, dS \geqslant 0 \tag{18}$$

for every admissible velocity field $\underset{\sim}{v}^*$ vanishing on S_u. In the special case of dead loading, $\partial T_j/\partial x_i = 0$ and the second term in (18) vanishes. Then the condition (18) with strict inequality takes the form proposed by Hill [3] as a sufficient condition for stability. There is also correspondence with Drucker's postulate of stability in the small of a time-independent system [4].

The sufficient condition for stability of a fixed equilibrium configuration is

---

The present approach has the advantage that the equilibrium configuration may also be considered in moving coordinate basis. Results are coordinate invariant.

$$W^d(\underset{\sim}{\chi}^*,t) = E(\underset{\sim}{\chi}^*,t) \geqslant c_1\, d(\underset{\sim}{\chi}^*,\underset{\sim}{\chi}^o,t) \quad \text{for} \quad d \leqslant c_2 \qquad (19)$$

for every perturbed process $\underset{\sim}{\chi}^*$ and all $t > t_1$, where c_1 and c_2 are fixed positive numbers. The proof follows directly from the stability definition. The condition (19) is also sufficient for dynamic stability against initial disturbances, in the sense of the definition used by Nguyen and Radenkovic [5].

Drucker [4] proposed a criterion of stability in the large of a time-independent system, which in the present notation becomes $W^d(\underset{\sim}{\chi}^*,t) > 0$ for every $\underset{\sim}{\chi}^*$ and all $t > t_1$. In general this is neither sufficient nor necessary for stability in the present sense.

We also mention that Krawietz [12] used the measure for initial disturbances which is equivalent to (7) for $W_{ext} \equiv 0$.

References

1. Klushnikov, V.D.: Stability of elastic-plastic systems /in Russian/. Nauka, Moscow 1980.
2. Rice, J.R.: The localization of plastic deformation. Proc. 14th IUTAM Congr. Theor. and Applied Mech. (ed., W.T.Koiter), North-Holland, Delft 1976, 207-220.
3. Hill, R.: A general theory of uniqueness and stability in elastic-plastic solids. J.Mech.Phys.Solids 6 (1958) 236-249.
4. Drucker, D.C.: On the postulate of stability of material in the mechanics of continua. J. Mec. 3 (1964) 235-249.
5. Nguyen, Q.S.; Radenkovic, D.: Stability of equilibrium in elastic plastic solids. Proc. IUTAM/IMU Symp. Appl. Methods of Functional Analysis to Probl.Mech., Marseille 1975, 403-414.
6. Hill, R.: Aspects of invariance in solid mechanics. Advances in Appl. Mech. 18 , Acad. Press, New York , 1978, 1-75.
7. Hutchinson, J.W.: Plastic buckling. Advances in Appl. Mech. 14, Acad. Press, New York 1974, 67-144.
8. Miles, J.P.: On necking phenomena and bifurcation solutions. Arch. Mech. 32 (1980) 909-931.
9. Petryk, H.: A stability postulate for quasi-static processes of plastic deformation. GAMM Tagung, Wurzburg 1981. (To be published in ZAMM).
10. Hill, R.: On the state of stress in a plastic-rigid body at the yield point. Phil. Mag. 42 (1951) 868-875.
11. Hill, R.: Uniqueness criteria and extremum principles in self-adjoint problems of continuum mechanics. J. Mech. Phys. Solids 10 (1962) 185-194.
12. Krawietz, A.: Stabilitat des Gleichgewichts Mechanischer Systeme. ZAMM 61 (1981) H4, T46.

Dynamic Stability of Mechanical Systems with Hereditary Properties

R. PYRZ

Institute of Machine Design Fundamentals
Warsaw Technical University, Warsaw

Summary

In the paper a new Lyapunov functional is constructed for the systems with hereditary properties. Stability of two problems is considered. The first one is a plastic stability of tensioned round specimen. The boundedness of the strain gradient is obtained. Dynamic stability of viscoelastic rotating shaft is the second problem. It is taken into consideration conservative as well as nonconservative uniformly distributed load.

Introduction

It is important for the description of many real, physical processes to take into consideration their memory effect i.e. the influence of all past states in which processes have been passed from the initial to the current instant. Phenomena governed by such systems are called hereditary phenomena.

Investigations of the stability of hereditary systems give rise to the consideration of two fundamental problems. Firstly, there is the need for the development of more potent analytical procedures for the better representation of the mechanical behaviour of structural materials under a variety of service conditions. This demands a deeper study of constitutive equations. Then, there is the requirement for the application of more refined methods of analysis to investigate the stability of the solutions of integral and integro-differential equations, which in majority cases describe hereditary phenomena.

Generaly we can describe the rheological systems in the following manner

$$y = H(t)y + \int_0^t K(t-\tau)\,G[y(\tau)]\,d\tau , \tag{1}$$

where y is an element of adequate Hilbert Space while H, K, G are an operators. Let us choose the Lyapunov functional in the form

$$V(y,t) = \langle y, Ny\rangle + \int_0^t \langle y(\tau), \int_0^\tau K(\tau-s)y(s)\,ds\rangle\, d\tau , \tag{2}$$

where $\langle\ ,\ \rangle$ is an inner product in Hilbert Space. The stability analysis of onedimesional rheological vibrating system using Lyapunov functional

(2) was described in the autonomous as well as nonautonomous case [1]. It is the aim of the present study to establish a condition for the stability of two mechanical systems. The first system is the non-uniform plastic tensile flow of rate-dependent material. The second one is the stability of rotating long shaft. These problems appear as a special case of (1) and we can apply the Lyapunov functional (2).

Plastic stability in tension

We consider a tension alluminium-alloy specimen of initial cross-sectional area $A_o(x)$, where coordinate x is measured along the axis of the specimen. The variations in A_o are assumed small. The specimen is subjected to an axial force P which increases in an arbitrary manner with time t. The hereditary properties of the specimen material is assumed to be governed by the equation

$$F(\varepsilon)=\sigma+\int_0^t K(t-s)\sigma(s)\,ds, \tag{3}$$

where ε and σ are true strain and true stress respectively. Stress is related to the modulus of elasticity [2]. The equation $F(\varepsilon)=\sigma$ determines the instantaneous straining diagram when the second term on right hand side of equation (3) is negligable. The function F and the integral kernel K are determined on constant load rate and creep experiments, see Fig.1 and Fig.2.

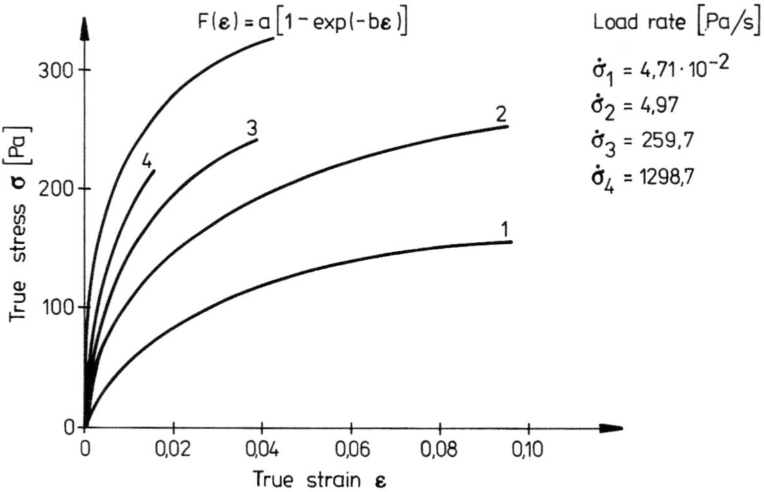

Fig.1. Constant load rate test

Let us note that we may write the following equation

$$\frac{\partial\sigma}{\partial x}=\sigma_x=(PA^{-1})_x=PA_o^{-1}\exp(\varepsilon)\varepsilon_x-PA_o^{-2}\exp(\varepsilon)(A_o)_x=\sigma\varepsilon_x-\sigma(\ln A_o)_x. \tag{4}$$

where we used known definition of true strain, true stress and incompressi-

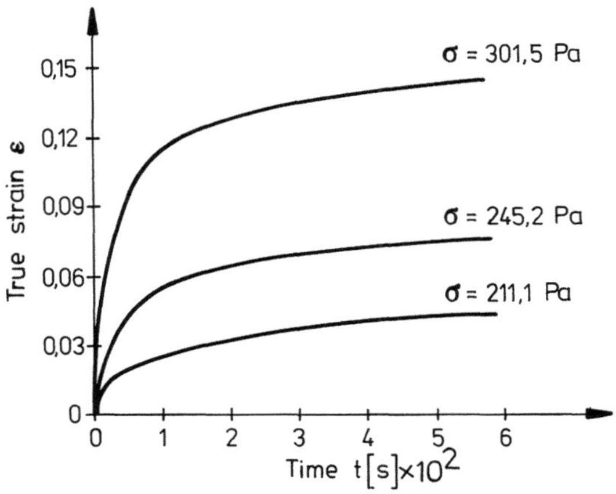

Fig. 2. Creep test in room temperature

bility condition. By the fact that A_o=const. we have

$$\sigma_x = \sigma \cdot y, \qquad y = \varepsilon_x . \tag{5}$$

Now we differentiate the equation (3) twice with respect to x and t obtaining

$$a\dot{y} = by + \int_o^t \sigma(s) A(t-s) y(s) ds, \tag{6}$$

where

$$a = E_\varepsilon - \sigma, \quad b = \dot{\sigma} + K(0)\sigma - E_{\varepsilon\varepsilon} \cdot \dot{\varepsilon}, \quad A(t) = \dot{K}(t).$$

We deal with equation (6) as a scalar version of (1) with weight function σ. We will investigate the stability of (6) in the sense of Lagrange i.e. we will determine conditions on the boundedness of the strain gradient y. Let us choose the Lyapunov function as a special case of (2)

$$V = ay^2 + a \int_o^t y(\tau) \int_o^\tau A(\tau - s) ds d\tau. \tag{7}$$

TheLyapunov function is positively determined if

$$a > o \ \text{ and } \ (-1)^i A^{(i)}(t) \quad 0, \ i=0,1,2. \tag{8}$$

The second term of (8) results from the estimation of positively definite kernel [3].
Consider the derivative of V along a solution of (6). All components of $\dot{V}$ may be estimated by $[\sup_{t \geqslant o} y(t)]^2$. Therefore we have the estimate for $\dot{V}$

$$\dot{V} \leqslant c[\sup_{t \geqslant o} y(t)]^2. \tag{9}$$

Time derivative (9) is negative semidefinite if

$$c=2b+\dot{a}+t^{2}\Big[a\sup_{t\geqslant 0}A(t)\Big]^{\bullet}+atA(0)+2\lambda_{\sigma}^{-\frac{1}{2}}+at^{3}(\max.\ ba^{-1})\sup_{t\geqslant 0}A(t)+ at(\max.\ a^{-1})\int_{0}^{t}(\lambda_{\sigma}\lambda)^{\frac{1}{2}}ds\leqslant 0, \qquad (10)$$

where λ and λ_{σ} are smallest eigenvalues of the kernels

$$A_{1}(t-\tau)=\int_{0}^{t}A(s-t)A(s-\tau)ds, \qquad A_{1\sigma}(t-\tau)=\int_{0}^{t}\sigma(s)A(s-t)A(s-\tau)ds. \qquad (11)$$

For the case of a constant croshead speed v we may take approximately $tvL_{o}^{-1}=\exp\varepsilon-1$, where L_{o} is an initial length. We find boundary (10) on the σ-ε plane, determining the stress from (3). The boundary points marked by small crosses and calculated from (10) is shown on Fig.3.

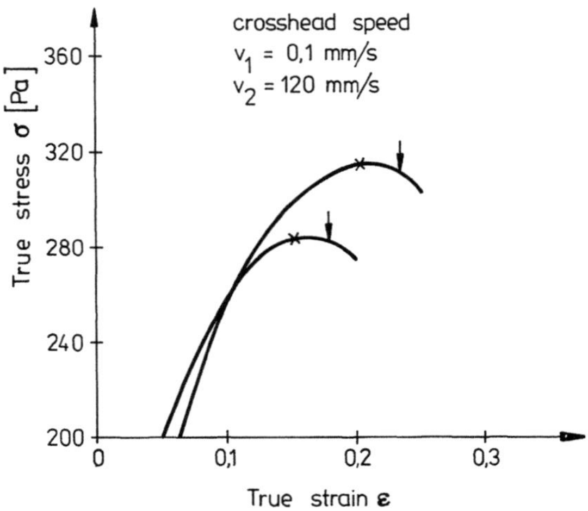

Fig.3. Stability points for alluminium-alloy specimen.

The arrows indicate experimentaly obtained necking points.
Detailed analysis and computations will be contained in the forthcoming paper [4]. It is interesting to note that in the case of rate independent material we neglect integral component of (6) and (7). The function $V=ay^{2}$ and its derevative give us the estimation $F_{\varepsilon}\geqslant\sigma$, a well known Considere condition for necking [5].

Stability of rotating shaft

As a second example a shaft rotating with angular velocity ω is considered. The shaft have mass μ per length, length l, the same bending stiffness EI in all directions normal to its axis and it is subjected to an axial force

$$P = S + \int_{x}^{l} q_{\alpha}(r)dr$$

The Boltzman viscoelastic constitutive equation with relaxation function R(t) is used for the shaft material. The shaft is damped by an external viscous damping with coefficient b, Fig. 4.

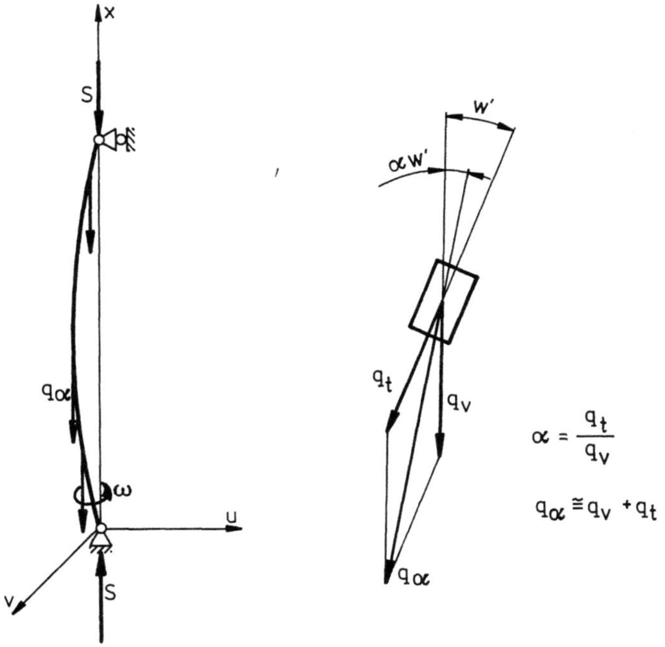

Fig. 4. The rotating shaft

The transverse displacements relative to the rotating axis with the same angular velocity as the shaft are governed by the matrix-operational equation

$$w = Aw + \int_0^t C(t-s)\, Bw(s)\, ds, \tag{12}$$

where $w = \mathrm{col}\{u, \dot{u}, v, \dot{v}\}$,

$$A = A_1 \frac{\partial^4}{\partial x^4} + A_2 \frac{\partial^2}{\partial x^2} + A_3 \frac{\partial}{\partial x} + A_4,$$

$$A_1 = -\begin{bmatrix} 0 & 0 & 0 & 0 \\ a_1 & 0 & 0 & 0 \\ 0 & 0 & 0 & 0 \\ 0 & 0 & a_1 & 0 \end{bmatrix}, \quad A_2 = -\begin{bmatrix} 0 & 0 & 0 & 0 \\ a_3 & 0 & 0 & 0 \\ 0 & 0 & 0 & 0 \\ 0 & 0 & a_3 & 0 \end{bmatrix}, \quad A_3 = -\begin{bmatrix} 0 & 0 & 0 & 0 \\ a_3+a_4 & 0 & 0 & 0 \\ 0 & 0 & 0 & 0 \\ 0 & 0 & a_3+a_4 & 0 \end{bmatrix}$$

$$A_4 = \begin{bmatrix} 0 & 1 & 0 & 0 \\ \omega^2 & -a_2 & -\omega a_2 & -2\omega \\ 0 & 0 & 0 & 1 \\ a_2\omega & 2\omega & \omega^2 & -a_2 \end{bmatrix}, \quad C(t) = \begin{bmatrix} 0 & 0 & 0 & 0 \\ 0 & R(t) & 0 & 0 \\ 0 & 0 & 0 & 0 \\ 0 & 0 & 0 & R(t) \end{bmatrix}, \quad B = \begin{bmatrix} 0 & 0 & 0 & 0 \\ 1 & 0 & 0 & 0 \\ 0 & 0 & 0 & 0 \\ 0 & 0 & 1 & 0 \end{bmatrix} \frac{\partial^4}{\partial x^4},$$

$$C_1(t) = \begin{bmatrix} 0 & 0 & 0 & 0 \\ R(t) & 0 & 0 & 0 \\ 0 & 0 & 0 & 0 \\ 0 & 0 & R(t) & 0 \end{bmatrix},$$

$a_1=\mu^{-1}EI$, $a_2=\mu^{-1}b$, $a_3(x)=\mu^{-1}P$, $a_4(x)=\mu^{-1}\alpha q_\alpha$

As w belongs to the space $L_r^2\{\Omega\times[0,\infty)\}$ we can introduce the inner product and the norm

$$\langle w^1,w^2\rangle = \int_0^t (w^1,w^2)ds, \qquad \|w\|^2 = \int_0^t |w|^2 ds ,$$

where

$$(w^1,w^2) = \int_\Omega \sum_{i=1}^r w_i^1 w_i^2 d\Omega, |w|^2 = \int_\Omega \sum_{i=1}^r w_i^2 d\Omega \quad \Omega=(x:0<x<1).$$

We'll find the stability condition for the null solution of the equation (12) in a Lyapunov-Movchan sense with respect to the norm $|w|$. As the Lyapunov functional we choose

$$V = \|w\|^2 + \langle w, w_{C1}\rangle , \tag{13}$$

where we introduce the notation

$$w_{C1} = \int_0^t C_1(t-s)w(s)ds.$$

Following [6] we have the estimate

$$\langle w, w_{C1}\rangle \geqslant \frac{1}{2} C_1(t)\|\int_0^t w(s)ds\|^2 \geqslant 0, \tag{14}$$

if $C_1(0)=0$, $(-1)^k C_1^{(k)}(t) \geqslant 0 \quad t>0, \ k=0,1,2.$

From the inequality (14) it follows the positive-definiteness of the functional V. The time-derivative of V along the solution of (12) is given by

$$\dot V=2\langle w,Aw\rangle +2\langle w,w_{CB}\rangle +\langle Aw,w_{C1}\rangle +\langle w_{CB},w_{C1}\rangle +\langle w_1 w_{C1}\rangle+\langle w,C_1(0)w\rangle . \tag{15}$$

Consider the conservative case of load P. Then $\alpha=0$, $q_\alpha=q=\text{const.}$ and noting $a_5=\mu^{-1}S$, $a_6=\mu^{-1}q$ we have

$$\begin{aligned} \dot V=&2(1+\omega^2)(\langle u,\dot u\rangle + \langle v,\dot v\rangle)+2\omega a_2(\langle \dot v,u\rangle - \langle v,\dot u\rangle)-2a_2(\|\dot u\|^2+\|\dot v\|^2) \\ &-2a_1(\langle \dot u,\overset{IV}{u}\rangle + \langle \dot v,\overset{IV}{v}\rangle) -2a_5(\langle \dot u,u''\rangle + \langle \dot v,v''\rangle) -2a_6[\langle \dot u,(1-x)u''\rangle + \\ &+\langle \dot v,(1-x)v''\rangle] +2a_6(\langle u',\dot u\rangle + \langle v',\dot v\rangle) +2\langle \dot u,u_R^{IV}\rangle +2\langle \dot v,v_R^{IV}\rangle +\langle \dot u,u_R\rangle + \\ &+\langle \dot v,v_R\rangle +\langle u,u_{\dot R}\rangle +\langle v,v_{\dot R}\rangle + R(0)(\|u\|^2+\|v\|^2) . \end{aligned} \tag{16}$$

The dots and prime indicate partial derivatives with respect to time t and coordinate x. We take into consideration pinned-pinned boundary conditions. Using the Schwarz and the Steklov inequalities and some special estimates [7],[8] we obtain

$$\dot{V} \leqslant \left(\frac{1}{2} a_2 \omega^2 - 2R(0) f^4\right) \left(\|u\|^2 + \|v\|^2 \right) +$$

$$\left[(1+\omega^2) + \lambda^{-\frac{1}{2}} + 2tM(t) f^4 \sup_{t\geqslant 0} R(t) + (1-M(t)) f^2 \left(a_5 - f^2 a_1 + f^2 g^{-2} a_6 \right)\right] \left(|u|^2 + |v|^2\right) \quad (17)$$

where $f=\pi/1$, $g^2=18{,}961^{-3}$, λ is the smallest eigenvalue of (11) and we supposed that

$$\left(|u''|^2 + |v''|^2 \right)_{t=0} \leqslant \left(|\sup u''|^2 + |\sup v''|^2 \right)_{t>0} \leqslant M(t) \left(|u''|^2 + |v''|^2 \right)$$

The time derivative of V is negative semidefinite if

$$a_2 \omega^2 \leqslant 4R(0) f^4 \quad (18)$$

$$(1+\omega^2) + \lambda^{-\frac{1}{2}} + 2tM(t) f^4 \sup_{t>0} R(t) \leqslant (M(t)-1) f^2 \left(a_5 - f^2 a_1 + f^2 g^{-2} a_6 \right).$$

For the nonconservative case $\alpha=1$. The functional V still has the form (13). Its time derivative is given by

$$\dot{V} = \dot{V}_{16} - 2a_6 \left(\langle u', \dot{u} \rangle + \langle v', \dot{v} \rangle \right) \quad (19)$$

A simple computation estabilishes the fact that (19) is negative semidefinite if the inequality $(18)_2$ is fulfilled and

$$a_2 \omega^2 + 2a_6 \leqslant 4R(0) f^4 \quad (20)$$

More detailed information for above estimations one will find elsewhere [8].

References

1. Pyrz R.: Mechanical behaviour and stability of motion for the onedimensional rheological system. Mech. Research Comm. 7 1980 395-399.

2. Rabotnov Yu.N.: Suvorova J.V.: The nonlinear hereditary-type stress -strain relation for metals. Inst.J.Solids Structure 14 1978 173-185.

3. Londen S.-O.: On an integral equation in a Hilbert space. SIAM J. Math. Anal. 8 1977 950-970.

4. Pyrz R.: Deformation stability of rate dependent material by Lyapunov second method. To appear.

5. Jonas J.J.: Holt R.A.: Coleman C.E.: Plastic stability in tension and compresion. Acta Metallurgica 24 1976 911-918.

6. Barbu V.: Malik M.A.: Semilinear integro-differential equations in

Hilbert space. J.Math.Anal.Appl. 67 1979 452-475.

7. Tylikowski A.: Dynamic stability of rotating shafts. Ing.Archiv 50 1981 41-48.

8. Pyrz R.: Stability of rotating shaft with hereditary characteristic. To appear in Ing.Archiv.

Existence of Solutions to Unstable Problems in Linear Elastodynamics

R. J. KNOPS

Department of Mathematics
Heriot-Watt University
Edinburgh EH14 4AS
Scotland

1. Introduction

We study existence of solutions to the initial boundary value problem of linearised homogeneous anisotropic elasticity for elasticities that are symmetric but not sign-definite. We show that a solution exists in a weak sense for data that is dense in L_2. Only a brief account is given since a more detailed description will be presented elsewhere.

The analysis may be motivated in several ways, the most obvious of which occurs perhaps in the investigation of instability. This involves examining the time evolution of initially small perturbations and thus questions of existence immediately arise if a purely formal treatment is to be avoided. Let us recall that, existence apart, a sufficient condition for stability of the null solution is that the elasticities are positive-definite, while a sufficient condition for instability (and hence a necessary condition for stability) is that the elasticities are indefinite. As regards existence, Marsden and Hughes [7] (see also Fichera [2]) have shown that a broad class of solutions exist to the initial boundary value problem provided the elasticities are strongly-elliptic. Hence, stability and instability can both be properly discussed in the case of strong-ellipticity, but when this condition is not satisfied there appears to be little that is so far known about existence.

Our result allows for strong-ellipticity to fail and thus helps understanding of existence in this case. Nevertheless, the kind of existence we establish could be regarded as too restrictive and that consequently it should be generally assumed that there is non-existence. Such a conclusion need not affect the instability classification, since the definition of instability can be extended from its usual form to include situations where there is non-existence. Alternatively, and perhaps more importantly, we may use the conclusion in the different context of constitutive hypotheses to endorse strong-ellipticity as a reasonable constitutive postulate.

Broadly similar remarks apply to other circumstances where a knowledge of existence in the absence of strong-ellipticity is desirable; examples include phase transitions, equilibrium shocks and buckling. For instance, in buckling, it is of interest to discover how the body can move at criticality from an unstable to a stable equilibrium position. In a small neighbourhood of the unstable position, the motion is presumably governed by the linearised equations which, if they are reliably to reflect physical behaviour, must admit solutions of appropriate generality. In conjunction with the existence result proved here, this kind of argument may again be used to support strong-ellipticity as a constitutive assumption, especially as it is known that this condition allows buckling. Conversely, the density aspect of our result could well provide the basis for an adequate description of the immediate post-buckling motion, in which event there would be less reason for the adoption of strong-ellipticity as a constitutive assumption.

However, it is not the intention to pursue this discussion which is introduced merely to motivate the questions of existence treated here.

In the following sections, we state the equations of linearised elastodynamics together with the accompanying initial and boundary conditions. Selected results based upon logarithmic arguments are next presented and used to show that solutions do not exist for general data. On the other hand, eigenmodal solutions exist demonstrating that the problem is not empty. Finally, the main theorem on existence is stated and its proof is briefly outlined.

Scalar, vector and tensor quantities are not, in general, typographically distinguished, while the precise meaning of various inner products will become clear from the context. Other notation is explained as required. Exceptionally, we use the standard summation and comma conventions.

2. Basic equations

Consider a linear homogeneous anisotropic elastic material occupying a bounded region Ω in three-dimensional euclidean space whose boundary $\partial\Omega$ is smooth enough to permit application of the divergence theorem. The cartesian components u_i of the vector displacement $u(x,t)$, $x \in \mathbb{R}^3$, are governed by the following equations of motion:

$$\rho \ddot{u}_i = (c_{ijkl} u_{k,l})_{,j} + f_i, \quad (x,t) \in Q = \Omega \times [0,T), \tag{2.1}$$

where $[0,T)$ is the maximal interval of existence, $\rho(x) (\geqslant \rho_0 > 0)$ is the density, f is the body-force per unit volume, a superposed dot denotes time differentiation, and $c_{ijkl}(x)$ are the elasticities. We assume that the system (2.1) is formally self-adjoint so that the elasticities possess the symmetry

$$c_{ijkl} = c_{klij}. \tag{2.2}$$

Furthermore, we suppose the elasticities are indefinite in the sense that the sign of the quadratic form

$$c_{ijkl} \xi_{ij} \xi_{kl} \tag{2.3}$$

depends upon the choice of ξ_{ij}.

We write equations (2.1) more succinctly as

$$\rho \ddot{u} = Cu + f \qquad (x,t) \in Q, \tag{2.4}$$

where the differential operator C is formally given by

$$Cu = \mathrm{Div}\, \mathbb{C}\, \nabla u, \tag{2.5}$$

in which Div denotes the usual tensor divergence operator, the gradient operator is denoted by ∇ so that

$$\nabla u = (\partial u_i / \partial x_j), \tag{2.6}$$

and $\mathbb{C}$ is the elasticity tensor. Thus, corresponding to (2.2), we have

$$\mathbb{C} = \mathbb{C}^T, \tag{2.7}$$

where a superscript T denotes the transposed tensor.

We adjoin to the differential equation (2.1) the boundary and initial conditions which respectively are

$$u = 0, \quad (x,t) \in \partial\Omega \times [0,T), \tag{2.8}$$

$$u(x,0) = u_0(x), \quad \dot{u}(x,0) = u_1(x), \quad x \in \Omega, \tag{2.9}$$

where u_0, u_1 are prescribed functions. For simplicity, we treat only the displacement boundary value (2.8), but other conditions may be included.

When $f = 0$, conservation of energy follows in the usual way from (2.1),(2.2) and the symmetry condition (2.2) for classical solutions. However, for weak solutions the computation is invalid and so instead we separately postulate an energy inequality in the form (again for $f = 0$)

$$E(t) \leqslant E(0), \tag{2.10}$$

where

$$E(t) \equiv \tfrac{1}{2}(\dot{u},\dot{u}) + \tfrac{1}{2}C(u,u) \tag{2.11}$$

and we have introduced the inner product given by

$$(u(\cdot),v(\cdot)) = \int_\Omega \rho u(x,\cdot)v(x,\cdot)dx, \tag{2.12}$$

together with the quantity $C(u,v)$ defined by

$$C(u,v) = -\int_\Omega Cuvdx \quad (= \int_\Omega c_{ijkl}u_{i,j}v_{k,l}dx). \tag{2.13}$$

We thus recognise $\tfrac{1}{2}C(u,u)$ as the strain-energy.

We employ (2.10) in the next section to establish some results based upon logarithmic convexity arguments. Further details of this method may be found in [4,5], with an extensive survey, also discussing related techniques, due to Payne [8].

3. Logarithmic Convexity Arguments and Some Consequences

We first show that the logarithm of the function $F(t)$ given by

$$F(t) = (u(t),u(t)) + \alpha(t+t_0)^2, \tag{3.1}$$

where α, t_0 are non-negative constants to be determined, is convex with respect to t. Then by means of well-known properties of convex functions, we derive information on the solution $u(x,t)$. We observe that the technique remains valid for suitably defined weak solutions (cp [5]) although we content ourselves with a discussion of classical solutions.

On differentiating (3.1) twice and using (2.1) to replace the inertia term followed by an integration by parts and use of (2.10) to eliminate the strain-energy we get

$$\ddot{F}(t) \geqslant 4[(\dot{u},\dot{u}) + \alpha] - 2(\alpha + 2E(0)). \tag{3.2}$$

With the help of Schwarz's inequality, we next obtain the inequality

$$F\ddot{F} - (\dot{F})^2 \geqslant -2(\alpha + 2E(0))F(t). \tag{3.3}$$

Now let us suppose that the elasticities are indefinite and that the initial data are chosen to give $E(0) \leqslant 0$. We may then set $\alpha + 2E(0) = 0$, and since $F(t) \neq 0$, we obtain from (3.3)

$$(\ln F)^{\cdot\cdot} \geqslant 0. \tag{3.4}$$

A convex function lies above the tangent at points of its curve, and hence from (3.4) we have

$$F(t) \geqslant F(0)\exp[t\dot{F}(0)/F(0)], \quad t \geqslant 0. \tag{3.5}$$

We may always select t_0 to ensure that $\dot{F}(0) > 0$ irrespective of the initial data, so that we see from (3.5) that solutions, when they exist, are bounded below in an L_2-sense by a function increasing exponentially in time. Thus, we may conclude that formally, at least, the null solution is unstable under the stated conditions on the elasticities and initial data. Furthermore, by considering data analogous to those in the classic Hadamard treatment of the Cauchy problem for Laplace's equation, it follows either directly or from (3.5) that the solution does not depend continuously upon the initial data. However, on recalling the further property that a convex function lies below the join of any two points on its curve, we may derive from (3.4) a second inequality:

$$F(t) \leqslant [F(0)]^{1-t/T}[F(T)]^{t/T}, \quad t \in [0,T]. \tag{3.6}$$

This shows that in the class of functions given by

$$\mathfrak{N} = \{u\colon (u(T),u(T)) \leqslant N^2, N \text{ constant}\} \tag{3.7}$$

the solution depends Hölder continuously upon the initial data as measured by $F(0)$. We add parenthetically that (3.6) may also be used to prove uniqueness of the solution, but that in so doing the energy inequality (2.10) is employed. A proof making no such assumption is based upon a Lagrange identity and is due originally to Brun [1] for classical solutions and to Levine [6] for weak solutions. This remark becomes important in the proof of existence.

Similar conclusions hold for initial data with zero or positive total energies provided the elasticities are indefinite and the initial strain energy non-positive. However, in these cases, the initial data must be further restricted to values making (u_0,u_1) positive. Details are presented in [5].

It is a simple exercise using (3.5) and (3.6) to show that the function (u,u) is non-decreasing with respect to t under initial data that, for instance, makes $E(0) \leqslant 0$ and $(u_0,u_1) \geqslant 0$. (An earlier version of this result contained in [3] is trivially inconclusive and should be discarded.) The result is the best possible in the sense that, by uniqueness, zero initial data implies the identical vanishing of the solution. The argument may be extended in an obvious manner to obtain conditions on the data for the solution to be either non-periodic or time-increasing, but we do not discuss details here.

Instead, let us use (3.5) to show that solutions do not exist for general data. We assume for the moment that the operator associated with our linear problem is closed and we take the elasticities to be indefinite. We may now use a scaling procedure, introduced by Marsden and Hughes [7] and Wilkes [9] to establish the non-existence of a contractive semi-group in our problem, to show that initial data can be chosen to ensure that (u_0,u_1) is arbitrarily large and that the initial total energy is negative. We thus have the situation in which (3.5) holds, and hence at any given finite value of time, the quantity $(u(t),u(t))$ is unbounded. We therefore conclude from the closed graph theorem that a solution does not exist for general initial data.

Of course, particular solutions do exist to our problems. There are, for example, the eigenmodal solutions given by

$$u(x,t) = e^{i\lambda t}v(x) \tag{3.8}$$

where for $f = 0$, the eigenfunction $v(x)$ satisfies

$$Cv + \rho\lambda^2 v = 0, \tag{3.9}$$

with

$$\lambda^2 = C(v,v)/(v,v),$$

and the initial data given by

$$u_0(x) = v(x), \quad u_1(x) = i\lambda u_0(x). \tag{3.10}$$

By uniqueness, these eigenmodal solutions cannot exist unless at least the initial data satisfies the compatibility relations $(3.10)_2$. Even if other kinds of special solutions exist it can reasonably be expected that they are also subject to a compatibility relation. The question now arises whether these relations are too restrictive to allow the formal instability computation discussed earlier to become meaningful or, whether for instance, in a buckling problem it is possible for a destabilising motion to be initiated at criticality. The existence result presented in the next Section provides a partial answer since it states that a solution exists for data dense in $L_2(Q)$.

4. Existence

We briefly describe in this section the existence of solutions to the initial boundary value (2.1),(2.8),(2.9) for indefinite elasticities satisfying the symmetry condition (2.2). The problem must be formulated in a weak sense and without loss we may suppose the initial data are zero but that the body force is non-zero. Let us consider the equations of motion in the form (2.4) and let us introduce the operator $A:\mathfrak{D}(A) \to L^2(Q)$, where A is given by

$$Av = \rho\ddot{v} - Cv, \qquad v \in \mathfrak{D}(A), \tag{4.1}$$

and $\mathfrak{D}(A)$ is defined explicitly in the sequel. It will then be seen that A is self-adjoint. We say that $u \in L^2(Q)$ is a weak solution to our problem if for $f \in L^2(Q)$ there holds

$$\int_Q uAv\,dx\,d\eta = \int_Q vf\,dx\,d\eta \qquad \forall\ v \in \mathfrak{D}(A). \tag{4.2}$$

We may use (4.2) to define the operator A^* adjoint to A and hence conclude that since A is self-adjoint, that in a distributional sense u satisfies the equation

$$Au = f. \tag{4.3}$$

We wish to discuss the existence of solutions to (4.3). The major difficulty stems from the indefiniteness of the elasticities which means that the equation is of no definite type (e.g., neither hyperbolic nor elliptic). We overcome this difficulty by projecting onto subspaces for which the equation is of definite type and in this way prove that A is closed. Then we have the following theorem:

Theorem The range of A^* is dense in $L^2(\Omega)$.

The proof of this theorem follows at once from standard results in adjoint operator theory combined with the uniqueness result mentioned in Section 3. The result itself is similar to others found in the literature of ill-posed problems.

Finally, we indicate briefly the decomposition used in the proof that A is closed. We suppose that $C:\mathfrak{D}(C) \to L^2(\Omega)$ is a self-adjoint densely defined operator and we let $P_\pm : L^2(\Omega) \to H_\pm$ be two orthogonal projection operators such that $L^2(\Omega) = H_+ \oplus H_-$. We suppose that $CP_\pm = P_\pm C$ and hence that $H_\pm$ are invariant subspaces with respect to C. We next use this invariance to ascribe sign-definite values to C in the following way:

$$-C(P_+v,v) \geq \kappa \| P_+v \|^2_{L^2(\Omega)}, \qquad \forall\, v \in \mathfrak{D}(C), \tag{4.4}$$

$$C(P_-v,v) \geq 0, \qquad \forall\, v \in \mathfrak{D}(C). \tag{4.5}$$

where κ is a positive constant.

We next introduce two operators $A_\pm$ defined by

$$A_\pm v = \rho\ddot{v} - Cv, \qquad v \in \mathfrak{D}(A_\pm),$$

where

$$\mathfrak{D}(A_+) = \{v \in W^{2,2}(\Omega) : v(t) \in H_+, v(T) = \dot{v}(T) = 0, v|_{\partial\Omega} = 0\},$$

$$\mathfrak{D}(A_-) = \{v \in C^0([0,T]:X) \cap C^1([0,T];L^2(\Omega)) : v(t) \in H_-,$$
$$v(T) = \dot{v}(T) = 0,\ v|_{\partial\Omega} = 0\},$$

and X is the subspace of $L_2(\Omega)$ equipped with the norm

$$\|v\|^2_X = \|v\|^2_{L^2(\Omega)} + C(v,v), \qquad v \in H_-.$$

By means of standard procedures in the theory of partial differential equations it may be shown that A_+, A_- are closed. Finally, we define the operator A by

$$\mathfrak{D}(A) = \{v \in L^2(Q) : P_+v \in \mathfrak{D}(A_+),\ P_-v \in \mathfrak{D}(A_-)\}$$

and

$$Av = A_+P_+v + A_-P_-v.$$

The closedness of A now follows from that of $A_\pm$.

Acknowledgement: The author expresses his thanks to Professor C.M. Dafermos for helpful discussions during the preparation of this paper.

References

1. Brun, L.: Méthodes énergétiques dans les systèmes evolutifs linéaires. Deuxiéme Partie: Théorèmes d'unicité. J.Mécanique 8 (1969) 167-192.

2. Fichera, G.: Existence theorems in elasticity. Handbuch der Physik. Vol VIa/2. Springer 1974, 347-389.

3. Knops, R.J.: Instability and the ill-posed Cauchy problem in elasticity. Mechanics of Solids. The Rodney Hill 60th Anniversary Volume. (eds.) H.G. Hopkins and M.J. Sewell. Oxford: Pergamon 1981, 357-382.

4. Knops, R.J., Payne, L.E.: Hölder stability and logarithmic convexity. Instability of Continuous Systems. IUTAM Symposium Herrenalb 1969. Berlin: Springer-Verlag 1971, 248-255.

5. Knops, R.J., Payne, L.E.: Growth estimates for solutions of evolutionary equations in Hilbert space with applications in elastodynamics. Arch. Rational Mech. Anal. 41 (1971) 363-398.

6. Levine, H.A.: An equipartition of energy theorem for weak solutions of evolutionary equations in Hilbert space. The Lagrange identity method. J. Differential Equations 24 (1977) 197-210.

7. Marsden, J.E., Hughes, T.J.R.: Topics in the mathematical foundations of elasticity. Nonlinear Analysis and Mechanics: Heriot-Watt Symposium II. Research Notes in Mathematics 27. London: Pitman

8. Payne, L.E.: Improperly Posed Problems in Partial Differential Equations. Regional Conference Series in Applied Mathematics. Philadelphia: SIAM 1975.

9. Wilkes, N.S.: On the non-existence of semi-groups for some equations of continuum mechanics. Proc. Roy. Soc. Edinburgh Sect. A86 (1980) 303-306.

Asymptotic Behaviour of Some Linear Evolution Equations

J. CARR

Department of Mathematics,
Heriot-Watt University,
Riccarton,
Currie.
EH14 4AS

1. Introduction

Let A be a linear operator on a Hilbert space H with discrete spectrum $\sigma(A)$ and suppose that A generates a continuous semigroup $\exp(At)$. In this paper we study the problem of finding sufficient conditions on A such that the spectrum of A determines the asymptotic behaviour of solutions of the equation

$$\dot{x} = Ax \tag{1.1}$$

The above problem arises naturally when studying the linearised stability of partial differential equations by means of an eigenvalue analysis. It also arises in the study of local dynamic bifurcation theory [3].

For special semigroups (e.g., analytic semigroups, unitary groups), the spectral mapping theorem

$$\sigma(\exp(At)) = \overline{\exp(t\sigma(A))} \tag{1.2}$$

is true. However, (1.2) does not hold in general [11].

The following initial boundary value problem provided motivation for this study:

$$u_{tt} + bu_{tx} + u_{xxxx} + au_{xx} = 0, \quad 0 < x < 1, \ t > 0, \tag{1.3}$$

$$u(0,t) = u_x(0,t) = u_{xx}(1,t) = u_{xxx}(1,t) = 0, \ t \geq 0.$$

where $a,b \geq 0$. The above equation represents the linearised equations of a system modelling the plane motions of a tube carrying fluid. Using ideas based on centre manifold theory, it is shown in [1] that for certain parameter values, a Hopf bifurcation occurs. The present study verifies the assumptions needed on the linear part of the equation in order to apply centre manifold theory.

2. Riesz Basis

A natural way in which to study solutions of (1.1) is to make an eigenfunction expansion. Let $\sigma(A) = \{\lambda_n\}$, $Ax_n = \lambda_n x_n$, $\|x_n\| = 1$ and suppose that the $\{x_n\}$ form a basis. Then the general solution to (1.1) can be written $\exp(At) = \sum a_n e^{\lambda_n t} \phi_n$. However, in order to make further progress we must also assume that any permutation of the $\{x_n\}$ is also a basis. This is equivalent to the $\{x_n\}$ forming a Riesz basis, that is $\{Sx_n\}$ is a complete orthonormal basis for some bounded invertible linear operator S on H. In this case $B = SAS^{-1}$ is normal and $\exp(At) = S\exp(Bt)S^{-1}$. Thus if we could prove that the eigenfunctions of A formed a Riesz Basis then the spectral mapping theorem would be true. Such results for certain differential operators are well known, see for example [5,6,7]. One of the methods for obtaining such results uses the following result:

Lemma [8]. Let $\{P_n\}$ be a sequence of projections and let $\{E_n\}$ be a complete self-adjoint sequence of projections. Furthermore, assume that

$$\dim P_o = \dim E_o < \infty,$$

$$\sum_{n=1}^{\infty} \|E_n(P_n - E_n)u\|^2 \leq C\|u\|^2 \quad \text{for all} \quad u \in H$$

where $0 \leq C < 1$. Then there exists a bounded invertible linear operator S such that $P_n = S^{-1}E_nS$.

To apply the above lemma, suppose that $A = A_1 + A_2$ with A_1 normal and A_2 small in some sense relative to A_1. Then if E_n and P_n are the projections onto the eigenspaces associated with A_1 and A_2 respectively, we have that

$$E_n - P_n = \frac{1}{2\pi i} \int_{\Gamma_n} [R_\lambda(A_1 + A_2) - R_\lambda(A_1)]\,d\lambda \tag{2.1}$$

where Γ_n is a contour surrounding the relevant eigenvalues. Under certain assumptions regarding the distribution of eigenvalues of A_1 and the size of A_2, (2.1) can be used to verify the hypothesis of the above Lemma.

In order to apply this to (1.3) we must first formulate it as an equation on Hilbert space. Let $Q_1 f = f''''$, $D(Q_1) = \{f \in H^4(0,1) : f(0) = f'(0) = f''(1) = f'''(1) = 0\}$, $Q_2 f = f''$, $Q_3 f = f'$, $H = D(Q_1^{\frac{1}{2}}) \times L^2(0,1)$. Then we write equation (1.3) as

$$\begin{bmatrix} \dot{u} \\ u_t \end{bmatrix} = (A_1 + aA_2 + bA_3) \begin{bmatrix} u \\ u_t \end{bmatrix}$$

where

$$A_1 = \begin{bmatrix} 0 & I \\ -Q_1 & 0 \end{bmatrix}, \quad A_2 = \begin{bmatrix} 0 & 0 \\ -Q_2 & 0 \end{bmatrix}, \quad A_3 = \begin{bmatrix} 0 & 0 \\ 0 & -Q_3 \end{bmatrix}.$$

It is easy to check that A_1 is skew-adjoint, $A_1 + bA_3$ is dissipative and that A_2 is bounded. Thus $A = A(a,b) = A_1 + aA_2 + bA_3$ generates a semigroup. If $b = 0$, it can be shown that the above theory is applicable [4,5]. The final result is $A(a,0)$ is similar to a skew-adjoint operator plus an operator with finite dimensional range which justifies an eigenvalue analysis of the initial value problem. However, because A_3 is in some sense too big, a similar result for $b > 0$ only holds for very small b. Thus we need a different method in order to discuss the case $b > 0$. In the remaining sections we describe some recent work of Harald Röh which shows that for $b \geq 0$, $A(a,b)$ is similar to a normal operator plus an operator with finite dimensional range [9]. In particular, it shows that solutions of the dissipative equation (1.2) exist for all $t \in \mathbb{R}$, that is $A(a,b)$ generates a group.

3. Results

Let A be a maximal dissipative operator with $\sigma(A) \subseteq \{z \in \mathbb{C} : \operatorname{Re} z < 0\}$.

Theorem Assume

(i) $\dim D(A)/\{x \in D(A) : \text{Re}\,(Ax,x) = 0\} < \infty$ (3.1)

(ii) $\sigma(A)$ consists of uniformly separated simple eigenvalues $\{\lambda_n\}$ with $|\text{Re}(\lambda_n)|$ bounded.

Then

(a) the eigenfunctions $\{x_n\}$ of A form a Riesz Basis in their closed linear hull.

(b) $\{x_n\}$ complete in H iff $|\text{Re}\, z_n| \, \| R(z_n, G) \|$ is bounded for some sequence $\{z_n\}$ in the resolvent set of A with $\text{Re}(z_n) \to -\infty$.

(c) $\sigma(\exp\,(At)) = \overline{\exp(t\sigma(A))}$

(the point zero should be added to the right hand side of (c) iff the $\{x_n\}$ are not complete).

Remark A similar result to the above holds when the λ_n are not simple [9].

The proof of the above theorem uses the functional calculus developed for bounded linear contractions [10]. Contraction operators are related to maximal dissipative operators by means of the Cayley transform

$$T = \phi(A) = (A + I)(A - 1)^{-1}$$

Moreover, any contraction T not having 1 as an eigenvalue is the Cayley transform of a maximal dissipative operator. These relations between A and T enable us to transform results concerning contractions into results about dissipative operators.

In this paper we shall only give an outline of the proof of the above theorem and refer to [9] for the details. We first introduce some notation. Let H^∞ denote the space of bounded analytic functions on $|z| < 1$. Then if $u(z) = \sum_{k=0}^{\infty} c_k z^k$ is in H^∞, $U(T)$ is a bounded linear operator, where $U(T)$ denotes the radial limit

$$U(T) = \lim_{r \to 1} \sum_{k=0}^{\infty} c_k r^k T^k .$$

Finally, $u \in H^\infty$ is an inner function if $|u(z)| = 1$ almost everywhere on the unit circle.

The fact that A has no eigenvalues on the imaginary axis implies that T is completely non unitary, that is there is no reducing subspace for which T is unitary. This together with (3.1) implies that $u(T) = 0$ for some non-zero inner function u. The minimal function m_T is the largest common inner divisor of all such u and it plays a similar role to that of the minimal polynomial of a matrix. The minimal function has the form

$$m_T(z) = e^{\gamma\phi(z)} \prod_{n=1}^{\infty} \frac{a_n - z}{1-\bar{a}_n z} \frac{\bar{a}_n}{|a_n|}$$

where ϕ is the linear fractional transform $\phi(z) = (z+1)(z-1)^{-1}$, $a_n = \phi(\lambda_n)$ and $|\alpha| = 1$. The constant γ is nonegative and in fact $\gamma = 0$ iff the eigenspaces of A span H.

It is straightforward to prove that the $\{x_n\}$ form a Riesz Basis in their closed linear hull iff

$$G_{\sigma_1 \cup \sigma_2} = G_{\sigma_1} \oplus G_{\sigma_2} \tag{3.2}$$

for any disjoint subset of the integers σ_1, σ_2 where

$$G_\sigma = \{x \in H : m_\sigma(T)x = 0\}$$

$$m_\sigma(z) = \Pi \left(\frac{a_n - z}{1 - \bar{a} z} \right) \frac{\bar{a}_n}{|a_n|}$$

where the product is taken over all n in σ.

Suppose there exists u,v in H^∞ such that

$$u(z)m_{\sigma_1}(z) + v(z)m_{\sigma_2}(z) = 1 \quad \text{on} \quad |z| < 1. \tag{3.3}$$

Then in exactly the same was as in matrix theory, this implies that (3.2) holds. In matrix theory, (3.3) is a statement about polynomials which follows from the Euclidean algorithm. In our case (3.3) follows from assumption (ii) which in terms of the eigenvalues of T is

$$\Pi \frac{|a_k - a_n|}{|1 - a_k \bar{a}_n|} \geqslant \delta > 0 \quad \text{for all} \quad k, \tag{3.4}$$

where the product is taken over all $n \neq k$. By a result of Carleson [2], (3.4) implies that the $\{a_n\}$ form an interpolating sequence, that is for each bounded sequence of complex numbers $\{b_n\}$, there exists $u \in H^\infty$ such that $u(a_n) = b_n$ for all n. By (3.4) $|m_{\sigma_1}(a_n)|^{-1} \leqslant \delta^{-1}$ for all n in σ_2. Hence there exists $u \in H^\infty$ such that $u(a_n) m_{\sigma_1}(a_n) = 1$ for all n in σ_2 and

$$v(z) = [m_{\sigma_2}(z)]^{-1}[1 - u(z) m_{\sigma_1}(z)] \in H^\infty$$

so that (3.3) is satisfied.

4. Applications

We first apply the results of the previous section to equation (1.3) with $a = 0$. A simple computation shows that if $u(x,t)$ is a solution of (1.3) then the energy satisfies

$$\frac{d}{dt}\int_0^1 [\,|u_t|^2 + |u_{xx}|^2]\,dx = -\frac{b}{2}|u_t(1,t)|^2.$$

This verifies condition (i) of Theorem 1 for $A(0,b)$. To check condition (ii) we have to study the eigenvalues of $A(0,b)$. Clearly $\lambda \in \mathbb{C}$ is an eigenvalue of $A(0,b)$ iff there exists nontrivial f satisfying

$$f''''(X) + \lambda b f'(X) + \lambda^2 f(X) = 0 \tag{4.1}$$

$$f(0) = f'(0) = f''(1) = f'''(1) = 0 \tag{4.2}$$

By studying the asymptotic behaviour of solutions of (4.1) for large λ and using this information in (4.2), it can be shown that the eigenvalues λ_n enjoy the asymptotic representation

$$\lambda_n = \pm i\Pi^2(n - \tfrac{1}{2})^2 - b \pm \frac{b^2}{16} i + O(n^{-1})$$

and are simple for n large enough. Also simple estimates on the resolvent shows that $A(0,b)$ satisfies the completeness hypothesis in Theorem 1. Thus $A(0,b)$ is similar to a normal operator (modulo possibly a finite dimensional part). By applying perturbation theory it follows that $A(a,b)$ enjoys the same status.

References

1. Bajaj, A.K.; Sethna, P.R.: Lundgren, T.S.; Hopf bifurcation phenomena in tubes carrying a fluid. SIAM J. Appl. Math. 39 (1980) 213-230.

2. Carleson, L.: Interpolation by bounded analytic functions and the corona problem. Ann. of Math. (2) (1962) 547-559.

3. Carr, J.: Applications of centre manifold theory. Applied Math. Sciences, Vol. 35, Springer Verlag. Berlin, New York, 1981

4. Carr, J; Malhardeen, M.Z.M.: Beck's Problem. SIAM J. Appl. Math. 37 (1979) 261-262.

5. Carr, J.; Malhardeen, M.Z.M.: Stability of nonconservative linear systems. Springer Lecture Notes in Math. Vol. 799, Springer Verlag. Berlin, New York 1980.

6. Clark, C.: On relatively bounded perturbations of ordinary differential operators. Pacific J. Math. 25 (1968) 59-70.

7. Dunford, N.; Schwartz, J.T.: Linear Operators, Part III, Wiley-Interscience, New York 1971.

8. Kato, T.: Similarity for sequences of projections. Bull. Amer. Math. Soc. (1967) 904-905.

9. Röh, H.: Dissipative operators with finite dimensional damping. Submitted for publication.

10. Sz.-Nagy, B.; Foias, C.: Harmonic Analysis of Operators on Hilbert Space. North-Holland 1970.

11. Zabczyk, J.: A note on C_0-semigroups. Bull. Acad. Polon. Sci. Ser. Sci. Tech. 23 (1975) 895-899.

Numerical Nonlinear Analysis of Structural Instability

Z. WASZCZYSZYN

Institute of Structural Mechanics
Technical University of Cracow, Poland

Summary

The main problems concerning the application of the finite element method to the nonlinear analysis of elastic and elastic-plastic structures are presented. After recapitulation of basic definitions and relations, equilibrium and stability of conservative systems is considered. The computation of equilibrium paths and critical points on the critical zone is shortly discussed. The problems of the elastic-plastic stability analysis is outlined.

Introductory Remarks

In recent years a great deal of attention has been devoted to the formulation of foundations of the nonlinear theory of stability. Unified approach based on the energetic considerations and the perturbation technique enable us to investigate a number of phenomena related to structural instability of elastic structures [1]. A big progress can be also noticed in the stability analysis of elastic-plastic structures [2]. Analytical approaches have proved to be useful rather for simple structures. Much more efficient are so called 'computer methods' which explore either finite differences or finite elements and standard procedures of the linear algebra.

Since the early 70-ies the finite element method (FEM) has been applied also to the nonlinear problems of stability [3] . It has been found out that the standard FEM algorithms have to be modified or supplemented [4].

The aim of the paper is to point out the new possibilities of the FEM application to the analysis of equilibrium and critical states of elastic structures under conservative loads. The main attention is focused on the appropriate formulation to consider also multiple-parameter loads, on computation of equilibrium paths (surfaces) and critical points (zones) in the load-configuration space.

The incremental FEM equilibrium equations related to the total Lagrangian formulation are used as the starting point. FE approximation problems of the displacement field and details of algorithms are not discussed in the paper.

Basic Definitions and Relations

The FE approximation of the displacement field $\underline{u}=\underline{u}(\underline{x})$ is introduced

$$\underline{u} = \underline{N} \cdot \underline{q} \qquad \text{and/or} \qquad \Delta\underline{u} = \underline{N} \cdot \Delta\underline{q} \;, \tag{1}$$

where $\underline{N}(\underline{x})$ is the shape function and $\underline{q}$ is the one-column matrix (vector) of generalized displacement of the FE nodes:

$$\underline{q} = \{q_i\} \equiv \{q_1,\ldots,q_N\} \in R^N \;. \tag{2}$$

In what follows the increments $\Delta\underline{u}$ and $\Delta\underline{q}$ are used.

External surface and point loads, $\underline{p}(\underline{u},\underline{\lambda})$ and $\underline{G}(q,\underline{\lambda})$ correspondingly, are reduced to the equivalent node forces on the basis of FEM procedure [5]:

$$\left.\begin{matrix} \underline{p}(\underline{u},\underline{\lambda}) \\ \underline{G}(\underline{q},\underline{\lambda}) \end{matrix}\right\} \xrightarrow{\text{FEM}} \underline{P}(\underline{q},\underline{\lambda}) \in R^N \;, \tag{3}$$

where $\underline{\lambda}$ is the vector of independent load parameters

$$\underline{\lambda} = \{\lambda^k\} \equiv \{\lambda^1,\ldots,\lambda^M\} \in R^M \;. \tag{4}$$

The position vector $\tilde{\underline{q}}$ in the load-configuration space R^{N+M} is introduced [6,4]:

$$\tilde{\underline{q}} = \{\underline{q},\underline{\lambda}\} \equiv \{q_\alpha\} \in R^{N+M} \;. \tag{5}$$

Equilibrium and Stability of Conservative Systems

Potential energy of the system is summed over all finite elements and nodes :

$$V = U + W \equiv \sum_{e=1}^{m} U_{(e)}\left(\underline{q}_{(e)}, \underline{a}_{(e)}\right) - \underline{q}^T \underline{P} \;, \tag{6}$$

where $U_{(e)}$ is the strain energy of a finite element. Internal imperfections of all finite elements are collected as components of the vector:

$$\underline{\ni} = \{ \ni_s \} \in R^S . \qquad (7)$$

The system is called Conservative System (CS) if stationary condition of the potential energy is equivalent to the principle of virtual work:

$$\begin{array}{ccc} \delta V = 0 & \longrightarrow & \delta U = - \delta W \\ & & \updownarrow \qquad \updownarrow \\ & & \delta L_{int} = \delta L_{ext} . \end{array} \qquad (8)$$

For independent displacement variations $\delta \underline{q}$ the relations (8) give the set of equilibrium equations:

$$V_i (\underline{\tilde{q}}; \underline{\ni}) \equiv U_i (\underline{q}; \underline{\ni}) - P_i (\underline{q}, \underline{\lambda}) = 0 \qquad (9)$$

for $i = 1, \dots, N$. Expanding the functions V_i into the Taylor series in the vicinity of $\tilde{q}$ and conserving linear terms with respect to the increments $\Delta \underline{\tilde{q}}$, the following set of equations is obtained

$$(U_{ij} - P_{ij}) \Delta q_j = P_i^k \Delta \lambda^k + (P_i - U_i) . \qquad (10)$$

In Eg.(8) the components of the following matrices are used:

$$U_{ij} - P_{ij} \equiv K_{ij} = \frac{\partial^2 V_i}{\partial q_j^2} \bigg|_{\underline{\tilde{q}}}, \qquad P_i^k = \frac{\partial P_i}{\partial \lambda^k} \bigg|_{\underline{\tilde{q}}} , \quad P_i - U_i \equiv R_i , \qquad (11)$$

where $i, j = 1, \dots, M$ and $k = 1, \dots, M$. The linearized incremental eqation (10) is usually written in the following matrix form:

$$\underline{K} \Delta \underline{q} = \underline{P}' \Delta \underline{\lambda} + \underline{R} , \qquad (12)$$

where the stiffness tangent matrix $\underline{K}$, the reference load matrix $\underline{P}'$ and the unbalanced force vector $R_i = P_i - U_i$ depend on the formulation and especially on the reference configuration [7]. A short discussion of these matrices on the basis of the total Lagrangian formulation has been given in [8].

If the system is in equilibrium state, then $\underline{R} = \underline{0}$ and the Cramer formula leads to the following set of equations:

$$D \Delta q_i = d_i^k \Delta \lambda^k , \qquad (13)$$

where the abbreviate notation is used:

$$D = \det \left| K_{ij} \right| , \qquad d_i^k = \frac{\partial D}{\partial K_{ij}} P_j^k . \qquad (14)$$

The necessary condition of the critical state implies the following relation:

$$D = 0 \implies d_i^k \Delta \lambda^k = 0 \qquad (15)$$

which can be adopted as the basis of a classification of critical points on the critical zone.

The above mentioned approach has been suggested by E.Riks [9] for one-parameter load and generalized on the case of multiple-parameter loads in [10] accordingly to the Husseyin classification of critical point [6] .

Computation of Equilibrium Paths and Critical Points

Various methods for calculation of equilibrium paths were reviewed in [4]. The main problems are the convergence of iterative procedures in the vicinity of critical points and starting with the evaluation of postcritical paths. If an independent parameter τ is selected from the components of the vector $\tilde{\underline{q}} = \{\underline{q}, \lambda\} \in R^{N+1}$ as the load parameter $\tau \equiv \lambda$ or the displacement $\tau \equiv q_j$ the convergence falls down in the neighbourhood of points L and C, even in the case of a smooth path - ef.Fig.1.

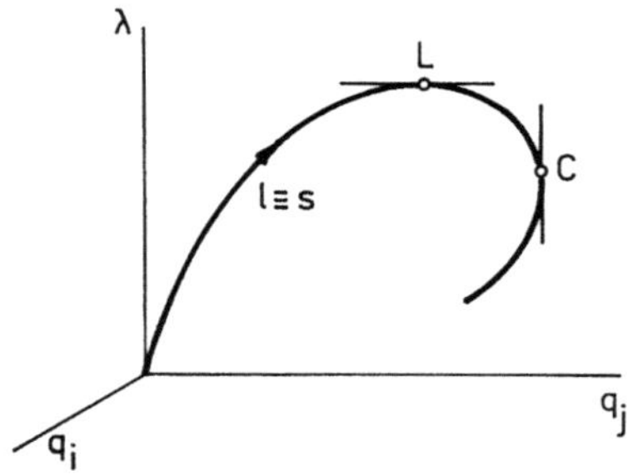

Fig.1 Equilibrium path in load configuration space.

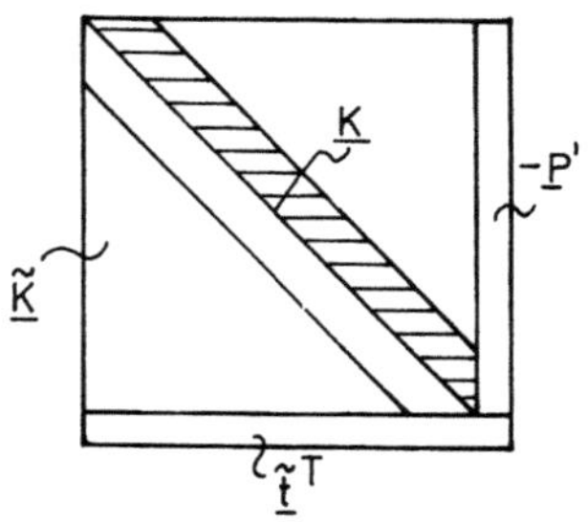

Fig.2 Structure of matrix $\tilde{\underline{K}}$.

This disadvantage is eliminated if all components of the vector $\tilde{q}$ are treated equivalently. In such an approach the enlarged set of incremental equations has to be considered:

$$K_{ij}\Delta q_j - P_i'\Delta\lambda = R_i \,,$$
$$t_j\Delta q_j + t_{N+1}\Delta\lambda = \Delta\tau \,, \tag{16}$$

where $i,j = 1,\ldots, N$. Eqs (16) are written down for the one - parameter load $P_i(\lambda)$ and $P_i' = \partial P_i/\partial\lambda$.

The last equation $(16)_2$ relates the increment $\Delta\tilde{q}$ to the control vector $\tilde{t}$ according to the choice of the control parameter τ . Three possibilities are worth attention:

$$\begin{aligned} &\text{i)}\ \tau\equiv\lambda && \tilde{t} = \{0,\ldots, 0, 1\} \,, \\ &\text{ii)}\ \tau\equiv q_j && \tilde{t} = \{0,\ldots, 0, 1, 0,\ldots, 0\} \,, \\ &\text{iii)}\ \tau\equiv s && \tilde{t} \approx \Delta\tilde{q} / |\Delta\tilde{q}| \,. \end{aligned} \tag{17}$$

In the first two cases the control vector is paralled to an appropriate coordinate axis in R^{N+1} . The third case corresponds to so called arc parameter s /cf.Fig.1/ introduced in [9], and the control vector $\tilde{t}$ is then close to the tangent vector of the equilibrium path.

In what follows, Eqs (16) will be called the enlarged set of equations. They can be written in the matrix form

$$\tilde{K}\,\Delta\tilde{q} = \tilde{R} \,. \tag{18}$$

Structure of the enlarged matrix $\tilde{K}$ is shown in Fig.2. The halfband of the stiffness matrix K can be stored for computation if $K^T = K$. In general the matrix $\tilde{K}$ is non-symmetric but this disadvantage can be easily overcome in algorithms [4] . The enlarged residual vector $\tilde{R}$ has the following components

$$\tilde{R}^{(o)} = \{0, \Delta\tau\}, \quad \tilde{R}^{(r)} = \{R^{(r)}, 0\} \quad \text{for } r\geq 1 \,, \tag{19}$$

where r denotes succesive iteration steps in the Newton-Raphson method — cf. Fig.3 . During computation the enlarged matrix $\underline{\tilde{K}}$ is to be modified, i.e. $\underline{\tilde{K}}^{(r)} = \underline{\tilde{K}}(\underline{q}^{(r)})$. If the modification is performed only at the beggining of the iteration, i.e. $\underline{K}^{(r)} \equiv \underline{K}^{(1)}$, then the modified NR method takes place. More detailed description of appropriate algorithms has been done in [4].

The most important property of the enlarged matrix $\underline{\tilde{K}}$ is its nonsingularity at limit critical points. In the critical state determinants of the matrices equal

$$\det|\underline{K}| = 0 \quad \text{and} \quad \det|\underline{\tilde{K}}| \begin{cases} \neq 0 \text{ for L} \\ = 0 \text{ for B.} \end{cases} \tag{20}$$

Relation (20) is explored in the algorithms to distinguisch limit and bifurcation points, L and B correspondingly.
In subsequent points m of the equilibrium path values of a scalar function $S(\Upsilon)$ can be computed. If zero points of this function

$$S(\underline{\tilde{q}}, \Upsilon) = 0 \tag{21}$$

correspond to critical points $\underline{\tilde{q}}^{c} = \underline{q}(\Upsilon_c)$, the function $S(\Upsilon)$ can be adopted to define the critical state.

As the function S the stability determinant $D = \det|\underline{K}|$ or the basic eigenvalue $\omega = \min(\omega_i)$ of the stiffness matrix $\underline{K}$ — cf. [9] , can be used. Other functions suitable for numerical calculations are discussed in [4] .

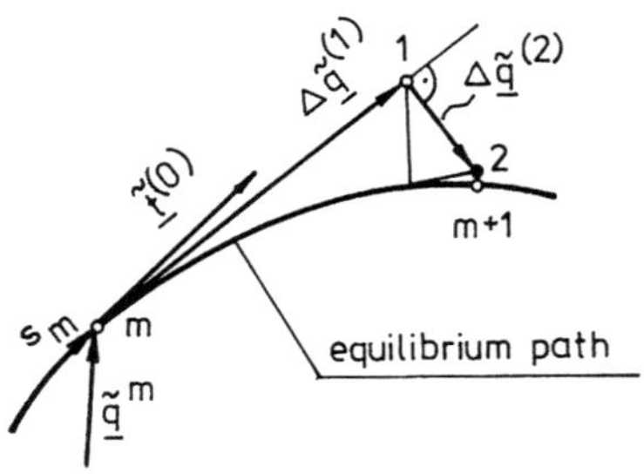

Fig.3 Computation of subsequenent points on equilibrium path

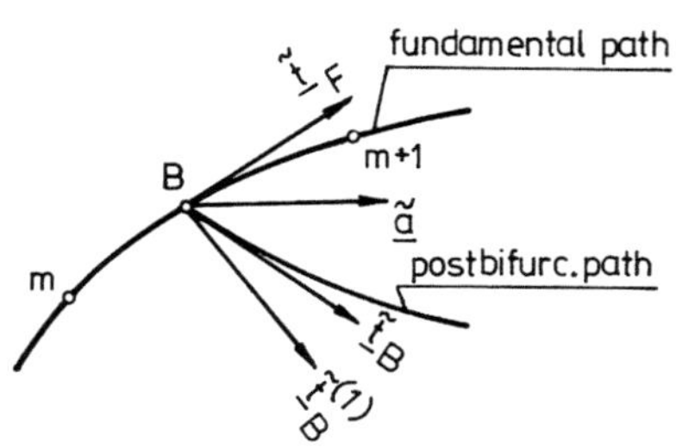

Fig.4 Tracing of post-bifurcation path

After the bifurcation point B is found on the fundamental path of the tangent vector $\underline{\tilde{t}}_F$, the postbifurcation path can be traced. In [4, 11] iterative procedures have been suggested to evaluate the first approximation of the vector $\tilde{\underline{t}}_B$ tangent to the postbifurcation path in the singular critical point (Fig.4). The idea of the approach lies in evaluation of the vector $\tilde{\underline{t}}_B^{(1)}$ as a linear combination of the vector $\tilde{\underline{t}}_F$ and the basic eigenvector $\underline{a}(\omega)$ of the stiffness matrix $\underline{K}(\tilde{\underline{q}}^B)$.

Application of the above mentioned concepts have been presented in [11] where local and overall buckling of plane elastic trusses has been considered.

Computation of Critical Zone and Stability Boundary

Having applied Eqs (16) the critical zone can be evaluated on the equilibrium surface as a set of critical points for different fixed values of load parameters λ^l changing only one parameter λ^m for $l = 1,\dots,m-1,m+1,\dots,M$ where $m=1,\dots,M$ subsequently.

Such an approach is rather unacceptable for $M > 2$. Certain progress depends on computation of contour lines on the critical zone. The algorithm is based on two variable parameters, e.g. λ^1 and λ^2, fixing other load parameters $\lambda^l = \text{const.}$ for $l = 3,\dots,M$ and solving the following set of equations in the space R^{N+2} - cf. [12]:

$$
\begin{aligned}
&K_{ij}\Delta q_j - P_i^1\,\Delta\lambda^1 - P_i^2\,\Delta\lambda^2 = R_i, \\
&t_j\Delta q_j + t_{N+1}\Delta\lambda^1 + t_{N+2}\Delta\lambda^2 = \Delta\tau, \qquad (22)\\
&D_j\Delta q_j = -D(\underline{q}\,;\underline{\partial}).
\end{aligned}
$$

The last equation is related to the zero value of the stability determinant D. Coefficients of these equations D_j are the partial derivatives of D:

$$
D_j \equiv \frac{\partial D}{\partial q_j} = \frac{\partial D}{\partial K_{rs}}\,K_{rsj}\,. \qquad (23)
$$

As it is shown in (23), derivatives of the stiffness matrix $\underline{K}$ have to be computed. Analitically it is possible for a small number N of degrees of freedom or for systems whose matrix $\underline{K}$ is of narrow band.

Eqs (22) can be easily used for computation of general and singular critical points. As a result of computations the position vector is obtained:

$$
\tilde{\underline{q}}(\tau) = \{\underline{q}(\tau), \underline{\lambda}(\tau)\}\,, \qquad (24)
$$

where Υ is the parameter of a contoure line on the critical zone.
The subset

$$\underline{\lambda}(\Upsilon) = \{\lambda^1(\Upsilon), \quad \lambda^2(\Upsilon) ; \quad \lambda^1 = \text{const.}\} \qquad (25)$$

defines a contour line of the stability boundary as a result of projection of $\tilde{\underline{q}}(\Upsilon)$ on the load parameter plane.

Special critical points if they exist for the ideal structures disappear if initial imperfections $\ni_s$ are taken into account. This extends the application of Eqs (22) .

Simple example of evaluation of the critical zone for a two-member truss of four degrees of freedom has discussed in [10] .

Instability of Elastic-Plastic Structures

Elastic-plastic structures are in fact nonconservative because of dependence of the displacement field on the history of the loading process. This is related peticularly to multiple - parameter loads. Nevertheless, the static criterion of instability can be used especially in cases of absence or small regions of local unloadings [2, 12] .

The algorithms which have been worked out for elastic material have to be modified to take into account new phenomena related to elastic - plastic material. The most important modications correspond to the following problems: i/ Dependence of the 'linear' part of the tangent matrix on the current stresses $\underline{K}_o(\underline{\sigma})$, ii/ Introduction of a current refference configuration to consider influence of the loading process, iii/ Taking into account of local unloadings and crossing of the yield surface.

Numerical aspects of differences between elastic and elastic-plastic systems can be outlined as enlargement of the needed computer memory, more arithmetic operations and worse convergence of iterative procedures.

Some basic problems related to application of algorithms on the basis of presented concepts modified to the analysis of elastic-plastic arches and cylindrical panels have been discussed in [13] .

References

1. Koiter, W.T.: Current Trends in the Theory of Buckling. Symp. IUTAM Buckling of Structures, Cambridge/USA. Springer-Verlag 1976, 1-16.

2. Tvergaard V.: Buckling Behaviour of Plate and Shell Structures. Theoret. and Appl. Mech., Ed. W.T.Koiter. Noth-Holland Publ. Co. 1976, 233-246.

3. Gallagher R.H.: Finite Element Representations for Thin Shell Instability Analysis. Symp. IUTAM Buckling of Structures, Cambridge/USA. Springer-Verlag 1976, 40-51.

4. Waszczyszyn Z.: Numerical Problems of Nonlinear Stability Analysis of Elastic Structures (in Polish). Modern Methods of the Structural Stability Analysis. Wrocław: Ossolineum 1981.

5. Zienkiewicz O.C.: The Finite Element Method. Ed.3. Mc Graw-Hill 1978.

6. Huseyin K.: Nonlinear Theory of Elastic Stability. Leyden: Noorhoff Intern. Publ. 1975.

7. Bathe K.J.; Ozdemir H: Elastic-Plastic Large Deformation, Static and Dynamic Analysis. Comp. and Struct. 6 (1976) 81-92.

8. Waszczyszyn Z.: Application of the Finite Element Method to the Structural Stability Analysis (in Polish). Proc. 5-th Polish Conference of Comp. Methods in Struct. Mechanics. Karpacz 1981.

9. Riks E.: An Incremental Approach to the Solution of Snapping and Buckling Problems. Int. J.Solids /Structures 15 (1979) 529-551.

10. Waszczyszyn Z.; Pytel E.; Nguyen-Cao-Duong: Numerical Analysis of Nonlinear Problems of Instability of Elastic Trusses under Multiple-Parameter Loads (in Polish). Eng. Transactions (in press).

11. Cichoń Cz.: Stability Analysis of Elastic Trusses (in Polish). Eng Transactions (to be published).

12. Yokoo Y.; Nakamura T.; Uetani K.: The Incremental Perturbation Method for Large Displacement Analysis of Elastic-Plastic Structures. Int. J.Num. Met. Eng. 10 (1976) 503-525.

13. Nguyen-Cao-Duong, Waszczyszyn Z.: Stability of Elastic-Plastic Arches and Cylindrical Panels (in Polish), Eng. Transctions (to be published).

Stochastic Buckling of Shells with General Imperfections

I. ELISHAKOFF* and J. ARBOCZ**

*Technion - Israel Institute of Technology,
Haifa, Israel

**Delft University of Technology,
The Netherlands

Summary

The paper is concerned with the effect of random axisymmetric and asymmetric imperfections on the buckling of circular cylindrical shells under axial compression. The initial imperfections are considered as random functions of the axial and the circumferential coordinates. In contrast to earlier works, the probabilistic properties are not assumed but calculated from experimental measurements of the shell profiles. The Fourier coefficients needed for application of the Monte Carlo Method are simulated by a special numerical procedure, which "creates" a large number of shells. The reliability function representing the probability of the buckling stress exceeding the specified stress is calculated. The reliability function permits evaluation of the allowable stress for the whole ensemble of shells produced by a given manufacturing process, defined as the stress level for which the desired reliability is attained.

Introduction

Application of the available theories of imperfection sensitivity to buckling load calculations necessitates prior knowledge of the initial imperfections which are present in the structure under consideration. These, however, vary from case to case, as their shape and magnitude derive from the specific fabrication process, itself subject - by its very nature - to a large number of random variables.

In the present work we consider axisymmtric as well as general nonsymmetric initial imperfections as random functions and solve the resulting stochastic stability problem by the Monte Carlo Method. This treatment may be regarded as the first step in a combined theoretical, experimental and numerical effort with a view to a probabilistic design code for axially compressed cylindrical shells with general nonsymmetric random initial imperfections. Unlike earlier investigations, the statistical information needed in the analysis is not assumed but drawn from available measurement data.

Case Capable of Exact Solution

As an illustration of the stochastic analysis of initial-imperfection sensitivity, let us first consider an example which can be solved in "closed" form - specifically, a long isotropic shell with an axisymmetric imperfection in the shape of the classical axisymmetric buckling mode. The nondimensional critical load versus imperfecton amplitude relationship was first derived by Koiter in 1945 [1] in his general theory. Here, we treat the initial imperfection amplitude $\overline{X}$ as a random variable with given probability density function as shown in Fig. 1. The buckling load is then also a random variable. What is sought is the reliability of the shell, for the present purpose as the probability of the random buckling load Λ exceeding the specified load λ:

$$R(\lambda) = \text{Prob}\ (\Lambda > \lambda) \qquad (1)$$

According to Fig. 1, Λ exceeds λ if the initial imperfection $\overline{X}$ lies between the values, $\overline{\xi}_1$ and $-\overline{\xi}_1$. Thus reliability is represented by the shaded area and clearly depends on the shape of the probability density curve, namely:

$$R\ (\lambda) = \int_{-\overline{\xi}_1}^{\overline{\xi}_1} f_{\overline{X}}\ (\overline{\xi})\ d\overline{\xi} \qquad (2)$$

The reliability function (shown in Fig. 2) may be used directly for design purposes. Suppose we stipulate that the reliability must exceed some specified high level, say 0.95. Now, if we subject the structure to loads less than or equal to the allowable load, then the reliability will be greater than or equal to 0.95, and the probability of failure smaller than or equal to 0.05.

It is also possible to determine the mean buckling load, which can be shown to equal the area under the reliability curve. However, it appears that the "allowable load" associated with the specified "required reliability" is a more meaningful design criterion than the mean buckling load.

Probabilistic Properties of Initial Imperfection

We have just treated the simple case of a structure represented by a single random variable. The question arises: "How does one describe the random initial imperfections in general?" If we have N realizations of the initial imperfection functions

$$\overline{W}(x)^{(m)} = \overline{w}(x)^{(m)}/t = \sum_i A_i^{(m)} \phi_i(x) \text{ , } (m = 1,2,\ldots,N) \tag{3}$$

we can desire two deterministic functions: the mean imperfection function $m_{\overline{W}}(x) = E[\overline{W}(x)]$ and the autocorrelation function

$$R_{\overline{W}}(x_1, x_2) = E[\overline{W}(x_1) - m_{\overline{W}}(x_1)][\overline{W}(x_2) - m_W(x_2)] \tag{5}$$

Without loss of generality, the mean imperfection function may be set at zero. The autocorrelation function yields information on the probabilistic relations between two sections of the random function, so that if the mean fuction is zero, at $x_1 = x_2$ the autocorrelation function equals the variance and suffices for characterizing the manufacturing process.

If $\phi_j(x)$ is a complete set on (o,L), where L is the shell length, the autocorrelation function may be represented as a double Fourier series [2]

$$R_{\overline{W}}(x_1, x_2) = \sum\sum \sigma_{jk}\, \phi_j(x_1)\, \phi_k(x_2) \text{ , } \mu_j^2 = \int_0^L \phi_j^2(x)\, dx \tag{6}$$

$$\sigma_{jk} = \mu_j^2 \mu_k^{-2} \int_0^L \int_0^L R_{\overline{W}}(x_1, x_2)\, \phi_j(x_1)\, \phi_k(x_2)\, dx_1\, dx_2 \tag{7}$$

where σ_{jk} forms the variance-covariance matrix of the initial imperfections. It can be shown [3] that when $L \to \infty$, i.e. the shell becomes infinite - then, instead of the covariance the generalized spectral density is obtained as the double Fourier transform of the autocorrelation function $R_W(x_1,x_2)$. When the initial imperfections are homogeneous in the axial direction, i.e. depend on $\xi = x_2 - x_1$, the conventional spectral density is obtained. This latter concept was applied by Amazigo

[4] to an infinite shell with random axisymmetric imperfections and yielded expressions for the main buckling load. Our own objective is the reliability function.

The Monte Carlo Method

Since a closed-form solution of the reliability function appears unfeasible for a finite shell, we shall resort to the Monte Carlo Method. Application of this method comprises three steps: first, we simulate the random initial imperfections, i.e. "create" a large number of shells with generally differing initial imperfection profiles; next, we determine the buckling load for each shell on the basis of accumulated deterministic data; finally, we derive the sought reliability function.

How can imperfect shells be "created"? Briefly, the procedure is as follows: Having N realizations of the initial imperfection functions

$$\bar{W}(x)^{(m)} = \bar{w}(x)^{(m)}/t = \sum_i A_i^{(m)} \phi_i(x) \ , \ (m = 1,2,\ldots,N) \tag{7}$$

we calculate, with the aid of "ensemble averages", first the estimated mean of the Fourier coefficients $A_i^{(m)}$

$$\tilde{A}_i^{(e)} = \sum_{m=1}^{N} A_i^{(m)}/N \tag{8}$$

and then the estimated variance-covariance matrix

$$\sigma_{jk}^{(e)} = \sum_{m=1}^{N} \sum_{n=1}^{N} [A_j^{(m)} - \tilde{A}_j^{(e)}] [A_k^{(n)} - \tilde{A}_k^{(e)}] (N-1)^{-1} \tag{9}$$

This is a non-negative symmetric matrix, and thus can be decomposed as a product of lower and upper triangular matrices

$$[\sigma_{jk}^{(e)}] = [C] \ [C]^T \tag{10}$$

The "created" initial imperfection vector is then obtainable [2,3]

$$A = [C] \{b\} + \{\tilde{A}^{(e)}\} \quad (11)$$

where $\{\tilde{A}^{(e)}\}$ is the estimated mean vector while the b's are random numbers generated by the computer and representing the realizations of a normal random variable with zero mean and unit variance. Taking, for example, 1000 different b's we obtain 1000 different A's, that is, different shells with the A's as Fourier coefficients of the initial imperfections. With these "created" initial imperfections we carry out repeated deterministic buckling analysis, which yields the buckling-load histogram of the group of shells under consideration. Defining the reliability R (λ) as the probability of the buckling load exceeding a prescribed value (see also Eq. (1)), we can calculate $R(\lambda)$ from the histogram through the frequency (i.e. fraction-of-the-ensemble) interpretation.

What is the accuracy of this method? For an infinite cylindrical shell with cosine-shaped axisymmetric iperfections, using Koiter's Eq. (5.2) from Ref. [5], as a nonlinear transfer funtion, the results of Monte Carlo simulation for 1000 shells practically coincide with analytical results obtained in terms of error functions [6] (Fig. 3).

Reliability of Shells with Axisymmetric Imperfections

Before proceeding further, we wish to reemphasize the differences between the approaches used by earlier investigators and the present one. The former worked with infinite shells and assumed that the random process is stationary and the ergodicity assumption valid. We work with finite shells, employ nonstationary processes and replace the ergodicity assumption with ensemble averages. The most important difference, however, is that while all other investigators to date (with the single exception of Makaroff [7]) assumed arbitrary autocorrelation functions for the initial imperfections, we make use of experimental surveys data stored in the Imperfection Data Bank of the Delft University of Technology [8] and directly derive the variance-covariance matrix of the imperfections, to serve as input in the Monte Carlo Method.

The proposed procedure was applied to a group of machined seamless shells (B-shells). The "simulated" autocorrelation function from 100 "created" shells practically coincides with its "measured" counterpart. The variance shown in Fig. 4 varies with the axial coordinate, so that the measured initial imperfections are neither stationary nor ergodic. Next 100 realizations of the B-shells were "created" by the simulation procedure described earlier, and used in constructing the histogram of the nondimensional buckling loads. Finally, the reliability function was determined from the histogram. The mean buckling load due to the axisymmetric imperfections alone came out at 0.816, significantly higher that its experimental counterpart, which equals 0.592; this indicates that for correct prediction of the reliability function the asymmetric imperfections have to be included as well.

Reliability of Shells with Asymmetric Imperfections

In representing the general imperfections, we used the following imperfections modes: $A_{2,0}$, $A_{4,0}$, $A_{26,0}$, $C_{1,2}$, $C_{1,6}$, $C_{1,8}$, $C_{2,3}$, $C_{2,11}$, $C_{24,3}$, $C_{24,11}$, $C_{25,2}$, $C_{25,6}$, $C_{25,8}$, $C_{25,10}$.
Here $A_{2,0}$ stands for an axisymmetric cosine Fourier coefficient, with 2 half-waves in the axial direction and (naturally) none in the circumferential directions, and $C_{1,10}$ for an asymmetric Fourier coefficient with a single sine half-wave in the axial direction and 10 full waves in the circumferential direction. As pointed by Arbocz and Babcock [9] the measured data becomes less reliable for the higher harmonics in view of the limited number of meshpoints used (41 x 49). In these circumstances, the simulation procedure was confined to the 8 lower-order modes, while the Donnell-Imbert imperfection model was used [9] for the 7 higher-order ones. To facilitate computation of the variance-covariance matrices, the following new notation was adopted for the initial imperfections:

$$\bar{W}(x,y) = \frac{\bar{w}(x,y)}{t} = \sum_{i=1}^{2} A_{io} \cos i_i \frac{\pi x}{L} + \sum_{r=1}^{6} C_r \sin k_r \frac{\pi x}{L} \cos \ell_r \frac{y}{R} + D_r \sin k_r \frac{\pi x}{L} \sin \ell_r \frac{y}{R} \tag{12}$$

The approximate autocorrelation function then reads:

$$R_{\bar{W}}(x_1, y_1; x_2, y_2)$$

$$\approx \sum_{i,j=1}^{2}\sum K_{A_{io}A_{jo}} \cos(i\pi x_1/L)\cos(j\pi x_2/L) \tag{13}$$

$$+ \sum_{r,s=1}^{6}\sum K_{C_rC_s} \sin(k_r\pi x_1/L)\sin(k_r\pi x_1/L)\cos(\ell_r y_1/R)\cos(\ell_s y_2/R)$$

$$+ \sum_{r,s=1}^{6}\sum K_{D_rD_s} \sin(k_r\pi x_1/L)\sin(k_s\pi x_2/L)\sin(\ell_r y_1/R)\sin(\ell_s y_2/R)$$

with the variance-covariance $K_{A_{io}C_s}$, $K_{A_{io}D_s}$ and $K_{C_rD_s}$ disregarded. Significantly (as can be seen from Fig. 5) in this general nonsymmetric case, the approximate total variance is not constant, either with the same conclusion as above regarding the initial imperfections. Also, as can be seen from Fig. 6, the approximate total variance for 500 simulated B-shells is almost undistinguishable from its "measured" counterpart for the group of B-shells shown in Fig. 7.

The deterministic analyses were performed with the aid of the so-called Multi Mode Analysis developed by Arbocz and Babcock [9], yielding the reliability function shown in Fig. 7. Inclusion of the asymmetric imperfections reduced the mean buckling to 0.739, still considerably higher than the experimental value of 0.592 or the deterministic value of 0.663 obtained on the exclusive basis of the Donnell-Imbert model [9]. Further refinement of the nonsymmetric random imperfection model can be expected to yield still lower results.

Conclusion

The results obtained may be summed up as follows:

1. It was demonstrated that the Monte Carlo Method can be used successfully to yield reliability functions for shells with axisymmetric as well as asymmetric imperfections.
2. It was found that for finite shells nonstationary statistics must be used (i.e., ergodicity is ruled out).
3. With the aid of the simulation procedure developed by Elishakoff [2], measured initial imperfections were used directly to generate input data for the Monte Carlo Method.

It is hoped that these preliminary results will encourge investigators throughout the world to compile extensive experimental information on initial imperfections, classified according to the manufacturing process. Such initial Imperfection Data Banks will make it possible to associate statistical measures with the different methods of fabrication. As outlined in this paper, the variance-covariance matrices and mean vectors can be used effectively to generate input for the Monte Carlo Method, which in turn yields the corresponding reliability functions. It is felt that by this means the imperfection-sensitivity concept can eventually be incorporated as a routine in design procedures.

References

1. Koiter, W.T.: On the stability of elastic equilibrium"(in Dutch), Ph.D. Thesis, Delft University of Technology, Amsterdam: H.J. Paris 1945. English translations: a) NASA TTF-10, 1967, b) AFFDL-TR-70-20, 1970.

2. Elishakoff, I.: Simulation of space-random fields for solution of stochastic boundary-value problems. J. Acoust. Soc. Amer. 65 (1979) 399-403.

3. Elishakoff, I.; Arbocz, J.: Reliability of axially compressed cylindrical shells with random axisymmetric imperfections. Rep. LR-306, Dept. Aerospace Eng., Delft University of Technology: 1980.

4. Amazigo, J.C.: Buckling under axial compression of long cylindrical shells with random axisymmetric imperfections. Quart. Appl. Math. 26 (1969) 537-566.

5. Koiter, W.T.: The effect of axisymmetric iperfections on the buckling of cylindrical shells under axial compression. Amsterdam: Koninklijke Nederlandse Akademie van Wetenschappen, Series 13, 65, No. 15 (1963) 265-279.

6. Roorda, J.; Hansen, J.S.: Random buckling behavior in axially loaded cylindrical shells with axisymmetric imperfections. J. Spacecraft 9 (1972) 88-91.

7. Makaroff, B.P.: Statistical analysis of stability of imperfect cylindrical shells, 7th All-Union Conference on the Theory of Plates and Shells, Proceedings, Dnepropetrovsk (1976) 387-391 (in Russian).

8. Arbocz, J.; Abramovich, H.: The initial imperfection data bank at the Delft University of Technology - Part 1, Rep. LR-290, Dept. Aerospace Eng., Delft University of Technology: 1978.

9. Arbocz, J.; Babcock Jr., P.D.: Prediction of buckling loads based on experimentally measured initial imperfections, Proceedings, IUTAM Symp. on Buckling of Structures, Heidelberg: Springer, 1976, 291-311.

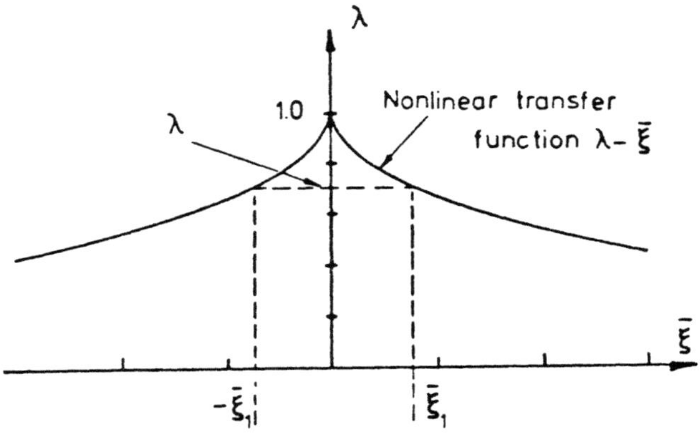

a. Critical buckling load λ vs imperfection amplitude $\bar{\xi}$

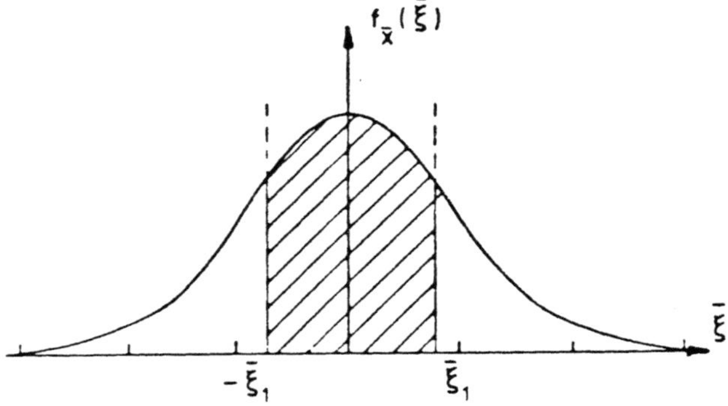

b. Probability density of the initial imperfection amplitude

Fig 1 Illustration of the reliability calculation

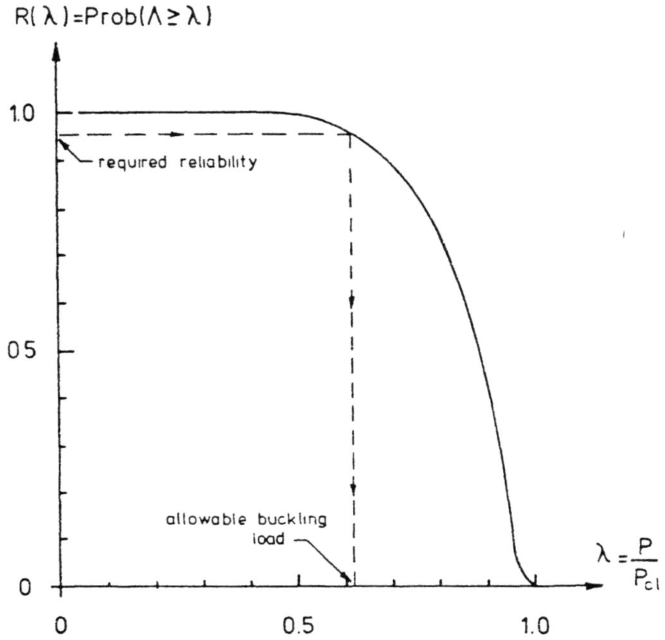

Fig 2 Plot of the reliability function R(λ)

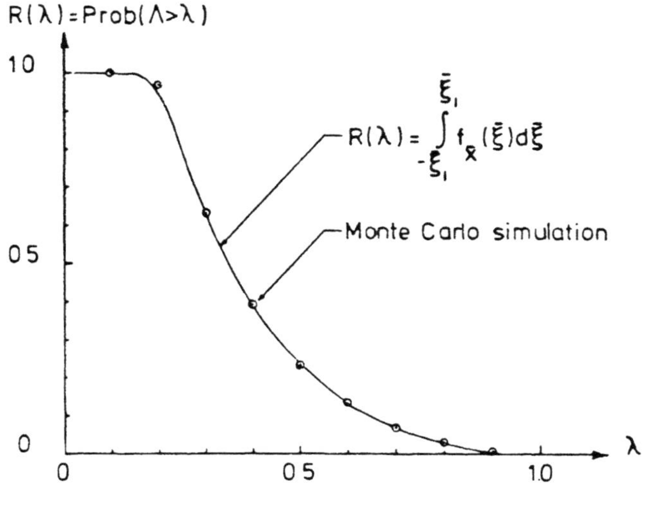

Fig.3 Comparison of analytical reliability function with results of the Monte Carlo method

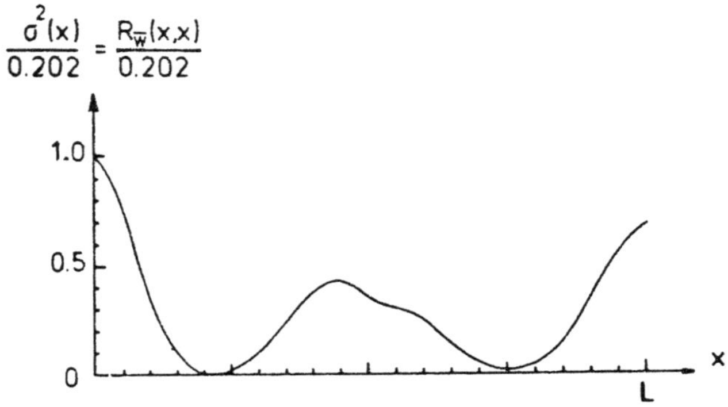

Fig. 4 Variance of the measured initial imperfections (Group of B-shells)

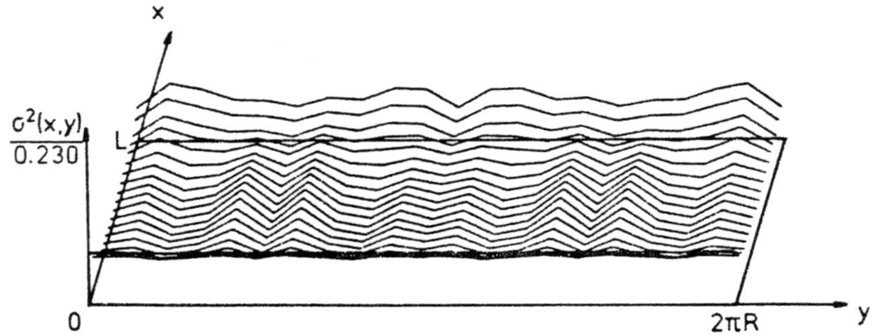

Fig. 5 Estimated approximate variance (Group of B-shells)

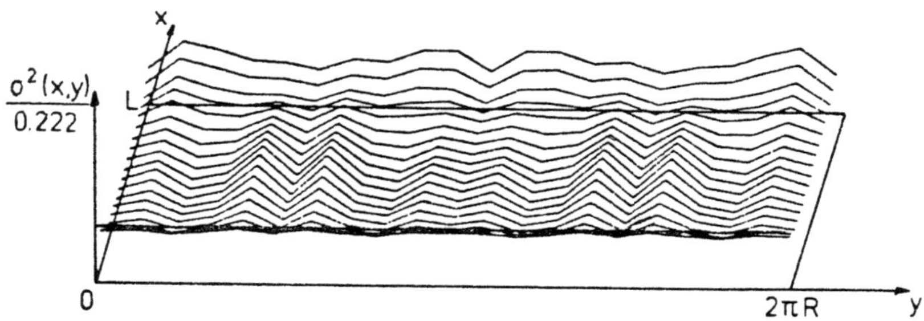

Fig 6 Simulated approximate variance (Group of 500 equivalent B-shells)

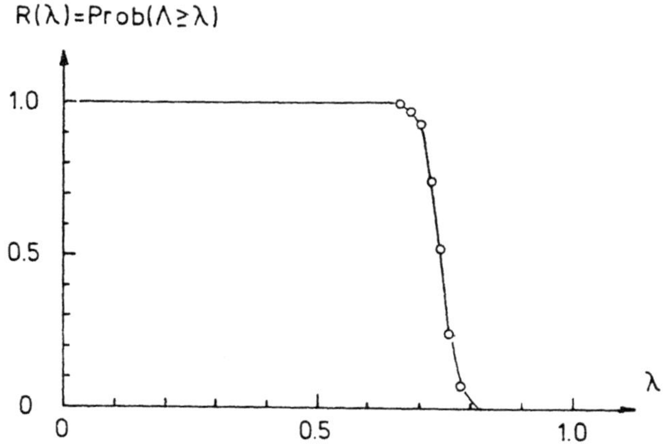

Fig 7 Calculated reliability function for the asymmetric imperfection model (Group of B-shells)

Reduced Molecular Chaos and Flow Instability

Joseph A. Johnson III

Department of Physics
The City College, CUNY
New York, N.Y. (USA)

Lin I and Raghu Ramaiah

Department of Physics
Rutgers University
New Brunswick, N.J. (USA)

Summary

A test for the applicability of a reduced molecular chaos hypothesis to unstable boundary layer flow has been derived along with approximate Reynolds number independence for a dimensionless frequency $\lambda_0 \delta^*/U_0$. A prediction of enhanced turbulence, from this hypothesis, in the presence of reaction distortion is qualitatively confirmed by Argon glow discharge instabilities with a negative ion admixture. Furthermore, the phase-coherent effects observed in unstable ionizing shock waves show mode-mode coupling consistent with the reduced molecular chaos approach.

Considerable uncertainty now exists concerning the sufficiency of the Navier Stokes equations as a basis for understanding turbulence. For at least two reasons, this (in a sense) should not be surprising. First, the very structure of the fluid mechanical description contains difficulties[1] (i) the equations are nonlinear and, as a result, are nonclosed for turbulent flow; (ii) furthermore, the continuous introduction of new unknowns (higher order moments) for a statistical characterization means that the number of unknowns continuously exceeds the number of equations. Secondly, second moment closure schemes (which, in one form or another, are the cornerstone for current phenomenological models) are all frame-dependent[2]. Since (by contrast) Reynolds stresses and turbulent correlations are frame indifferent tensors, the empirical approaches violate form invariance and cannot provide the complete description of an unstable fluid system. Furthermore, the evolution of the mechanical description of a fluid system from the molecular level (Liouville's equations) to the hydrodynamical level involves a considerable amount of averaging in momentum and co-ordinate space. It is easy to imagine that, for some physical events indeed, the details of the averaging procedure might prove to be important in ways beyond the predictive power of our present-day understanding of nature. In particular, the limitations in Boltzmann's approach arising from the assumption of molecular chaos are well known[3]. For example, persistence in velocity correlations is predicted but with an incorrect power law dependence[4]; amplified thermodynamic fluctuations cannot be treated without the implementation of ad hoc assumptions[5]; well established diffusion phenomena are anamolous to the conventional framework of kinetic theory[6]. In these several cases, one sees an influence of phenomena at the molecular scale on phenomena at the macroscopic and continuum scale. New points of view on turbulence also arise from a re-examination of its premises in kinetic theory.

The theoretical and philosophical foundations for this were laid some time ago by the speculations of Grad[7]. A rederivation of the Boltzmann equation is required using less restrictive assumptions for the "memory" during molecular collisions. The analytical techniques have been provided in the papers by Klimontovich[8]. With these, Tsuge derives a kinetic theory for turbulence from a reduced chaos hypothesis[9]. He finds that a new Boltzmann equation is achieved; the classical Navier-Stokes and Fourier laws are then reproduced along with new "turbulent" corrective terms for the stress arising from the relative velocity correlations between colliding particles and, for heat flux, from the stress convected by the turbulent component of the flow. In both cases, the corrections vanish when a new nonempirical binary chaos function term (the function Ψ in Tsuge's presentations) is negligible. Others have shown[10] that the Tsuge apporoach leads to the possible coupling of molecular relaxation and turbulence in flow with $\Psi \neq 0$. As a consequence of the several efforts which evolved from the reduced chaos approach, two unexpected conclusions have been reached. First, Orr-Sommerfeld solutions (normally restricted to weakly turbulent, i.e., amplitude and phase coherent, unstable flow) seem to be applicable to some turbulent systems with phase-coherent manifestations (turbulent bursts) even when the flow otherwise displays "normal" (i.e. amplitude in-coherent) fluctuations[9,11,12]. Secondly, flows with coherent structures show an accompanying enhancement of reaction processes beyond that which might be explained by rate theory[13,14,15]. Here, we shall extend this re-examination of kinetic theory by more precisely determining the applicability of the Orr-Sommerfeld solutions and by more precisely testing the kinetic (chemical) implications of a phase-coherent interpretation for turbulent bursts.

We have sought to determine the systematic trends which are predicted by this approach for the stream function $\Psi(x,y,t) = \varphi(y) \exp\{i(\alpha x - \lambda t$ where: α = wavenumber; λ = frequency; $D = d/dy$; $U(y)$ is the mean flow; and $Re \equiv U_o \delta^*/\nu$ is the Reynolds number based on free stream velocity, boundary layer displacement thickness, and boundary layer kinematic viscosity.[16] The perburbed flows are described by the Falkner-Skan equations: $F'''(\eta) = -FF'' + \beta(F'^2-1)$ where $U(\eta) = F'(\eta)$; $U''(\eta) = F'''(\eta)$; $\eta = y\{U_o/(x\nu(2-\beta))\}^{1/2}$ and the boundary conditions are $F(0) = F'(0) = \lim (F'-1)=0$.

Numerical solutions ranging from neutral velocity profiles, β =0 (the Blasius boundary layer), to nearly separated flow (β =-.1988) have been obtained using standard procedure[12]. For all $\beta< 0$, the velocity profile has a point of inflection and is therefore inherently unstable. For each β, a set of curves for turbulent intensity versus Reynolds number is produced. Each curve is in turn characterized by: (1) a nondimensional frequency, $fr(= \lambda\nu/U_o^2)$; (2) a peak Reynolds number Re_p; (3) a minimum Reynolds number Re_n; and (4) a ratio I_{max}/I_{min} where I_{max} and I_{min} are the maximum and minimum values respectively of the turbulent intensity at $Re=Re_p$ and $Re=Re_n$. For example, for β =-0.15, the curves in Figure 1 for $U(\eta)$ and log (I) are derived.

With these, we have learned that the Reynolds number window, (Re_p-Re_n), for disturbances which will be amplified increases with increasing Re_p for all boundary layers with

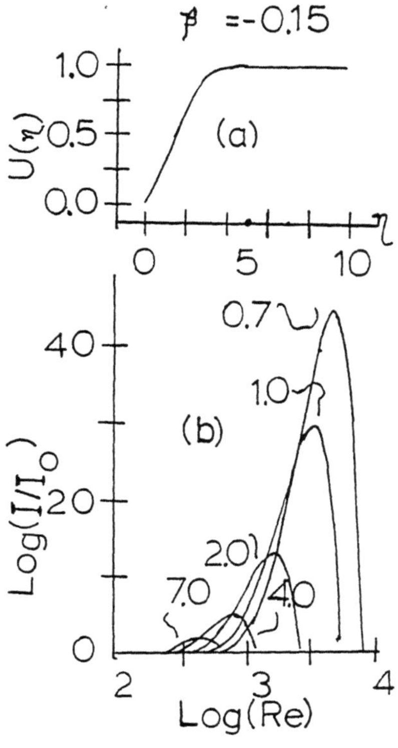

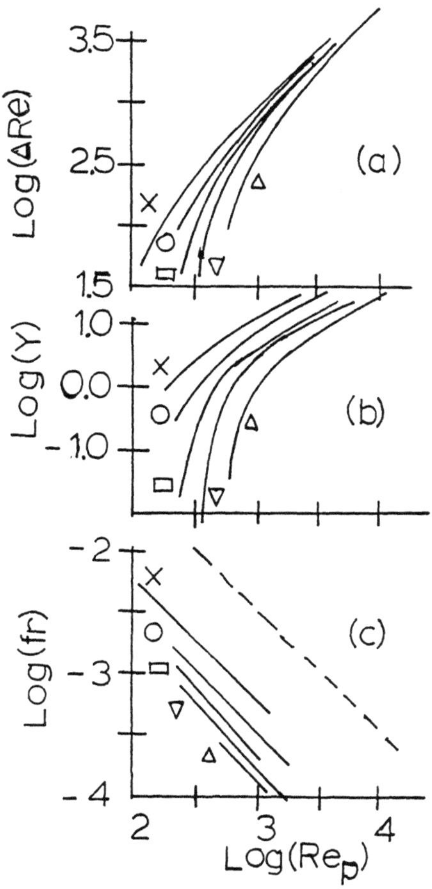

Figure 1. (Left) Orr-Sommerfeld Solutions: An Illustration. (a) Falkner-Skan Velocity Profile; β =-0.15. (b) Turbulent Intensity vs Reynolds Number: β =-0.15. In (b), the profiles are indicated with their associated values of 10^4(fr) = $10^4(\lambda\nu/U_o^2)$. The Reynolds number at the maximum of each profile ($I=I_{max}$) is Re_p for that profile; at $I/I_o=1$, ($I=I_{min}=I_o$) the corresponding Reynolds number is designated as Re_n.

Figure 2. (Right) Orr-Sommerfeld Solutions: Systematic Trends. (a) Reynolds Number Window ($\Delta Re=(Re_p-Re_n)$) vs Reynolds Number. (b) Maximum Turbulent Intensity vs Reynolds Number. (c) Non-dimensional Frequency fr vs Reynolds Number. Each profile, with specific fr and β, has values of $Y=\log(I_{max}/I_{min})$, Re_p and Re_n determined for it; the curves shown above are smoothed through these characterizing parameters. The symbols designate β as follows: X, β =-0.1988; O, β =-0.15; □, β =-0.10; ∇, β =0.05; Δ, β =0. The dashed line represents fr $\propto Re^{-1.0}$.

adverse pressure gradients; however, the distinctions based on β among these boundary layers diminish. We find that the maximum turbulent intensity, (I_{max}/I_{min}), for an amplified disturbance increases dramatically with increasing Re_p and with increasing strength of the adverse pressure gradient (i.e., with decreasing β). We find that the nondimensional frequency fr has approximately the same power law dependence on Re_p when $Re_p \lesssim 10^3$ for all the boundary layers treated. These three behaviors are shown in Figure 2. Furthermore for $\lambda\nu/U_o^2 \propto Re^{-1.0}$, as is implied in Figure 2 (c), and $\lambda\nu/U_o^2 = (\lambda\delta^*/U_o Re)$, one obtains $frRe = \lambda\delta^*/U_o \propto C_1$ where C_1 is constant $C_1 = C_1(\beta)$. In Figure 3, $(\lambda\delta^*/U_o)\cdot(1/C_1)$ vs Re_p and C_1 vs β are plotted; the theory predicts that the frequency λ, when made dimensionless by the ratio of displacement thickness to free stream velocity, should be relatively independent of Reynolds number.

For incompressible flow and x as the stream wise direction, the y-y fluctuation correlation becomes $Q(x,x) = \overline{((\Delta u_2)\ (\Delta u_2)_n)} = \int A(\lambda,\gamma)\,|Y(\eta,\lambda)|^2 \exp(2I_\beta(R,\lambda,\gamma)\,d\lambda d\gamma$. The weight function $A(\lambda,\gamma)$ depends strongly on the nature of the mechanism initiating the observed instability. Near transition, for the case $A(\lambda,\gamma) = A(\gamma)\delta(\lambda-\lambda_o)$, we can use $M \equiv Q(x,x)^{1/2}/Q(x_o,x_o,\lambda_o)^{1/2} = \exp(I_\beta(R,\lambda_o,0))$ in which the function I_β has been expanded about its maximum at $\lambda = \lambda_o$ and terms higher than quadratic have been neglected. Calculations of M provide the amplification factors for fluctuations of microscopic origin which lead to macroscopic turbulence, fluctuations whose smallest possible values are given by Q_o.

This leads to the usual families of curves for exp (I_β) vs R (as, e.g., in Fig. 1). These curves show (for a given family), individually, three distinct regimes: a first regime with $d(\ln M)/dR > 0$, $d^2(\ln M)/dR^2 > 0$; a middle regime which is essentially linear in the semi-log plane, with $d(\ln M)/dR > 0$, $d^2(\ln M)/dR^2 < 0$. The third regime shows a maximum at $R = R_p(\lambda_o)$. The linear middle regime can be described by $\ln M = m(\lambda_o)\ (R - R_n(\lambda_o))$. Since each curve is now specified by λ_o and associated values for $m(\lambda_o)$, $R_n(\lambda_o)$ and $R_p(\lambda_o)$,

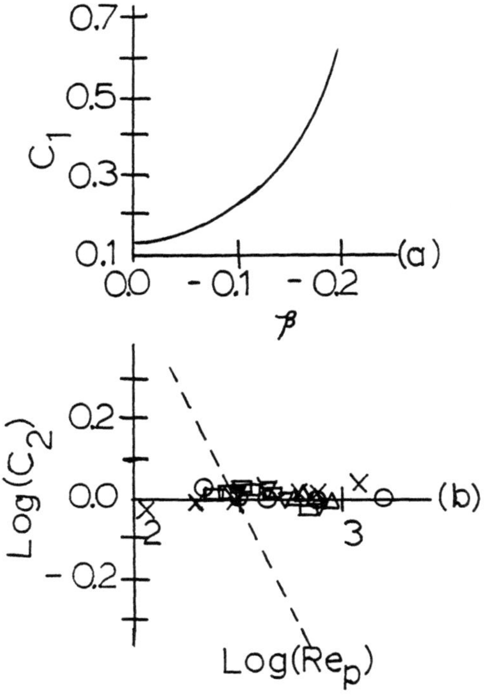

Figure 3. Orr-Sommerfeld Solutions: Systematic Trends. (a) C_1 vs β. The value of C_1 is given by $C_1 = (fr)\cdot(Re_{p,fr})$ for each β; the average is taken up to $Re_p \sim 10^3$. (b) C_2 vs Re_p. Here $C_2 = (frRe/C_1) = (\lambda\delta^*/U_o)\cdot(1/C_1)$ and the symbols designate β as in Figure 2. The dashed line represents $C_2 \propto Re^{-1.0}$.

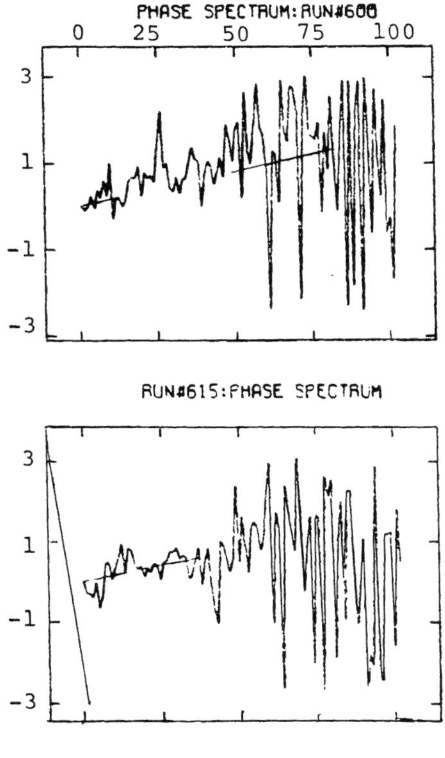

Figure 4. $K-\omega$ Plots, Glow Discharge. The upper trace is for the case p_{tube} = 70 mTorr Ar, 0% SF_6. The lower trace is for 70 mTorr Ar, 6.5% SF_6. (Units: in x, one unit = 5 kHz; in y, one unit = 0.65 cm^{-1}.)

this is the same as $M_\ell \equiv \exp(m(R - f(R_p(\lambda_o))))$ where $R_n = f(R_p)$ is achieved by eliminating λ_o. The expression for $A(\lambda,\gamma)$ can be recast to give a general formulation of M which is correct in the linear regime. Let's use $\Delta\lambda/\lambda_o \equiv (1/n)/\lambda_o$ as a gauge for the impact of a "natural" protuberance and: $A \equiv nA$ for $\lambda_o - (1/2n) < \lambda < \lambda_o + (1/2n)$; $A \equiv 0$ for $\lambda > \lambda_o + (1/2n)$, $\lambda < \lambda_o - (1/2n)$. This gives $A \underset{n\to\infty}{=} \{A\delta(\lambda-\lambda_o)\}$, as required, and allows the explicit computation of M in the linear regime. For the range of Reynolds numbers normally encountered, the solutions are consistent with $f(R_p) = c_1(\lambda_o) + c_3(\lambda_o)\ln(\lambda_o \nu/u^2 c_2(\lambda_o))$ and $M_\ell \underset{\lambda\to\lambda_o}{\sim} \{\exp(m(R-c_1))\}\{\lambda\nu/u^2)/c_2^{-mc_3}$ where c_1, c_2, c_3 and m are slowly varying functions of λ_o routinely available, e.g., from results such as those shown in Figures 2 and 3. We can now integrate over λ: $M_{\ell,n} \sim \{\exp(m(R - c_1))\}\{(\lambda_o \nu/u^2)/c_2\}^{-mc_3}\{1 + h_c\}$; where (up to 5th order in $\Delta\lambda/\lambda_o$) $h_c = \{(mc_3(1+2mc_3))/24\}(c_2/n\lambda_o)^2$. The term h_c is, by inspection, a measure of the uncertainty in the use of Tsuge's approach; this derivation therefore requires $h_c \ll 1$ or (with $\Delta\lambda = 1/n$)

$$\frac{\Delta\lambda}{\lambda_o} \ll \{24/(mc_3c_2^2)\ (1 + 2mc_3)\}^{1/2} \qquad (1)$$

This condition is easily met by any system which is "naturally" unstable. Whenever the diffusion rates for different components of a multicomponent system are different, one and two dimensional instabilities can be induced; soliton-like particle clumps can develop if there is a balance between the nonlinear reaction mechanism and the normal dispersion diffusive effects[18]. Diffusion induced chemical turbulence has already been demonstrated experimentally[19]. Ionization waves and turbulence in the discharge tube have also been studied for many years[20]. However, in a glow discharge, the presence of negative ions should have a significant effect on the local nonlinear mechanisms[21]. Thus, the possible coupling of turbulence and reaction mechanisms which is implied by the reduced chaos approach should show a pronounced effect in the unstable plasmas in a glow discharge tube. We have tested for this possiblity.

Our measurements were performed in a Pyrex glass tube of 80 cm length and 2.5 cm inner diameter. Thermionically emitted electrons assisted the discharge process and we were able to obtain the positive column at 15 cm from the cathode. An array of Langmuir probes was situated 40 cm from the cathode. Two of three probes (biased into the ion saturation regime) separated by an angle of 16° and equidistant from the axis, monitor the plasma fluctuations the third being used to monitor overall average values of plasma temperature and density. A grid, 2 cm in diameter and made of fine mesh steel wire, was inserted into the positive column 12 cm from the probe array. Pulses (80V, 0.1 ms) were applied to the grid at a repetition rate of 0.2 kHz. Since these pulses served only to perturb the plasma weakly in their neighborhood, the primary effect of the grid was to ensure that a source of ionization instability was present at the probe site. The probe outputs were passed through a series of filters in order to eliminate fluctuations above the Nyquist frequency of the analog to digital converter and those arising from 60 cycle noise. The resulting signals were digitized and stored at a rate such that several runs could be obtained during which the plasma parameters were essentially constant. Familiar techniques[22] were used to determine the auto and cross power spectra, the phase spectra and the resulting dispersion relations (i.e., K vs ω in the usual notation).

Two typical glow discharge conditions and results are shown in Figures 4 and 5. The phase spectra for the cases with and without the SF_6 admixture (SF_6 being a source of negative ions) are not very different. The nearly linear behavior indicated in both cases corresponding to a phase velocity of $v_{(ph)} \simeq 8\ (10^5)$ cm/s suggests that the extent of nonlinearity is not influenced by the negative ions. However, Figure 5 shows that the nature of the nonlinear process has been changed substantially; the low frequency content of the power spectra is significantly altered by the introduction of SF_6. The data shown are representative of our complete set. At increased pressures, the overall instability level increases; however, the enhancement arising from SF_6 is

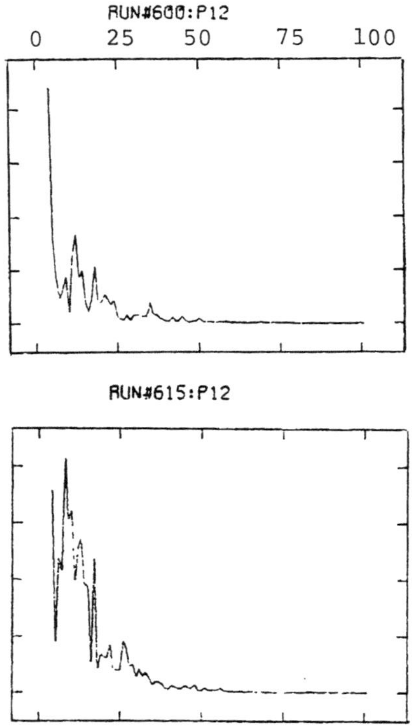

Figure 5. Cross-Power Spectra, Glow Discharge. (Left) The run conditions are given in Figure 4. (Units: in x, one unit = 5 kHz; y units are arbitrary (power) dimensions.)

Figure 6. (Below) Fluctuations in an Ionizing Shock Wave. Run conditions: p = 30 mTorr, M_s=41 (see reference 14). All x units are in cm^{-1}. Upper trace: log(P(k) vs k; y units are arbitrary. Middle trace: Bispectrum Coherence vs k, One Shock Tube Firing; y units are dimensionless.

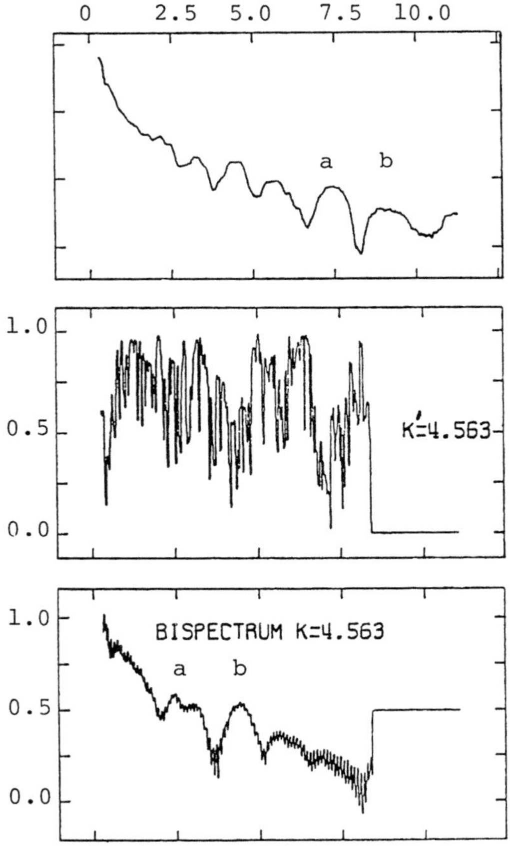

Figure 6 (cont'd). Lower trace: Bispectrum Coherence vs k, Four Shock Tube Firings; y units are dimensionless. Notice that k' is 4.563 cm^{-1}. Signals in the upper trace can be identified as having wavenumbers given by the sum of a signal k in the lower trace and the k' used for the bispectrum slice as indicated.

always noticeable, the degree of enhancement increasing with increased SF_6 concentration.

The overall effect might be accelerated in a system for which the reaction-diffusion mechanism predominates. If there are strong wave-wave interactions, then the three-wave resonance condition may be satisfied ($\omega_k = \omega_{k'} \pm \omega_{k''}$; $\vec{K} = \vec{K'} \pm \vec{K''}$) and we can use digital bispectrum techniques to study the coupled modes[23]. The ionizing collisional shock front satisfies these preconditions since turbulence has been found with coherent structures[14]. We have therefore tested the consequences of this expectation in our arc discharge shock tube data.

Our shock waves are produced in a cylindrical arc discharge stainless steel shock tube of 5 cm diameter and 185 cm length. The shock tube is filled with high purity room temperature argon and fired without the use of a diaphragm. The discharge is generated by triggering a 14.5 μF capacitor charged to 18.5 kV and typically takes about 50 μsec for completion. The shock wave's speed can be varied between 5 x 10^5 cm/sec and 3 x 10^6 cm/sec by varying the initial filling pressure between 250 mTorr and 25 mTorr. The test section is 160 cm from the discharge. Two piezoelecrtric pressure transducers with 15 cm separation are positioned fore and aft of the test section's window. At the window, a pair of lenses and pinhole diaphragm are used to collect white light emission (with a 3.5 mm x 1 mm spatial resolution). At the same location, two steel probes of 0.75 mm diameter are placed in the tube with varying separation along the axis of the tube. The probes are biased at negative voltage so as to collect ion current. The probe signals are sent simultaneously to a transient ADC recorder which has a 25 MHz bandwidth in the chopped mode. The digitized data from the probe current are used to obtain the plasma power spectrum via a Fast Fourier Transform. The bispectrum was determined by averaging over four shock tube firings at the same run conditions. The new quantity which is computed measures the phase coherence between modes of the system and takes on values close to 1 if the three wave condition is satisfied.

The relevance of the bispectrum test is shown in Figure 6. The averaged auto-power spectrum in Fig. 6(a) is relatively featureless. When a projection of the joint power spectrum (from the two probes) is shown on a specified plane for one of the two frequency co-ordinates without preconditions, Figure 6(b) results which is even less suggestive than Fig. 6(a). However, the use of the bispectrum analysis conditions produces Figure 6(c). Several prominent signals can be seen. More importantly, a comparison between Figures 6(a) and 6(c) shows that the existence of three wave resonance condition is at least qualitatively confirmed. This novel effect, which is expected from the reduced molecular chaos hypothesis, is a dramatic amplification of the traditional expectations of classical shock wave theory[24].

This work was supported in part by NASA Grant NSG-3280 to Rutgers University and by NSF Grant SER-810430 to The City College.

References

1. A.M. Yaglom, Am. Rev. Fluid Mech. 11, 505 (1979).

2. C. Speziale, Phys. Fluids, 23, 459 (1980).

3. G.E. Uhlenbeck, Acta Physica Austriaca, Suppl X, 107 (1973).

4. J.R. Dorfman in The Boltzmann Equation, E.G.D. Cohen et al. (ed) (New York: Springer-Verlag) p. 209 (1973).

5. M. Kac in The Boltzmann Equation, E.G.D. Cohen et al. (ed) (New York: Springer-Verlag) p. 379 (1973).

6. B.J. Adler in Transport Phenomena - 1973, J. Kestin (ed), (New York: American Institute of Physics) p. 1 (1973).

7. H. Grad in Handbook des Physik, S. Flugge (ed) (Berlin: Springer-Verlag) Vol. 12, p. 205 (1958).

8. Y.L. Klemontovich, The Statistical Theory of Nonequilibrium Processes in a Plasma, (Cambridge, Mass: MIT Press) p. 49 (1967).

9. S. Tsuge, Phys. Fluids, 17, 22 (1974).

10. J.A. Johnson III, S.C. Chen, Phys. Letts., 68A, 141 (1978).

11. J.A. Johnson III, W.R. Jones, J.P. Santiago, J. Phys. D: Appl. Phys. 13, 1413 (1980).

12. J.A. Johnson III, Appl. Phys. Lett., 37, 275 (1980).

13. D.R. White, Phys. Fluids, 4, 465 (1961).

14. J.A. Johnson III, J.P. Santiago, L.I., Phys. Lett., 83A, 443 (1981).

15. J.A. Johnson III, R. Ramaish, J.P. Santiago, Rev. Sci. Instr. 52, 936 (1981).

16. R. Jordinson, J. Fluid Mech., 43, 801 (1970).

17. L.M. Mack, J. Fluid Mech., 73, 497 (1976); see also H.J. Obrenski, M.V. Morkovin, M. Landahl, AGARDograph 134 (1969).

18. Y. Kuramoto, Prog. T heor. Phys. 63, 1885 (1980).

19. H. Yamazaki, Y. Ono, K. Hirahawa, J. Phys. Soc. Japan, 49, 335 (1978).

20. L. Pekarek, V. Krejii, Phys. Lett., 81A, 228 (1981); R.N. Franklin, Plasma Phenomena in Gas Dischargers, (Oxford: Clarendon Press), p. 159 (1975).

21. W.L. Nighan, J. Wiegand, Phys. Rev. A., 10, 922 (1974).

22. K.F. Shoenberg, Rev. Sci. Instr., 51, 1151 (1980).

23. Y.C. Kin, J.M. Beall, E.J. Powers, R.W. Miksad, Phys. Fluids, 23, 258 (1980).

24. J.N. Bradley, Shock Waves in Chemistry and Physics, (London: Methuen & Co., Ltd) p. 108 (1962).

Laminar-Turbulent Transition in Taylor-Couette Flow, Influence of Geometrical Parameters

G. COGNET ; A. BOUABDALLAH ; A. AIT AIDER

LEMTA - ERA CNRS n° 893

Institut National Polytechnique de Lorraine

Nancy - France

Summary

The Taylor-Couette flow shows a transition from the laminar to the turbulent regime which agrees with recent theoretical predictions : the cahotic features of the turbulence can be obtained after a small number of bifurcations. However the geometrical parameters and the end conditions of the annular gap have a great influence about the transition phenomena.
A first study points out the effect of the gap/radius ratio : d/R = 0.048 ; 0.10 ; 0.22 with the same height of the cylinders. The second one concerns a small gap geometry (d/R = 0.1) for which the aspect ratio is varied from H/d = 14 to 2. The critical values of the Taylor number and the characteristic frequencies of the flow are identified and their evolution is followed according the geometry.

Introduction

The understanding of the appearance and the development of the turbulence from the first instabilities remains an important problem in fluid mechanics even if numerous studies have covered this topic . Among the first contributors, Landau /1/ proposed to describe the laminar turbulent transition as the superposition of successive instabilities, each one being characterized by its frequency, until the flow is so perturbated that it can be considered turbulent. Recently Ruelle and Takens /2/ have theoretically shown that a small number of frequencies (3) is sufficient for a non linear dynamic system to reach a cahotic or "turbulent" behavior. Their predictions have found an experimental consistency from two well-known flows : the Rayleigh Bénard flow realized by a thermal convection between two parallel plates (Bergé /3/) and the Taylor-Couette flow between coaxial cylinders (Gollub and Swinney /4/).

From fine local measurements and spectral analysis carried out by Laser Doppler Velocimetry it has been shown in both cases, that after the appearance of stationnary convective rolls or cells, as the dynamic parameters are increased, the flows become periodic with a first and a second well defined frequencies (f_1 and f_2) ; then a spread band frequency is identified (f_B)

(essentially in Taylor-Couette flow), at last the spectra becomes completely turbulent like with the disappearance of the sharp frequencies. The transition and the critical values of the dynamic parameters are, however, very dependent of the geometrical factors and the boundary conditions. The most studies of the Taylor-Couette flow have been conducted with infinite height and small gap hypothesis. Experimentally only Cole /5/ and Benjamin /6/ have been interested in small height systems for the detection and the prediction of the first and the second bifurcation.

Several theoretical attempts have been performed recently (/7/ /8/ /9/) which consider finite length cylinders.

The study which is presented hereafter is divided in two parts.
One describes the influence of the gap width on the laminar turbulent transition for three gap/radius ratios (d/R = 0.048 ; 0.10 ; 0.22) with constant height H = 200 mm. The electrochemical method has been used for measurements and comparison with previous results (Bouabdallah and Cognet /10/).

The other relates to a small gap geometry (d/R = 0.1) with the aspect ratio which was varied from H/d ≑ 14 to 2 ; the experiments have been done by visualization.

In both cases the evolution of the flow is followed from the stable laminar regime up to the developed turbulent regime.

I. Laminar-turbulent transition - Influence of the gap width

Three flow geometries have been studied : with the same fixed outer cylinder R_o = 55 mm and three rotating inner cylinders R_i = 52.5, 50, 45 mm (annular gaps : d = 2.5, 5 and 10 mm), H = 200 mm. The intermediate geometry, d/R_i = 0.10, is considered as reference. The small electrochemical probes (∅ = 0.4 mm) were flush mounted to a part of the surface of the outer wall which could be shifted in the axial direction allowing the exploration of the flow by local wall shear measurements /10/. The evolution of the motion is described as the Taylor number T is slowly increased.

$$T = (R_i d\Omega/\nu)(d/R_i)^{1/2}$$

Ω angular speed of the rotating inner cylinder

ν kinematic viscosity of the fluid

1. Mean values

The results (figure 1) concern the azimuthal component of the wall shear , S, for the reference geometry. The axial exploration has covered, at least, one wave length according the z direction (two Taylor cells). The mean value, S, is reported from its maximum S_M (source position) to its minimum S_m (sink position). The main characteristics of the flow which are usually mentionned can be identified.

- Subcritical flow $T < T_c = 41$

The flow is only azimuthal. S is z-independent up to T_c. For this first critical value the Taylor vortices appear (in the following we also refer to the reduced Taylor number : $T^* = T/T_c$; $T_c^* = 1$).

- Taylor Vortex flow $T_c < T < T_c' = 48$ $(T_c'^* = 1,17)$

In this range the vortices grow in intensity. For fixed T, S has a periodical dependence with z but remains time independent.

- Wavy Taylor Vortex flow $T_c' < T < T_t = 710$ $(T_t^* = 17,3)$

This range corresponds to the appearance, the development then the decay of the azimuthal waves until their quasi disappearance for $T^* = T_t^* = 17,3$. The mean amplitude of the vortices $(S_M - S_m)$ continuously increases to reach its maximum for T_t^*.

- Turbulent Taylor Vortex flow $T > 710$ $(T^* > T_t^*)$

Well defined cells are still existing in the flow ; their amplitude decreases as the turbulence is growing. It must be noticed that S_M/S_m ratio exceeds 2 for $T^* \mp 100$ which confirms the remaining axial periodicity for speed higher than one hundred times the first critical value.

These results point out at least three critical Taylor numbers with the successive appearance of the Taylor cells (T_c), the azimuthal waves (T_c'), the disappearance of these waves and the development of the turbulence (T_t).

The azimuthal waves introduce a distortion in the cells but do not highly change the law of growth of their amplitude ; it is remarkable that this amplitude begins to decay and parallelly the turbulence of the flow significantly increases as soon as the travelling waves disappear.

The comparison with the other geometries can be established from the results which are presented on the figures 2 and 3. The evolution of S_M and S_m with T-number is similar in each case ; however it must be noticed that the transition towards turbulence is faster as the gap is smaller. The critical values of T are reported below, in the table 1.

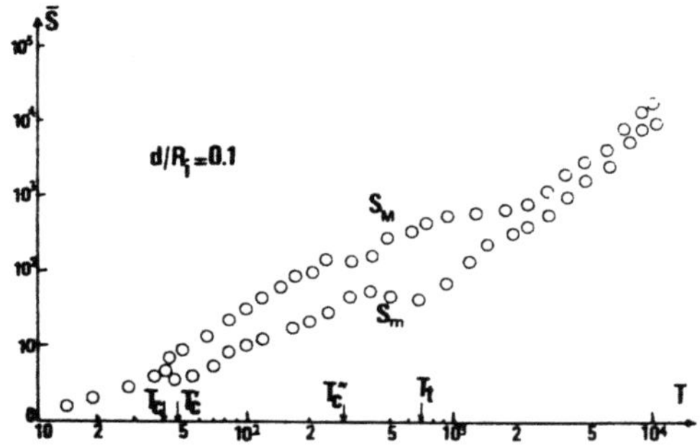

Figure 1

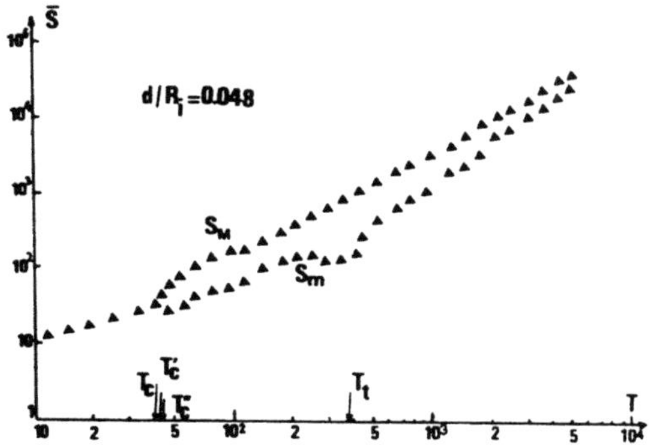

Figure 2

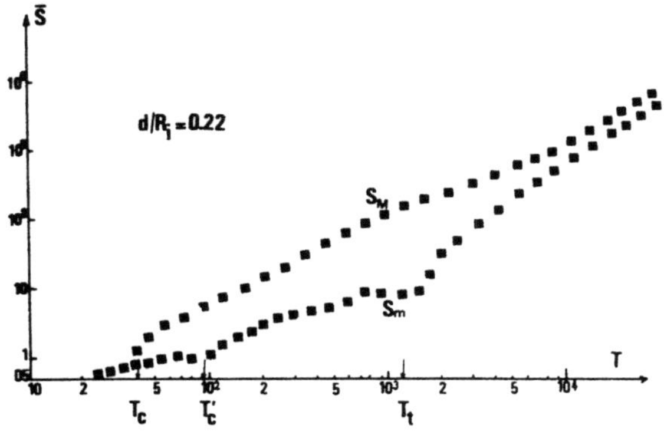

Figure 3

Mean wall shear in a maximum (S_M) and minimum (S_m) position vs Taylor number, for three gaps.

Author	d/R_i	H/d	$T_c'^*$	$T_c''^*$	T_t^*	upper (1) / lower (2) } end
Cognet /11/	0,10	14	1,20		15,6	(1) free surface (2) free end
Bouabdallah present work	0,048	80	1,07	1,17	9,3?	(1) fixed wall (2) free end
"	0,10	40	1,20	7,3	17,3	"
"	0,22	20	2,14	?	21,4	"
Ait Aider present work	0,10	14	1,28	4,0	16,4	(1),(2) fixed walls
Swinney and al /4/	0,14	20	1,20	10,1	21,9	(1),(2) fixed walls

Table 1. Critical values of the Taylor number according to the geometry

The cellular structure is very strong particularly in the large gap ; at the maximum of the cell amplitude ($T = T_t$) : $S_M \simeq 20\ S_m$; beyond T_t the turbulence slowly reduces this great difference of the flow between a maximum and a minimum position.

2. Motion analysis

The spectral analysis of the wall shear fluctuations, s, measured at the middle of a cell, allows a more complete and precise description of the transition.

The figures 4a, b, c, d, present the normalized power density spectra of s which point out the fundamental frequencies of the flow for the reference geometry as the Taylor number is slowly increased.

- In the range $T_c'^* < T^* < T_c''^* = 7,3$ the flow is purely periodical (fig. 4a) with a frequency, $f_1 = nF_o$, and its harmonics (n is the azimuthal wave number, F_o the frequency of the whole azimuthal wavy pattern). More than 98 % of the energy is concentrated in the fundamental. For $T^* = 6$ some noise

Fig. 4

$d/R_i = 0.1$

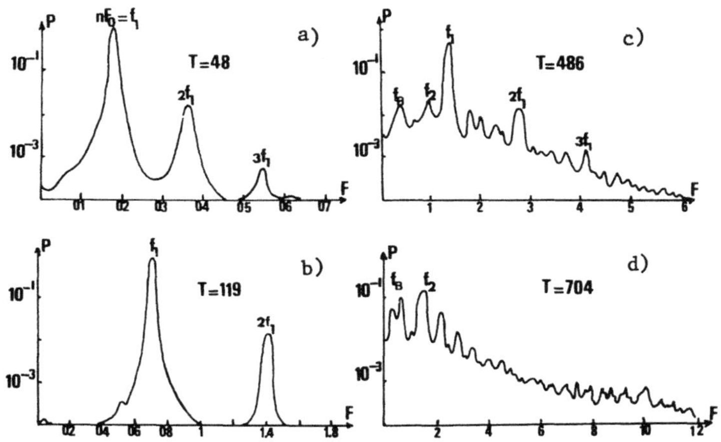

Fig. 5

$d/R_i = 0.048$

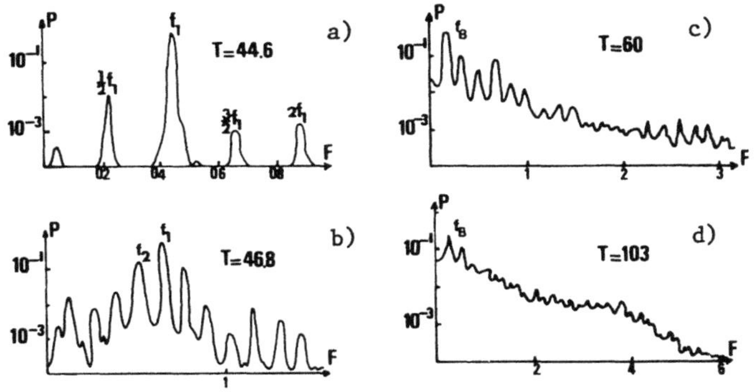

Fig. 6

$d/R_i = 0.22$

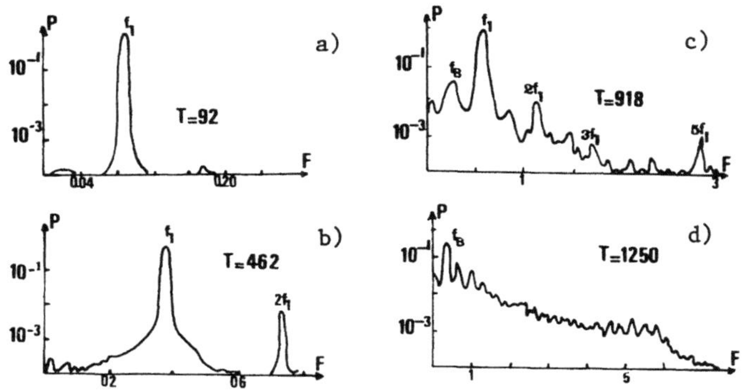

Normalized power density spectra of the wall shear fluctuations

appears in the spectrum but it represents less that three percents of the total energy.

- For $T^* = 7,3 = T''^*_c$, a new frequency is detected : f_2, distinct from f_1 ; it is accompanied by a less energetic frequency $f_m \simeq f_1/4$ which seams to be equal to F_o when n = 4. Many peaks defined by $(pf_1 + qf_2)$ frequencies, p and q being integers, can be identified in the spectra.

When T^* reaches 12, f_m becomes a large band frequency, f_B, and there is an increase of the noise. This seams to introduce the first "cahotic" element in the flow. (fig. 4 c)

As the Taylor number is still increased, f_1 remains the more important frequency but its energy decreases and it disappears for $T^* = T^*_t = 17,3$. This fundamental change shows the disappearance of the azimuthal wave. Beyond T^*_t only the large peak is still present, the spectrum is more extended and the noise is developed as in turbulent regime. (fig. 4 d)

The features of the transition, in that case, are similar to the description given by Swinney /4/ for a close geometry (Table 1).

The significance of f_2-frequency has been recently precised by GORMAN /12/ ; it corresponds to a modulation of amplitude of the azimuthal waves, which is clearly demonstrated by the pictures of figure 7 ; where the boundaries of the cells periodically appear flat or wavy.

The spectral analysis of the motion points out that the transition is highly dependent of the gap geometry. In the small gap apparatus it is more difficult to follow the successive bifurcations of the flow as they are concentrated near the first critical Taylor number (fig. 5 a, b, c, d). The azimuthal waves are detected for $T^* = T^*_c = 1.07$; the spectrum exhibits the f_1-frequency with the multiples $nf_1/2$, which are not observed in the reference geometry ; the f_2- frequency appears for $T^* = T'^*_c = 1.17$. A great number of sharp frequencies are identified. As soon as $T^* = 1.5$ the large peak, f_B, is dominant and the spectrum looks turbulent for $T^* = 2.5$; this does not correspond with the maximum of amplitude of the cell which is registered for $T^* = 9.3$ (fig. 2).

The large gap geometry is characterized by a wide transition range (fig. 6 a, b, c, d). The travelling waves appear for $T^* = T'^*_c = 2.14$ with the f_1-frequency. They do not seam to be submitted to any modulation as no second sharp frequency (f_2) is clearly detected. The broad peak f_B announces the occurence of turbulence but remains in the spectrum beyond the disappearance of the waves, for $T^* = T^*_t = 21.4$. An experiment of visualization has been conducted in that specific case without showing the typical modulation of wavy boundaries of the cells corresponding to the f_2-frequency.

So the results shows that the mechanism of transition can be affected according to the gap width even though the number of bifurcations between the laminar and the turbulent regime is effectively limited within two or three frequencies.

II. Influence of the aspect ratio : $A = H/d$

The aspect ratio is also an important parameter for the laminar-turbulent transition in Taylor Couette flow. Its influence has been experimentally studied by visualization using the classical suspension of Aluminium flakes in water. The flow geometry was similar to the reference one, but its size was twice ($d = 10$ mm ; $R_i = 100$ mm /11/). The acting height of the cylinders could be varied by the displacement of the annular upper end into the gap from $H = 140$ to 20 mm. The motion was analysed by direct measurements of the cell numbers (m), the wave number (n) and the wave frequency ($f_1 = nF_o$). A special equipment, with a video camera and a magnetic recorder fixed to a rotating plate, coaxial with the cylinders, has been settled to follow the flow in a frame linked to the azimuthal travelling waves and, in that way, to determine the frequency of their modulation (f_2).

Figure 8 presents the characteristic frequencies of the motion and their evolution with the Taylor number. The results corresponds to the greatest aspect ratio ($A = 14$) with a slowly increasing speed of the inner cylinder (Nrd/s). In spite of the great number of different flow states, each defined by a couple of values (m,n), they are representative of the general tendency all along the transition range. In that procedure the number of cells is not strongly affected ; as theoretically predicted for T_c : m=14 ; then the only change before the onset of turbulence is the disappearance of two cells (m = 12). The azimuthal wave number changes much more with T ; three waves are observed for T'_c then n increases steeply to reach 10 and decreases by stage to 8 and 6. The flow state (12,6), established for $T^* \simeq 9$, is stable until the turbulence. The f_1/N ratio follows this evolution but the variation of F_o/N is more significant; as it concerns the whole travelling pattern it is nearly independent of the state. After a steep decrease from

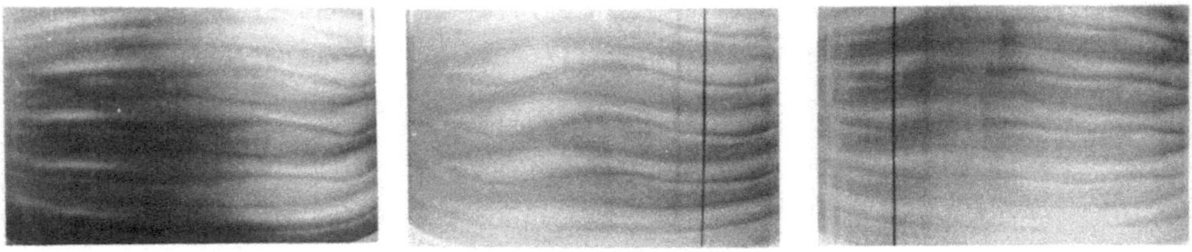

<u>Figure 7</u> : f_2 frequency ; modulation of the amplitude of the azimuthal wave.
A = 14 ; T* = 9,7 state (12-6) f_2/N = 0,62

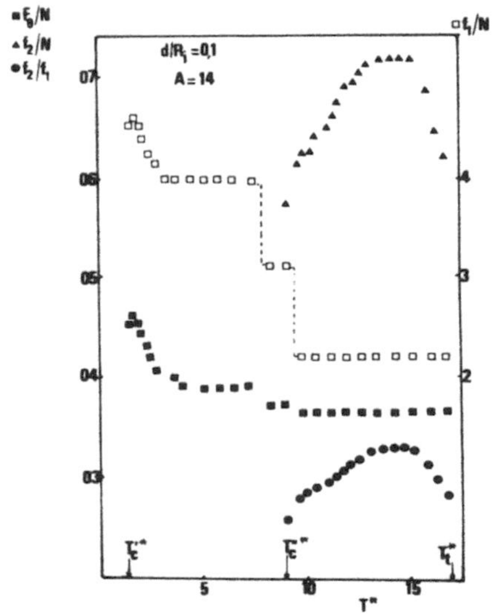

<u>Figure 8</u> :

Fo, f_1, f_2 frequencies vs Taylor number

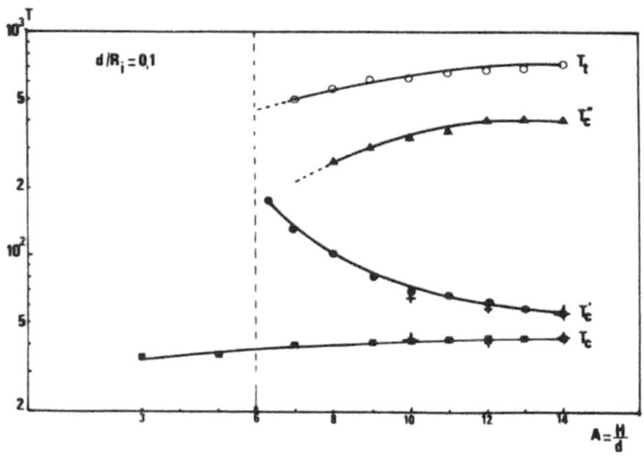

<u>Figure 9</u> :

Influence of the aspect ratio on the critical values of T

0.45 it keeps a nearly constant value until the vanishing of the waves ($T_t^* = 16.4$).

The f_2-frequency, only reported for the final (12,6) state, follows a quite different way. From T_c'' it increases, reaches a flat maximum then decreases just before T_t^*. The maximum value of f_1/f_2 has been interpreted /13/ as a "phase locking" of the two frequencies which could be at the origin of the wave instability and so an important stage leading to turbulence.

The evolution of the critical values of T when the aspect ratio is varied is given on figure 9. The appearance of the Taylor cells is not much affected by A variations. A small decrease of T_c is observed as A passes from 14 to 3. On the contrary the increase of T_c' is remarkable. Morever when $A < 6$ no wavy pattern can be obtained. These results are in agreement with Cole's data /5/ and confirm that the waves are much more sensitive to the geometrical parameters than the cells.

The values of T_c'', determined for the more stable states which precede for each case the onset of turbulence (cf. $T_c'' = 9$ for $A = 14$), decreases with A. It seams that the two curves $T_c'(A)$ and $T_c''(A)$ meet together at the boundary $A = 6$ but this has not been verified by experiment. T_t has been defined as the critical value of T at the vanishing of the waves which also corresponds with the beginning of the turbulent regime. The decrease of T_t with A is similar to the behaviour of T_c and T_c''.

Conclusion

The influence of the geometrical parameters (annular gap height of the cylinders-end conditions) on the laminar turbulent transition in Taylor-Couette flow is important and can explain the discrepancy between the different results quoted in the litterature. Particularly it appears that the stability range of the waves is strongly affected even by small changes in the geometry. The way towards turbulence is not unique. In front of this difficult problem it seams essential to distinguish the commun and specific elements of each geometry. The presence in the flow of several sharp frequencies is one of these commun characteristics. We need now to more clearly catch their physical meaning (f_2, f_B) and to get more precise information (measurement) about their interaction which is surely the key of a better understanding of the onset of turbulence.

REFERENCES

1. *L. LANDAU* C.R. (Dokl) Acad. Sci. URSS 44, 311 (1944)
2. *D. RUELLE et F. TAKENS* Comm. Math. Phys. 20, 167 (1971)
3. *P. BERGE et M. DUBOIS* Optics Communication 19, 129 (1976)
4. *J.P. GOLLUB et H.L. SWINNEY* Phys. Rev. Lett, 35, 14, 927 (1975)
5. *J.A. COLE* J.F.M. 75, 1-15 (1976)
6. *T.B. BENJAMIN* Proc. Roy. Soc. London A359, 27-43 (1978)
7. *I.C. WALTON* Proc. Roy. Soc. London A372, 201-218 (1980)
8. *J.T. STUART et R.C. DI PRIMA* Proc. Roy. Soc. London A 372, 357-365 (1980)
9. *P. HALL* Proc. Roy. Soc. London A372, 317-356 (1980)
10. *A. BOUABDALLAH et G. COGNET* "Laminar-turbulent transition" IUTAM Symposium Stuttgart (Springer-Verlag, 1980) p. 368-377
11. *G. COGNET* J. Mécanique 10, 65-90 (1971)
12. *M.A. GORMAN et H.L. SWINNEY* Phys. Rev. Lett. 43, 1871-1875 (1979)
13. *P.R. FENSTERMACHER ; H.L. SWINNEY ; J.P. GOLLUB* J.F.M. 94, 103 (1979)

Taylor-Görtler Vortices in Fully Developed or Boundary Layer Flows

P. HALL

Department of Mathematics, Imperial College, London SW7, U.K.

Our concern is with the stability of fluid flows over curved walls. The detailed asymptotic analysis required to derive the results discussed briefly below are given by Hall (1). The nonparallel nature of the basic flow in which Görtler vortices develop has effectively prevented the development of a self-consistent perturbation investigation of even the linearized stability equations. Görtler (2) and Hammerlin (3) have investigated the linear stability problem using the so-called 'parallel flow approximation' whilst, amongst others, Smith (4) has taken some account of nonparallel effect. Without exception the nonparallel nature of the basic flow has not been consistently taken into account in these investigations.

We shall restrict our attention here to time-independent perturbations and we define δ and R_E by

$$\delta = \ell R^{-1}, \qquad R_E = U_0 \ell \nu^{-1} \tag{1a,b}$$

where ℓ is a typical downstream lengthscale whilst R is a typical radius of curvature of the surface. We now consider the limit $R_E \to \infty$ with the Görtler number G defined by

$$G = 2R_E^{\frac{1}{2}}\delta \quad , \tag{2}$$

held fixed. Suppose that s,y and z are downstream, normal and spanwise variables scaled on ℓ, Δ and Δ respectively where Δ is the boundary layer thickness. If u,v and w denote the velocity components in the y,s and z directions then we perturb the basic flow such that,

$$u = U_0\{\overline{U}_0 + U\}R_E^{-\frac{1}{2}}, \quad v = U_0\{\overline{V}_0 + V\}, \quad w = U_0WR_E^{-\frac{1}{2}} \quad (3)$$

where $\overline{U}_0(Y,s)$ and $\overline{V}_0(Y,s)$ are the dimensionless velocity components of the basic flow. The linearized stability equations obtained in the limit $R_E \to \infty$ with G held fixed are

$$\begin{aligned} &U_{YY} + U_{ZZ} + P_Y + G\overline{V}_0V = [-(\overline{U}_0U)_y + \overline{V}_0U_s + V\overline{U}_{0s}] \\ &V_{YY} + V_{ZZ} + U\overline{V}_{0y} = [-\overline{U}_0V_y + \overline{V}_0V_s + V_0\overline{V}_{0s}] \\ &W_{YY} + W_{ZZ} + P_Z = [-\overline{U}_0W_y + \overline{V}_0W_s] \\ &-U_Y + W_Z = -[V_s] \end{aligned} \quad (4)$$

where P is the pressure perturbation corresponding to (U,V,W) and for definiteness we have taken the curvature of the wall to be constant.

The equations solved by Görtler are obtained by neglecting the terms in the square brackets above and there is clearly no justification for neglecting these terms. A similar criticism applies to the equations solved by Hammerlin (3), Smith (4), etc.

Suppose now that we consider small wavelength $(\sim 0(\epsilon))$ perturbations and expand G in the form

$$G = \epsilon^{-4}\{g_0 + e^{\frac{1}{2}}g_1 + \epsilon g_2 + \dots\} \quad . \quad (5)$$

Our previous experience, Hall (1), shows that such disturbances are concentrated in internal viscous layers located where $|U_0U_{0y}|$ is a maximum. This critical layer is of thickness $\epsilon^{\frac{1}{2}}$ and we therefore define η by

$$\eta = \epsilon^{-\frac{1}{2}}\{\xi-\xi^+\} \quad (6)$$

where $\xi = y/(2s)^{\frac{1}{2}}$ and ξ^+ is the location of the critical layer.

We now expand U,V,W,P in the form

$$\begin{aligned} U &= (U_0(\xi,s) + \epsilon^{\frac{1}{2}}U_1(\xi,s) + \dots)e^{iz/\epsilon}, \\ V &= \epsilon^{\frac{1}{2}}(V_0(\xi,s) + \epsilon^{\frac{1}{2}}V_1(\xi,s) + \dots)e^{iz/\epsilon}, \\ W &= \epsilon^{\frac{1}{2}}(W_0(\xi,s) + \epsilon^{\frac{1}{2}}W_1(\xi,s) + \dots)e^{iz/\epsilon}, \\ P &= \epsilon^{\frac{1}{2}}(P_0(\xi,s) + \epsilon^{\frac{1}{2}}P_1(\xi,s) + \dots)e^{iz/\epsilon}, \end{aligned} \quad (7)$$

and then $\bar{U}_0, \bar{V}_0$ in the form

$$\bar{U}_0 = \sqrt{2s}\,[U^+_{00} + \epsilon^{\frac{1}{2}} U^+_{01\eta} + \ldots] \quad ,$$
$$\bar{V}_0 = V^+_{00} + \epsilon^{\frac{1}{2}} V^+_{01\eta} + \ldots \quad . \tag{8}$$

The above expressions are then substituted into (4) and like powers of ϵ are equated. At first order we obtain

$$g_0 = \frac{\sqrt{2s}}{V^+_{00}V^+_{01}} \quad ,$$

whilst at second order we obtain

$$g_1 = 0 \quad .$$

At third order we obtain

$$\frac{\partial^2 V_0}{\partial \eta^2} + 2s\eta^2 \left[\frac{V^+_{02}}{V^+_{00}} + \frac{V^+_{03}}{V^+_{01}} \right] V_0 + g_2 \frac{V^+_{00}V^+_{01}}{3} \sqrt{2s}\, V_0 = 0$$

so that V_0 is determined in terms of parabolic cylinder functions. The function V_0 is then confined to the critical layer if

$$g_2 = g_{2\eta} = 6 \frac{\left\{ - \left(\frac{V^+_{02}}{V^+_{00}} + \frac{V^+_{03}}{V^+_{01}} \right) \right\}^{\frac{1}{2}}}{V^+_{00}V^+_{01}} \left(\tfrac{1}{2}+n\right), \qquad n = 0,1,\ldots$$

The most dangerous mode corresponds to n = 0, and for the Blasius profile the critical Görtler number is

$$G = \frac{5.91}{\epsilon^4} \{ s^{\frac{1}{2}} + 0.96\,\epsilon + - \} \tag{9}$$

The next term in this expansion can be calculated by proceeding to higher order and depends on nonparallel flow effects. This nonparallel flow effect term is not predicted by the reduced approximate equation of Görtler (1940), Smith (1956), etc. In the figure below we have compared the eigenfunction predicted by our above theory with that measured experimentally by Bippes and Görtler (1972). The argument between the curves suggests that our expansion procedure, valid for $\epsilon \to 0$, is useful at relatively large values of ϵ.

Finally we close by noting that the above theory has been extended by Hall (1) to take nonlinear and nonparallel flow effects simultaneously into account.

References

1. Hall, P. 1981a,b,c, submitted for publication.

2. Görtler, H., 1940, Tech. Memor. Nat. Adv. Comm. Aero. Res. Coun., London, No. 2779.

3. Hammerlin, G., 1956, Z.A.M.P., 7, 156.

4. Smith, A.M.O., 1955, Quart. Appl. Math., 13, 233.

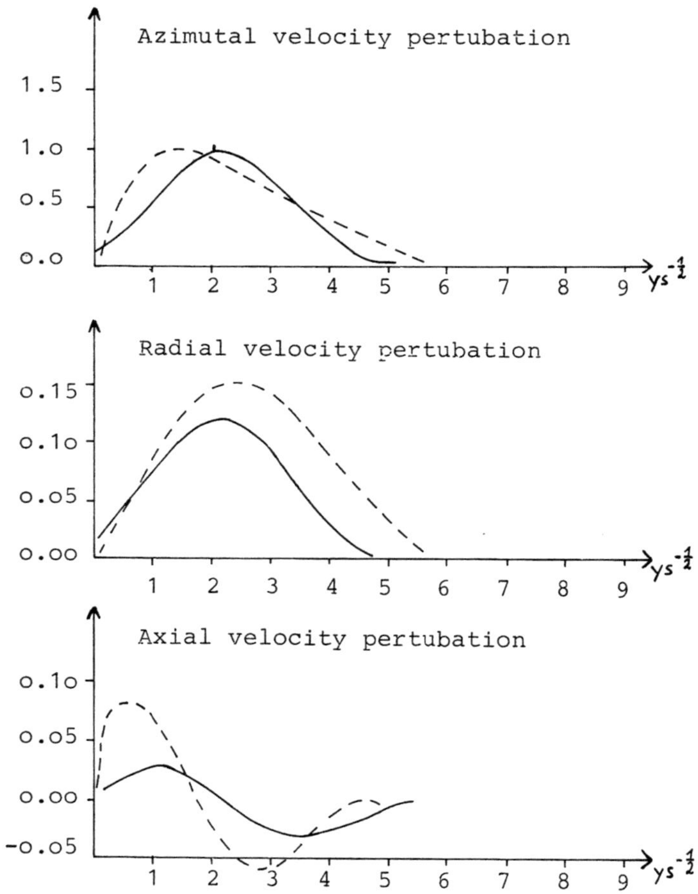

Nonlinear Development and Secondary Instability of Görtler Vortices

Y. Aihara and H. Koyama

Department of Aeronautics, Faculty of Engineering, University of Tokyo, Tokyo 113, Japan

Summary

Nonlinear development and secondary instability of Görtler vortices are investigated experimentally. The generation of horseshoe type vortices prior to the transition of the boundary layer is considered to be an inherent process in view of the nature of Görtler vortices in the linear region. The process is specifically attained through a three-dimensional distortion of the mean flow field due to nonlinear development of Görtler vortices and turbulent energy supply in the high shear layer near the edge of the boundary layer.
The entire process is considered to be partly common to one leading to the bursting phenomenon over a flat plate.

1. Introduction

The nonlinear development of Görtler vortices and the associated onset of fluctuations have been investigated for the explanation of the mechanism of early transition in the boundary layers along concave walls. Some proposals have been made concerning the transition, for instance, based on a three-dimensional distortion of Tollmien-Schlichting waves [1] or a meandering motion of Görtler vortices [2]. Our recent investigation reveals, however, that the more important and inherent process of transition is caused by a periodic formation of horseshoe type vortices interconnecting the pair of Görtler vortices [3], [4]. Therefore, it would be worthwhile to deal with our point of view about the transition process dependent upon the nature of Görtler vortices in the following section, before proceeding to the discussion of the present experiment.

2. Fundamental natures of Görtler vortices

2.1 Incompatibility of pressure field

The disturbances in velocity (u, v, w) and static pressure

p in the linear region of Görtler vortices are written as follows.

$$\begin{aligned} u &= \tilde{u}(y)\cos(\alpha z)e^{\beta x} \\ v &= \tilde{v}(y)\cos(\alpha z)e^{\beta x} \\ w &= \tilde{w}(y)\sin(\alpha z)e^{\beta x} \\ p &= \tilde{p}(y)\cos(\alpha z)e^{\beta x} \end{aligned} \tag{1}$$

where $\tilde{u}$, $\tilde{v}$, $\tilde{w}$ and $\tilde{p}$ are amplitude functions, α is the lateral wave number of Görtler vortices and β is the spatial amplification factor. x, y and z are the coordinates in streamwise direction, outward normal direction to the wall and lateral direction, respectively. As p is in phase with v, the extremum of induced pressure field does not appear at the center of the vortex but in the intermediate part between adjacent vortices. This feature differs from that of isolated vortices where the static pressure becomes minimum at the center of vortex, resulting in a balance between the centrifugal force and the pressure gradient force. Therefore, we may expect that the system of Görtler vortices needs some adjustment inevitably in relation to the incompatibility of the pressure field with continued amplification in the downstream direction.

A recent experiment concerning the effects of the streamwise pressure gradient on the secondary instability reveals actually that Görtler vortices are sensitive to the pressure field especially in the nonlinear region [4].

2.2 Vorticity distribution

Components of three-dimensional vorticities of Görtler vortices $(\omega_x, \omega_y, \omega_z)$ are obtained as follows.

$$\begin{aligned} \omega_x &= \frac{\partial w}{\partial y} - \frac{\partial v}{\partial z} = \left(\frac{d\tilde{w}}{dy} + \alpha\tilde{v}\right)\sin(\alpha z)e^{\beta x} \\ \omega_y &= \frac{\partial u}{\partial z} - \frac{\partial w}{\partial x} = (-\alpha\tilde{u}-\beta\tilde{w})\sin(\alpha z)e^{\beta x} \\ \omega_z &= \frac{\partial v}{\partial x} - \frac{\partial u}{\partial y} = \left(\beta\tilde{v} - \frac{d\tilde{u}}{dy}\right)\cos(\alpha z)e^{\beta x} \end{aligned} \tag{2}$$

As $|\tilde{u}| \gg |\tilde{v}|, |\tilde{w}|$ in the linear region of Görtler vortices [5], the following relations are derived for moderate α and β.

$$|\omega_x| \ll |\omega_y|, |\omega_z| \tag{3}$$

$$\omega_y \simeq -\alpha\tilde{u}\sin(\alpha z)e^{\beta x} \tag{4}$$

$$\omega_z \simeq -\frac{d\tilde{u}}{dy}\cos(\alpha z)e^{\beta x} \tag{5}$$

The inequality (3) shows that it is rather inadequate to look at Görtler vortices as longitudinal ones and the dominant components of vorticities are seen to be in normal and lateral directions.

A schematic diagram of three-dimensional vorticity is obtained as shown in Fig. 1, where a right hand system is employed. Inspection of the diagram suggests several possibilities of the vortex generation due to secondary instability of the system in the form of reasonable continuation of vortex axis. A periodic formation of horseshoe type vortices interconnecting the pair of Görtler vortices on both sides of section III (see Fig. 1) has been observed in the outer boundary layer (Fig. 2). This result is also explained reasonably in the following.

2.3 Horseshoe type vortices

Prior to the generation of horseshoe type vortices, the developed Görtler vortices change the lateral spacing in such a way that the ones on both sides of section III approach each other and make a pair. This brings about a change in the lateral distribution of streamwise velocity, i. e. , the accelerated part becomes wide and flat while the decelerated part does narrow and sharp. The change results in a further increase in $|\omega_y| \sim |\partial u/\partial z|$ on both sides of section III. The upwash v in section III increases with a reduction of the distance between the paired Görtler vortices, which renders the stationary velocity distribution in section III very unstable by forming a high shear layer near the edge of the boundary layer. The vortices shed from the high shear layer with ω_z are so bent that the vortices come into contact with the pair of Görtler vortices through an increased ω_y in the system (Fig. 1). Therefore, we can consider that the vortices thus generated are themselves necessary due to the secondary instability of Görtler vortices. Detailed study of the secondary instability is described in [3].

3. Apparatus

The experiments were carried out in a rectangular sectional curved channel, 300mm×300mm, made of transparent acrylic resin having a concave wall with a radius of curvature r=1m. The channel was attached to a low speed wind tunnel with a gap of 5mm so as to make the boundary layer develop from the leading edge of the channel. The streamwise length of the concave side was 2100mm (Fig. 3).

The main flow velocity, U_0, was defined at the outer edge of the boundary layer in section II, and was set at 1.7m/s throughout the present experiment. Turbulence level in the main flow was 0.13%. The static pressure distribution was satisfactorily uniform along the center line of the concave wall in the present experimental region. As for details of measurements, refer to [3] and [4].

4. Result and Discussion

A couple of Görtler vortices near the centerline of the concave wall was investigated in the range between x=425mm and 805mm. Reynolds number R_θ based on momentum thickness θ in section II, and Görtler parameter $G \equiv R_\theta (\theta/r)^{\frac{1}{2}}$ ranged from 1.70×10^2 to 5.78×10^2 and from 6.58 to 41.3, respectively.

4.1 Mean flow field

Normal distributions of u at x=425, 605 and 805mm are shown in Fig. 4 for a vortex located between z=4 and 12mm. In section I, the boundary layer thickness reduces and the velocity profile becomes more stable in the downstream direction. In section III, on the contrary, the boundary layer thickness increases and a characteristic velocity profile is formed consisting of an almost shearless region in the middle and a high shear layer near the edge of the boundary layer. The velocity profile in section II is deformed but its absolute value approaches that in section I in the early stage of nonlinear development of Görtler vortices corresponding to the change in lateral spacing of vortices (Fig. 4 (a)) as described in [3]. The tendency, however, is reversed in the downstream as shown in Figs. 4 (b) and (c). At x=805mm, there is almost no difference between sections II and III except near the wall and the boundary layer edge. The rapid changes in the mean flow field and the remarkable lateral nonuniformity in the boundary layer thickness can hardly be explained only by superposition of the induced flow due to Görtler vortices.

Normal distributions of v and w at x=605mm for the same vortex are shown in Fig. 5. According to linear theory, $\tilde{v}$ is of order of $\tilde{u}/10R_\theta$, but for developed Görtler vortices the absolute value of v becomes much larger keeping the normal distribution in the upstream region almost unchanged. This tendency was already recognized by Bippes and Görtler [2], and Aihara and Koyama [3]. A remarkable nonsymmetry of v in sections I and III (Fig. 5(a)), however, suggests some new considerations to account for the strong upwash through the most part of the boundary layer in the decelerated part of lateral periodicity in u.

The most reasonable explanation would be that a low speed air transported into the decelerated part by Görtler vortices forms a streamwise ridge on the wall so that the flow in the boundary layer suffers the local and three-dimensional displacement effects. The same sort of phenomenon is pointed out by Blackwelder and Eckelmann [6] for a pair of longitudinal vortices in the turbulent boundary layer along a flat plate. In our case, however, it is not clear at present whether a characteristic streamwise length exists or not for the ridge.

Figure 6 shows the lateral distributions of displacement thicknesses δ^* at three streamwise positions. δ^* varies periodically so that the displacement effect increases in section III and decreases in section I, and the effect becomes even stronger in the downstream direction except near section III at z=26mm. Stable regions where δ^* is decreasing in the downstream are rather restricted to the neighbourhood of respective sections I, whereas the state of section III seems to extend corresponding to the trends shown in Figs. 4(b) and (c). Mean velocity distributions in sections II and III are affected significantly by the formation of ridges. The ridge may be considered to promote the formation of a high shear layer. On the other hand, if we regard the amplitude of v in section I or w in section II as a measure of the intensity of Görtler vortices, then Görtler vortices seem to decay gradually in the downstream.

4.2 Fluctuations

As mentioned in [3], the spectra of fluctuations observed

in the present transitional region consist essentially of two components; the meandering motion of Görtler vortices with a low frequency and the secondary instability of Görtler vortices in the form of horseshoe type ones with a high frequency. Fluctuations are expressed with dash hereafter. As we are mainly interested in the secondary instability, we discuss the behaviours of both fluctuations separately with frequencies higher or lower than 8 Hz in the following.

As the growth of fluctuations is dependent upon the energy supply, we notice here two Reynolds stresses, $\overline{u'v'}$ and $\overline{u'w'}$. These are considered to be most important to the energy exchange between the mean flow and fluctuations in relation to the shears of mean flow, $\partial u/\partial y$ and $\partial u/\partial z$.

Results are discussed taking the data obtained at x=605mm as the most common. Figure 7 shows the distributions of u' with low and high frequencies. For f(frequency) < 8 Hz, the intensities of u' appear almost uniformly high through the boundary layer and are roughly unchanged in the downstream direction.

For f > 8 Hz, their relations to the mean shear of u are more obvious, and they are amplified continuously in the high shear layer in the downstream direction. These results suggest that the mechanism of growth is different for each fluctuation.

$\overline{u'v'}$ for f < 8 Hz is dominant over that for f > 8 Hz (Fig. 8), but as the turbulent energy supply from the mean flow is denoted by $-\overline{u'v'}\partial u/\partial y$, the former is inefficient to promote the meandering motion in relation to the distribution of u (Fig. 4(b)). On the other hand, $\overline{u'v'}$ for f > 8 Hz helps the extraction of energy from the mean flow in the high shear layer. The horseshoe type vortices are thus considered to depend on the formation of the high shear layer.

Figure 9 shows $\overline{u'w'}$ and the lateral turbulent transport for f < 8 Hz is again dominant. As $\partial u/\partial z$ is large in the boundary layer, the energy supply to the meandering motion, $-\overline{u'w'}\partial u/\partial z$, is mainly attained through the lateral movement of the row of Görtler vortices. The same sort of energy supply to horseshoe type vortices can be seen also in Fig. 9, probably due to the nonlinear interaction between two legs of horseshoe type vortices and the three-dimensionally distorted mean flow field.

Figure 10 shows a breakdown of the wave form. Spikes are generated from the high frequency waves, showing the importance of horseshoe type vortices to the boundary layer transition.

5. Concluding remarks

Experimentally investigated are a three-dimensional distortion of mean velocity field due to nonlinear development of Görtler vortices and the energy supply to two types of fluctuations. The results show that the horseshoe type vortices are generated and amplified as an inherent fluctuation and their development is important to the transition of the boundary layer.

References

1. Tani, I.; Aihara, Y.: Görtler vortices and boundary layer transition. Zeit. Angew. Math. Phys. 20 (1969) 609-618.

2. Bippes, H.; Görtler, H.: Dreidimensionale Störungen in der Grenzschicht an einer konkaven Wand. Acta Mech. 14 (1972) 251-267.

3. Aihara, Y.; Koyama, H.: Secondary instability of Görtler vortices - Formation of periodic three-dimensional coherent structure. Trans. Japan Soc. Aeron. Astron. 24 (1981)

4. Aihara, Y.; Sonoda, T.: Effects of pressure gradient on the secondary instability of Görtler vortices. AIAA 19th Aerospace Sci. Meeting, St. Louis, Missouri, AIAA-81-0197 (1981).

5. Görtler, H.: Über eine dreidimensionale Instabilität laminarer Grenzschichten an konkaven Wänden. Nachr. Acad. Wiss. Göttingen Math. Phys. Kl. 2, No. 1 (1940).

6. Blackwelder, R.F.; Eckelmann, H.: Streamwise vortices associated with the bursting phenomenon. J. Fluid Mech. 94 (1979) 577-594.

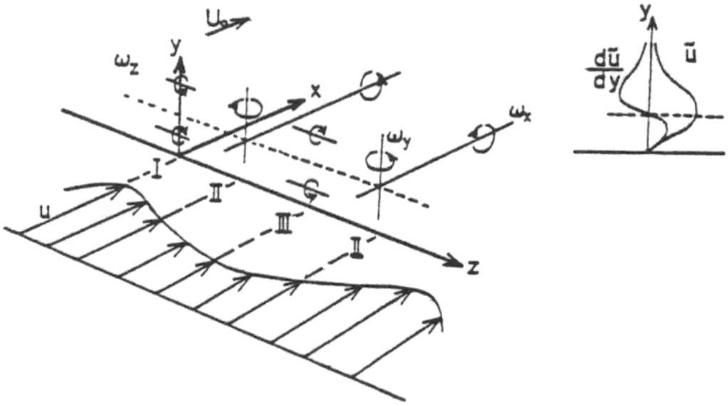

Fig.1 Vorticity of Görtler vortices

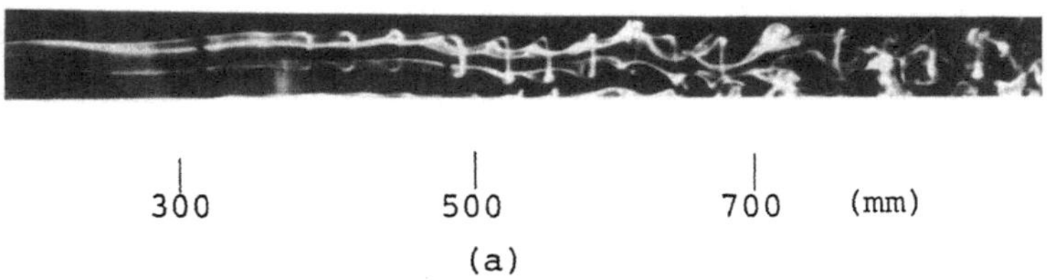

(b)

Fig.2 Flow visualizations of horseshoe type vortices
(a) Plan view (b) Side view

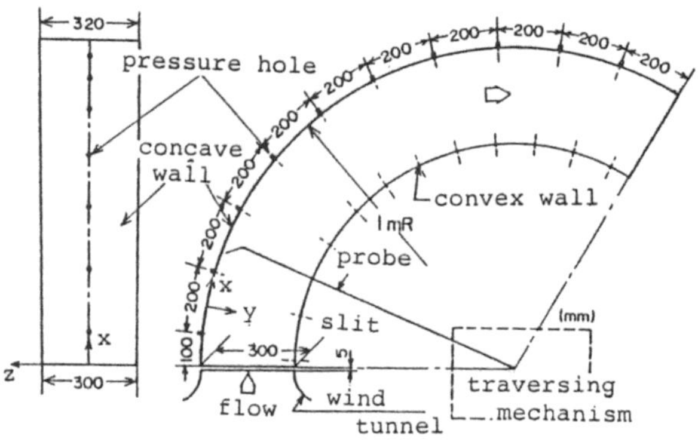

Fig.3 Curved channel

Fig.4 Streamwise mean velocity distributions
(a) x=425mm
(b) x=605mm
(c) x=805mm

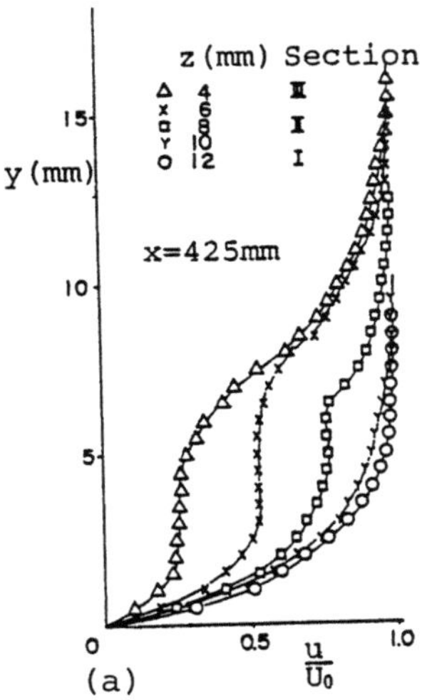

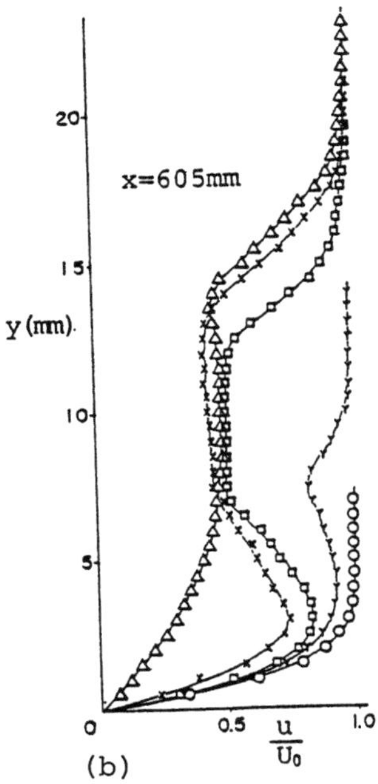

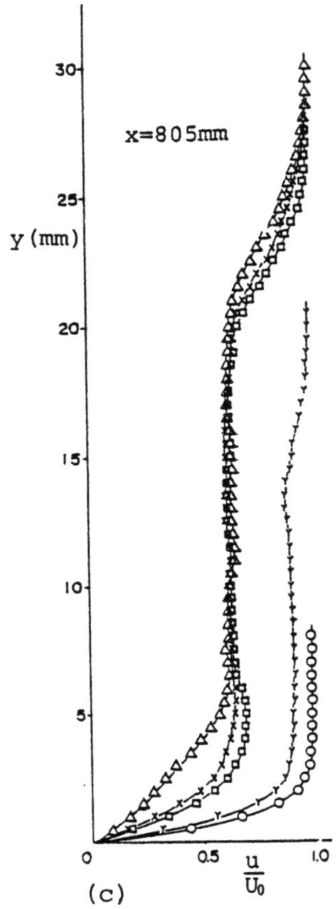

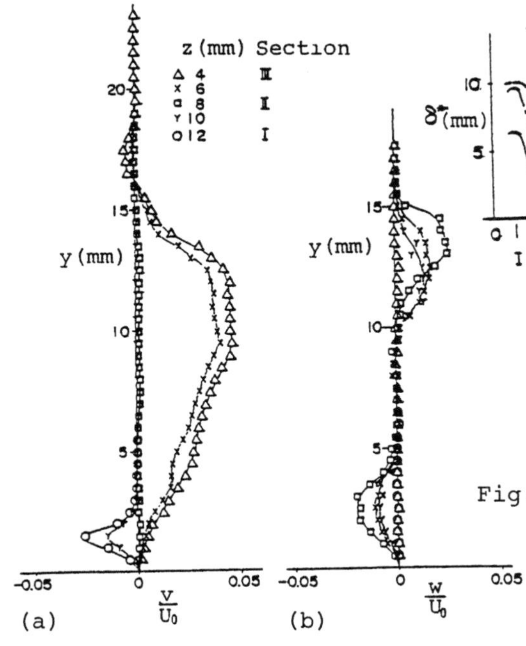

Fig.5 Mean velocity distributions in a cross section at x=605mm (a) v (b) w

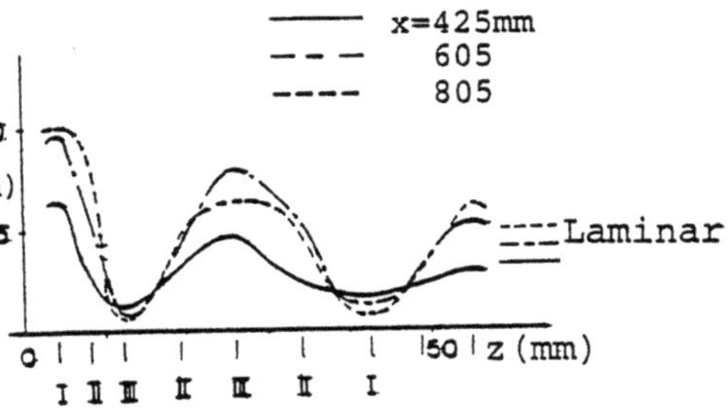

Fig.6 Lateral distributions of displacement thicknesses δ* at x=425, 605 and 805mm

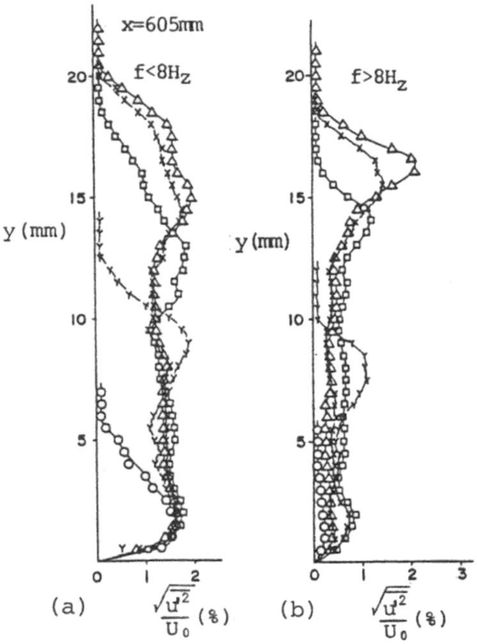

Fig.7 Velocity fluctuations u' at x=605mm (a) $f<8H_Z$ (b) $f>8H_Z$

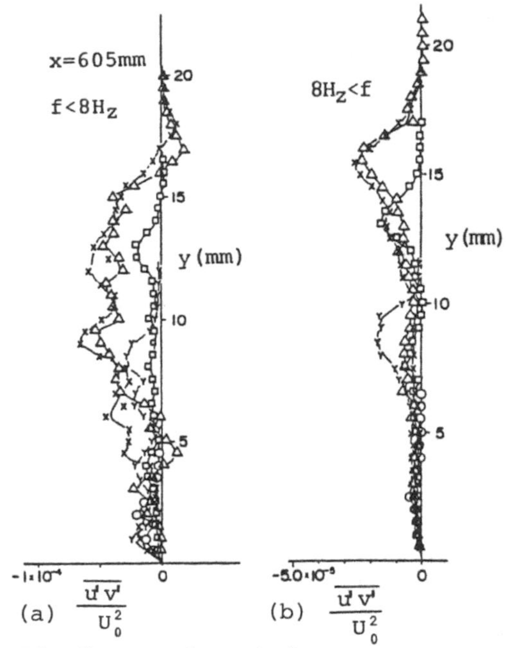

Fig.8 Normal turbulent transport $\overline{u'v'}/U_0^2$ at x=605mm (a) $f<8H_Z$ (b) $f>8H_Z$

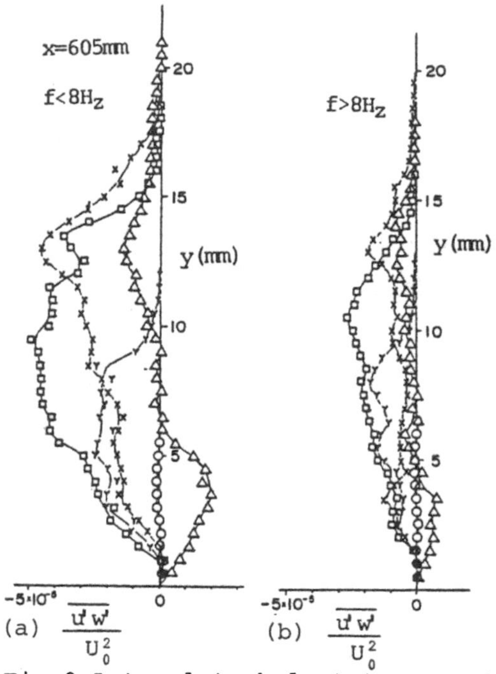

Fig.9 Lateral turbulent transport $u'w'/U_0^2$ at x=605mm (a) $f<8H_Z$ (b) $f>8H_Z$

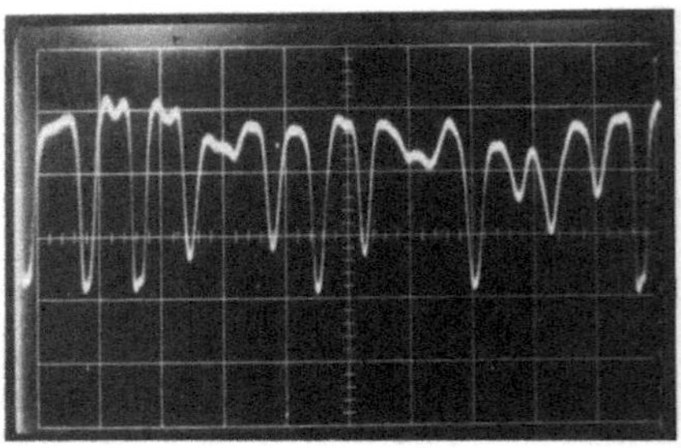

Stability of Nonparallel Developing Flow in Circular Ducts to Asymmetric Disturbances

V. K. GARG

Department of Mechanical Engineering
Indian Institute of Technology Kanpur, India

Summary

Linear spatial stability of the nonparallel developing flow in a rigid circular pipe and in a concentric annulus is described at several axial locations for asymmetric disturbances. The main flow velocity profile is obtained by an implicit finite-difference technique assuming uniform flow at entry to the duct. The method of multiple scales is used to account for all nonparallel effects. It is found that the nonparallel developing flow is most unstable to asymmetric disturbances with azimuthal wavenumber $n = 1$. Axisymmetric disturbances are, however, more unstable than asymmetric disturbances with $n \geq 2$ except in the near-entry region. The present results compare well with the experimental data of Sarpkaya [1] for pipe flow.

Introduction

Linear stability analysis of the developing flow in a circular pipe finds the flow to be more unstable to axisymmetric disturbances in comparison to the asymmetric disturbances in the near-entry region. This result is, however, obtained when the developing flow is assumed to be parallel though in the near-entry region it is certainly not so. While Shen et al.[2] did consider the effect of radial velocity component on the stability of developing flow in a pipe, they ignored other important nonparallel effects such as the streamwise variation of the eigenfunction, growth rate and wavenumber. It is now well known [3] that these nonparallel effects change considerably the stability characteristics of the developing flow for an axisymmetric disturbance. It is also known [4] that the Hagen-Poiseuille flow is least stable to asymmetric disturbances with $n = 1$. Results are therefore presented here for the stability of the nonparallel developing flow in a pipe and in a concentric annulus to asymmetric disturbances.

Analysis

Considering the flow of an incompressible, Newtonian fluid in the inlet region of a circular pipe or a concentric annulus of inner and outer radii a and b respectively, we nondimensionalize the variables with respect to the average velocity of flow in the duct, and radius a of pipe or (b-a)/2 for the annulus. The infinitesimal disturbance with dimensionless velocity components u, v and w, and dimensionless pressure p is considered superimposed on the main flow. Knowing that the developing flow velocity components U (axial) and V (radial) are slowly varying functions of X, the dimensionless axial distance from the inlet section, we introduce another independent variable X_1 such that $X_1 = \varepsilon X$, where ε is a small dimensionless parameter characterizing the nonparallelism of the flow. In the near-inlet region, ε is taken as $Re^{-1/2}$ [3], where Re is the Reynolds number of the flow. We take the disturbance to be of the form

$$u = [u_o(X_1,r) + \varepsilon u_1(X_1,r) + \varepsilon^2 u_2(X_1,r) + \ldots] e^{i\eta}, \qquad (1)$$

and similar expressions for v, w and p, with

$$\frac{\partial\eta}{\partial T} = -\omega, \quad \frac{\partial\eta}{\partial X} = k_o(X_1), \text{ and } \frac{\partial\eta}{\partial\theta} = n, \qquad (2)$$

where r is the dimensionless radial coordinate, T is dimensionless time, ω is the dimensionless frequency, the real part of k_o is the wavenumber and its imaginary part is the spatial growth rate for the parallel flow, and θ is the azimuthal coordinate. Using transformation relations between (X, T) and (X_1, η) planes, equating like powers of ε to zero, and denoting a derivative with respect to r by D, the continuity and linearized Navier-Stokes equations for the disturbance yield

order ε^o:

$$L_u(u_o) \equiv D^2u_o + \frac{Du_o}{r} - Au_o - Re\, v_o\, DU - ik_o p_o = 0,$$

$$L_v(v_o) \equiv Dv_o + \frac{v_o}{r} + ik_o u_o + \frac{inw_o}{r} = 0,$$

$$L_w(w_o) \equiv D^2w_o + \frac{Dw_o}{r} - (A + \frac{1}{r^2})\, w_o + \frac{2inv_o}{r^2} - \frac{inp_o}{r} = 0,$$

$$L_p(p_o) \equiv Dp_o + Av_o + ik_o Du_o + \frac{in}{r} Dw_o + \frac{inw_o}{r^2} = 0, \qquad (3)$$

where $A = \frac{n^2}{r^2} + k_o^2 + iRe\,(k_o U - \omega)$, and p_o stands for Re p_o, with the boundary conditions

$$\left.\begin{aligned} u_o &= v_o = w_o = 0 \quad \text{at } r = 1, \\ p_o &= u_o = v_o + iw_o = 0 \text{ at } r = 0 \text{ for } n=1, \\ p_o &= u_o = v_o = w_o = 0 \text{ at } r = 0 \text{ for } n>1, \end{aligned}\right\} \text{for pipe} \quad (4)$$

$$\left.\begin{aligned} \text{and} \quad u_o &= v_o = w_o = 0 \text{ at } r = r_1 \text{ and } r = r_2, \\ \text{where} \quad r_1 &= 2a/(b-a), \text{ and } r_2 = 2b/(b-a), \end{aligned}\right\} \text{for annulus} \quad (5)$$

order ε^1:

$$L_u(u_1) = Re\left(U \frac{\partial u_o}{\partial X_1} + VDu_o + u_o \frac{\partial U}{\partial X_1}\right) + \frac{\partial p_o}{\partial X_1} - i\left(2k_o \frac{\partial}{\partial X_1} + \frac{dk_o}{dX_1}\right)u_o,$$

$$L_v(v_1) = -\frac{\partial u_o}{\partial X_1},$$

$$L_w(w_1) = Re\left(U \frac{\partial w_o}{\partial X_1} + VDw_o + \frac{Vw_o}{r}\right) - i\left(2k_o \frac{\partial}{\partial X_1} + \frac{dk_o}{dX_1}\right) w_o,$$

$$L_p(p_1) = -Re\left(U \frac{\partial v_o}{\partial X_1} + VDv_o + v_o DV\right) - \frac{\partial^2 u_o}{\partial X_1 \partial r} + i\left(2k_o \frac{\partial}{\partial X_1} + \frac{dk_o}{dX_1}\right)v_o, \qquad (6)$$

with boundary conditions on u_1, etc. similar to those on u_o, etc.

The eigenvalue problem defined by eqns. (3) and (4) or (3) and (5) is the famous Orr-Sommerfeld problem for the parallel flow. Eqns. (6) show that the axial variation of the eigenvalue and eigenfunction, and radial component of main flow velocity determine the first order corrections over the parallel flow theory. For given values of ω, Re, $U(X_1, r)$ and $V(X_1, r)$, the solution of (3) and (4) or (3) and (5) may be expressed as

$$\begin{aligned} u_o &= B(X_1)\,\beta_1(r; X_1), \qquad v_o = B(X_1)\,\beta_2(r; X_1), \\ w_o &= B(X_1)\,\beta_3(r; X_1), \text{ and } p_o = B(X_1)\,\beta_4(r; X_1), \end{aligned} \qquad (7)$$

where β_1, β_2, etc. are the eigenfunctions. The amplitude function $B(X_1)$ is given by

$$\frac{dB}{dX_1} + i\, k_1(X_1)\, B = 0, \tag{8}$$

where k_1 is the ratio of quadratures involving products of eigenfunctions and their derivatives for the original and adjoint problems. The solution of (8) is

$$B(X_1) = B_o \exp\,[i \int k_1(X_1)\, dX_1], \tag{9}$$

where B_o is an arbitrary constant of integration. Thus to the first approximation

$$u = B_o \beta_1(r;\, X_1) \exp\,[i \int (k_o + \varepsilon k_1) dX - i\omega T + in\theta], \tag{10}$$

and similarly for v, w and p. The nonparallelism of the flow therefore makes the eigenvalue k_o a function of X_1, produces a correction $\varepsilon\, k_1(X_1)$ to k_o and causes the mode shape to vary both in the axial and radial directions.

Computational Procedure

The main flow velocity field was found using an implicit finite-difference technique [5,6]. This yields the velocity components U and V as functions of $\bar{X} = X/Re$ and r. Eqns.(3) were integrated using the fourth order Runge-Kutta method. Convergence to the eigenvalue, k_o, was achieved by means of Muller's technique [7]. Selective application of the Gram-Schmidt orthonormalization technique [8] was used to keep the solution vectors linearly independent during numerical integration.

With k_o and eigenfunctions known, a similar procedure was used to solve the adjoint problem

$$D^2u^* + \frac{Du^*}{r} + ik_o p^* - (A + k_o^2)\, u^* - \frac{nk_o w^*}{r} = 0,$$

$$Dv^* + \frac{v^*}{r} + ik_o u^* + \frac{inw^*}{r} = 0,$$

$$D^2w^* + \frac{Dw^*}{r} - (A + \frac{n^2+1}{r^2}) w^* - \frac{nk_o u^*}{r} + \frac{2inv^*}{r^2} + \frac{inp^*}{r} = 0,$$

$$Dp^* + Re\, u^* DU - Av^* - 2inw^*/r^2 = 0, \tag{11}$$

with the boundary conditions

$$\left.\begin{aligned} &u^* = 0 \text{ at } r = 0, \quad rp^* - nv^* \to 0 \text{ as } r \to 0, \\ &v^* + iw^* = 0 \text{ at } r = 0 \text{ for } n = 1, \\ &v^* = w^* = 0 \text{ at } r = 0 \quad \text{for } n > 1, \\ &u^* = v^* = w^* = 0 \text{ at } r = 1 \end{aligned}\right\} \text{for pipe} \quad (12)$$

and $\quad u^* = v^* = w^* = 0$ at $r = r_1$ and r_2 for annulus, (13)

the difference being that no iteration is necessary since the adjoint problem has the same eigenvalue as the original problem. For the pipe flow, it is instructive to note that the solution of the adjoint problem is singular at $r = 0$ for $n = 1$. In fact, for r small, we find

$$\left.\begin{aligned} u^* &= r^n (U_1 + U_2 r^2 + \ldots), \\ v^* &= r^{n-1} (V_1 + V_2 r^2 + \ldots), \\ w^* &= r^{n-1} (iV_1 + W_2 r^2 + \ldots), \\ \text{and} \quad p^* &= r^{n-2} (nV_1 + P_1 r^2 + \ldots), \end{aligned}\right\} \quad (14)$$

where all the coefficients in the above expansions are linear functions of U_1, V_1 and P_1.

Results and Discussion

With the inclusion of all nonparallel effects, the growth rate becomes a function of the disturbance property as well as of the axial and radial coordinates. Therefore, growth rates for the three components of disturbance velocity were found at various axial locations and for various diameter ratios for the annular flow. Of all the growth rates, it was found that g_w, the growth rate of the disturbance tangential velocity component, was the largest. Also, since g_w is a function of both the axial and radial coordinates, its maximum value occurring at r_w at a given axial location was selected in order to limit the number of neutral curves; r_w corresponds approximately to the edge of the

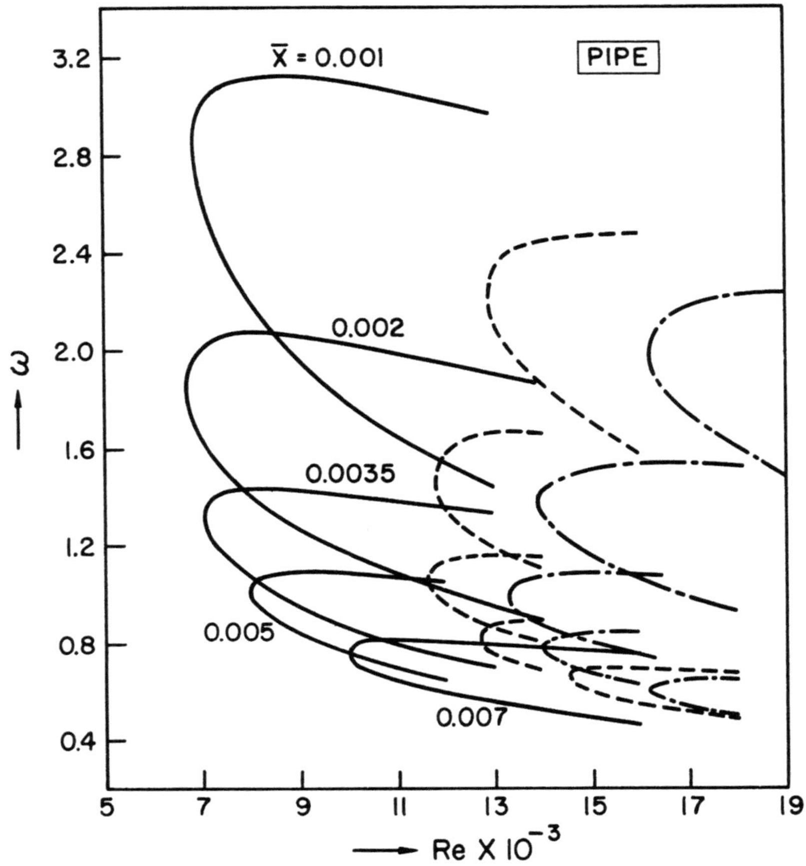

Fig. 1. Neutral curves at various axial locations for $n = 1$. ——, based on $g_w(\overline{X}, r_w)$; ----, based on $g_v(\overline{X}, 0)$; — - —, based on parallel flow theory.

boundary layer, which for annular flow is on either the inner or the outer wall depending upon the diameter ratio and axial location. For the pipe flow, the maximum value of g_v, the growth rate of the disturbance radial velocity component, was found to occur at the pipe axis. Figures 1 and 2 show the neutral curves corresponding to $g_w(\overline{X}, r_w)$ for the $n = 1$ disturbance at different $\overline{X}$ for the pipe and annular flow respectively. Also included in Fig. 1 are the neutral curves based on g_v $(\overline{X}, 0)$ and the parallel flow theory. In comparison to the results for the parallel flow theory, the nonparallel effects make the flow unstable at lower

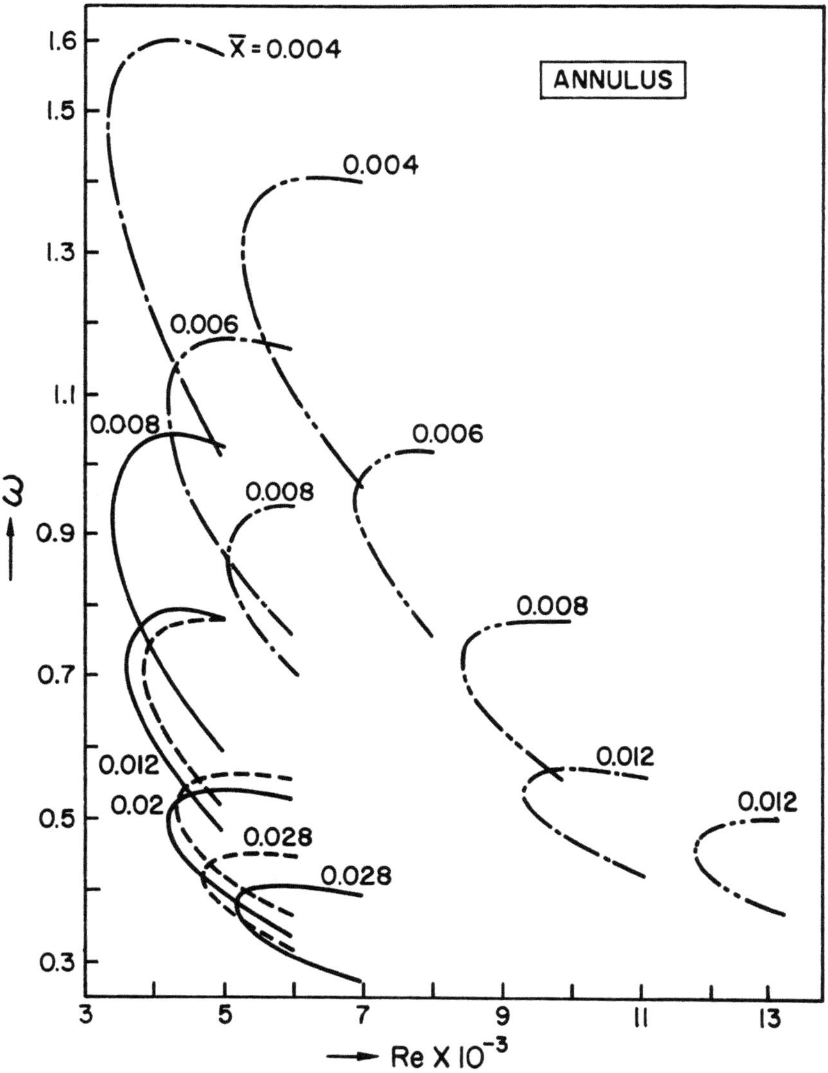

Fig. 2. Neutral curves based on $g_w(\bar{X}, r_w)$ for various γ and $\bar{X}$ and $n = 1$. ——, $\gamma = 0.01$; ---, $\gamma = 0.1$; — — —, $\gamma = 0.4$; — - - - —, $\gamma = 0.8$.

Reynolds numbers and for a wider range of frequencies.

Figures 3 and 4 show the axial variation of critical Reynolds number Re_c based on $g_w(\bar{X}, r_w)$, g_v $(\bar{X}, 0)$ and the parallel flow theory for the pipe and annular flow respectively. Fig. 3 also shows the axial variation of critical frequency ω_c and displays

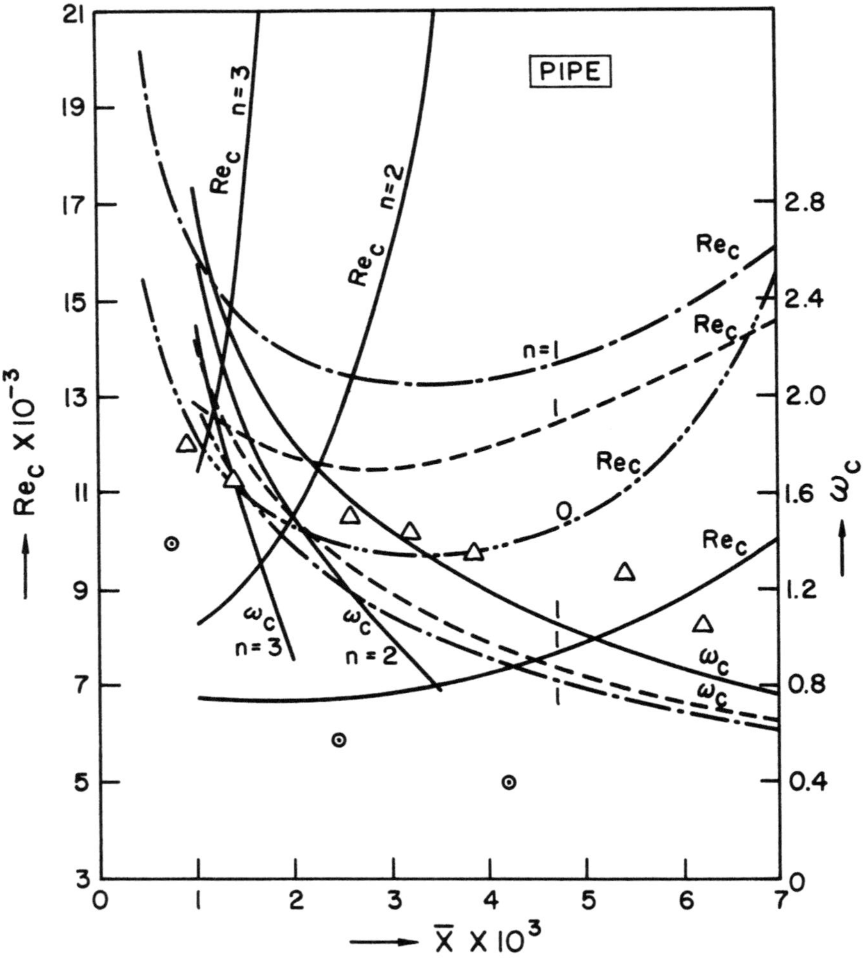

Fig. 3. Re_c and ω_c vs. $\bar{X}$. ——, based on $g_w(\bar{X}, r_w)$; ----, based on $g_v(\bar{X}, 0)$; — - —, based on parallel flow theory; — — — —, based on $g_\psi(\bar{X}, 0)$; ⊙, experimental Re_c; △, experimental ω_c.

the Re_c and ω_c values found experimentally by Sarpkaya [1] for n = 1 disturbance. For the pipe flow, the Re_c-$\bar{X}$ curves pass through a minima except for $n \geq 2$, and the present results for n = 1 disturbance are very close to the experimental data. Any discrepancy may be due to nonlinear effects in experimental data [1]. For the annular flow, the behaviour of Re_c-$\bar{X}$ curves appears to be quite different with the inclusion of nonparallel effects. For both the flows, Re_c based on the parallel flow theory is as

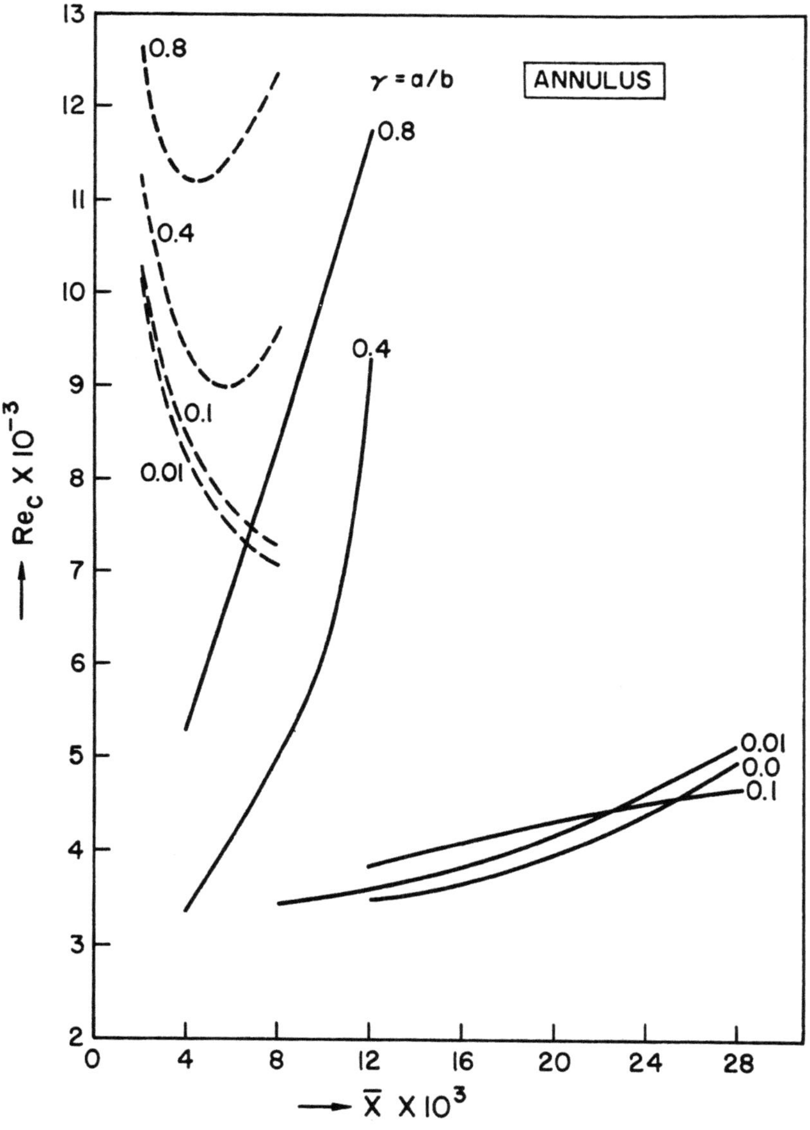

Fig. 4. Critical Reynolds number vs $\bar{X}$ for n = 1. —, based on $g_w(\bar{X}, r_w)$; ---, based on parallel flow theory.

much as two to three times that corresponding to $g_w(\bar{X}, r_w)$ for the n = 1 disturbance. Comparison of results for various asymmetric disturbances for pipe flow shows that n = 1 disturbance is the most unstable of all but axisymmetric disturbance (n = 0) is more unstable than asymmetric disturbance with $n \geq 2$ except in the near-entry region. It may be pointed out that in Fig.3, the Re_c-$\overline{X}$ curve for $n = 0$ corresponds to the growth rate of streamfunction ψ at the pipe axis.

Though not shown here, it was found that while the parallel developing flow in both the ducts is more unstable to axisymmetric disturbance than to the asymmetric disturbance with $n = 1$ in the near-entry region, the reverse is true for the entire entry region for the nonparallel flow. Thus, considering the developing flow in ducts to be parallel for stability analysis is highly improper.

References

1. Sarpkaya, T.: J. Fluid Mech. 68 (1975) 345.
2. Shen, F.C.T.; Chen, T.S.; Huang, L.M.: J.Appl.Mech. 43 (1976) 209.
3. Garg, V.K.; Gupta, S.C.: J.Appl.Mech. 48 (1981) 243.
4. Garg, V.K.; Rouleau, W.T.: J. Fluid Mech. 54 (1972) 113.
5. Hornbeck, R.W.: Appl. Sci. Res. A13 (1963) 224.
6. Gupta, S.C.; Garg, V.K.: Comp. Meth. Appl. Mech. Engng. 26 (1981).
7. Muller, D.E.: Math. Tables Other Aids Comp. 10 (1956) 208.
8. Garg, V.K.: Comp. Meth. Appl. Mech. Engng. 22 (1980) 87.

A Difficult Numerical Calculation Concerning the Stability of the Blasius Boundary Layer

A. DAVEY

School of Mathematics,
University of Newcastle upon Tyne,
England

Summary

At the Herrenalb 1969 Symposium on the Instability of Continuous Systems Tatsumi & Gotoh presented a paper on the structure of a small damped disturbance in the Blasius boundary layer when the Reynolds number is very large. They showed that for such a disturbance the eigenfunction of the Orr-Sommerfeld equation has the form of a concentrated wavepacket in the neighbourhood of the critical layer.

We explain how to use the Compound Matrix Method to calculate such an eigenfunction and we present a solution when the Reynolds number is 10^6 which is in excellent agreement with the theory of Tatsumi & Gotoh.

§1. Introduction

Mack [1], (see Figure 3 on page 508), has found numerically, at least for the particular wavenumber for which he did his calculations, that there is a damped eigenvalue of the Orr-Sommerfeld equation for the Blasius boundary-layer velocity profile which approaches an 'inviscid' limit as the Reynolds number becomes infinite. By an 'inviscid' limit we mean that the limiting value may be obtained by solving Rayleigh's equation using an appropriately chosen contour below the real axis. Tatsumi & Gotoh [2], (see especially Figure 3), have shown analytically that for such an eigenmode the structure of the eigenfunction of the Orr-Sommerfeld equation takes the form of a concentrated wavepacket in the neighbourhood of the critical layer.

In the last paragraph of Lin [3] he mentions that such a solution with finite damping closely resembles the structure of turbulence. "The fluid motion exhibits a highly oscillatory behaviour over a finite part of the field .... the increasingly refined spatial structure of the fluid motion is just enough to counterbalance the vanishing of viscosity and to maintain the effects of the viscous terms in the Navier-Stokes equations of motion."

When the Reynolds number is large the Orr-Sommerfeld equation is not an easy equation to solve numerically, (because the characteristic values are widely separated), and it would seem to be especially difficult to calculate an eigenfunction which takes the form of a concentrated wavepacket, perhaps bearing some resemblance to a 'turbulent spot'. We felt that it would be interesting therefore to use the Compound Matrix Method, which is a relatively new method for solving Orr-Sommerfeld type problems, to try and produce one of these eigenfunctions; partly because this type of problem is a good test problem for any new numerical method and partly to verify the work of Tatsumi & Gotoh.

§2. Statement of the Orr-Sommerfeld problem

The simplest formulation of the Orr-Sommerfeld problem for the Blasius boundary layer, using the parallel flow approximation, is

$$\{D^2 - \alpha^2 - i\alpha R(f' - c)\}\{D^2 - \alpha^2\}\varphi + i\alpha R f''' \varphi = 0, \tag{1}$$

where

$$D \equiv \frac{d}{d\eta} \equiv ', \tag{2}$$

and

$$f''' + f f'' = 0, \tag{3}$$

$$f(0) = f'(0) = 0, \; f' \to 1 \text{ as } \eta \to \infty. \tag{4}$$

The boundary conditions for φ are

$$\varphi(0) = \varphi'(0) = 0, \tag{5}$$

and

$$\varphi \sim Ae^{-\alpha\eta} + Be^{-p\eta}, \text{ as } \eta \to \infty, \tag{6}$$

where

$$p^2 = \alpha^2 + i\alpha R(1 - c), \; \Re(p) > 0. \tag{7}$$

Since numerical calculations have to be done on a finite interval it is necessary to translate (6) into a suitable form for some large value of $\eta = \eta_1$, say, where η_1 is usually of order 5 or 10. A good way to do this is to find two linearly independent operators which force the coefficients of $e^{\alpha\eta}$ and $e^{p\eta}$ in the general solution for φ when η is large, both to be zero, the simplest two such operators are $(D + \alpha)(D + p)$ and $D(D + \alpha)(D + p)$ and so for the outer boundary condition we use

$$\left.\begin{array}{l}(D+\alpha)(D+p)\varphi = 0,\\ D(D+\alpha)(D+p)\varphi = 0,\end{array}\right\} \text{ at } \eta = \eta_1. \tag{8}$$

In the above 'virgin' formulation calculation of the neutral stability curve for the usual primary mode yields a 'nose' at $\alpha = .24965$ and $R = 426.585$ because the length scale is the boundary layer thickness δ where

$$\delta = \sqrt{\frac{2\nu x}{U}}, \tag{9}$$

ν being the kinematic viscosity of the fluid, x measures distance downstream from the leading edge and U is the free-stream speed. In our opinion the best way to do numerical calculations is to use the above-mentioned virgin co-ordinates, and then to scale the results if one wishes to use a different length scale.

For example, the standard notation is to use the displacement thickness δ_1 of the boundary layer as the length scale, where, to the number of significant figures shown,

$$\delta_1 = 1.2167806216\delta. \tag{10}$$

So if α_s, R_s are the wavenumber and Reynolds number respectively in the standard notation then

$$\alpha_s = \frac{\delta_1}{\delta}\alpha, \; R_s = \frac{\delta_1}{\delta}R, \tag{11}$$

and the 'nose' of the neutral stability curve is at $\alpha_s = \alpha_c$, $R_s = R_c$, where

$$\alpha_c = .30377, \; R_c = 519.060. \tag{12}$$

In the notation used by Mack [1] his wavenumber α_M and his Reynolds number R_M are related to our virgin α, R by

$$\alpha_M = \frac{\alpha}{\sqrt{2}}, \quad R_M = \frac{R}{\sqrt{2}}, \tag{13}$$

so that in Mack's notation the critical Reynolds number is about 300 or so. The calculations which we report in this paper are all for $\alpha_M = 0.179$, as used by Mack, and the eigenfunction which we show in Figure 1 is for $R_M = 10^6$.

Before we explain in the next section how to use the Compound Matrix Method to solve (1), (5), (8) let us first explain briefly how we calculate the Blasius function $f(\eta)$. For any particular set of data values we use a Runge-Kutta type of integration scheme from the large value of η, namely η_1, to $\eta = 0$ and we integrate the Compound Matrix equations and the equation for $f(\eta)$ simultaneously. To determine suitable starting values for f, f', f'' when $\eta = \eta_1$ we make use of the asymptotic expansion of $f(\eta)$ for large η and set

$$
\begin{aligned}
f &= \eta - a + be^{-\frac{1}{2}(\eta - a)^2}\{(\eta - a)^{-2} - 3(\eta - a)^{-4} + 15(\eta - a)^{-6} \\
&\quad -105(\eta - a)^{-8} + 945(\eta - a)^{-10}\}, \\
f' &= 1 - be^{-\frac{1}{2}(\eta - a)^2}\{(\eta - a)^{-1} - (\eta - a)^{-3} + 3(\eta - a)^{-5} \\
&\qquad -15(\eta - a)^{-7} + 105(\eta - a)^{-9}\}, \\
f'' &= be^{-\frac{1}{2}(\eta - a)^2},
\end{aligned} \tag{14}
$$

where a, b are constants. Now of course the values of a, b are known from the asymptotic theory for $f(\eta)$, in fact $a = \delta_1/\delta$ and $b \doteq .33054$. However, the accuracy of the calculation may be improved, given η_1 and the mesh to be used, by allowing a, b to vary slightly from their known values and doing some preparatory runs selecting a, b via a Newton-Raphson iteration procedure so as to ensure that $f(0) = f'(0) = 0$. In practice we found that for different η_1 and different meshes a could be kept fixed at δ_1/δ, (see(10)), but that b might vary by about 10^{-7}.

§3. The Compound Matrix Method of Solution

Since we originally did this work (August 1977 - December 1978, regrettably unpublished) an excellent paper by Ng & Reid [4] has appeared which explains how to use the Compound Matrix Method to solve the Orr-Sommerfeld equation for the Blasius problem. Consequently we only give below the essential details of the method, for more information the reader should consult Ng & Reid's paper.

Let $\underset{\sim}{\varphi} \equiv [\varphi, \varphi', \varphi'', \varphi''']^T$ and suppose that $\underset{\sim}{\varphi}_1$ and $\underset{\sim}{\varphi}_2$ are two linearly independent solutions of (1) which satisfy the outer boundary conditions (8) at $\eta = \eta_1$, for example $\underset{\sim}{\varphi}_1$, $\underset{\sim}{\varphi}_2$ could be chosen to be orthonormal at $\eta = \eta_1$. If we integrate (1) from $\eta = \eta_1$ towards $\eta = 0$, using a Runge-Kutta type of routine, then because the general solution of (1) is dominated by a single term of order $e^{-p\eta}$ as η decreases $\underset{\sim}{\varphi}_1$, $\underset{\sim}{\varphi}_2$ become almost parallel. A consequence of this is that when c is iterated upon to satisfy the boundary conditions (5) at $\eta = 0$ there is a considerable loss of significant figures at $\eta = 0$ when we try to evaluate the secular determinant. The point is that as the integration proceeds $\underset{\sim}{\varphi}_1$, $\underset{\sim}{\varphi}_2$ span the subspace of all those solutions which satisfy (1), (8) more and more badly and so a special method such as orthonormalization or the Riccati method has to be used. An alternative method is to calculate instead the 'normal' to this subspace, that is $\underset{\sim}{\varphi}_1 \wedge \underset{\sim}{\varphi}_2$, and this is the essence of the Compound Matrix Method.

The components of this 'normal' are, in lexical order,

$$y_1 = \varphi_1\varphi_2' - \varphi_1'\varphi_2, \quad y_4 = \varphi_1'\varphi_2'' - \varphi_1''\varphi_2',$$
$$y_2 = \varphi_1\varphi_2'' - \varphi_1''\varphi_2, \quad y_5 = \varphi_1'\varphi_2''' - \varphi_1'''\varphi_2', \tag{15}$$
$$y_3 = \varphi_1\varphi_2''' - \varphi_1'''\varphi_2, \quad y_6 = \varphi_1''\varphi_2''' - \varphi_1'''\varphi_2'',$$

and they satisfy the quadratic identity

$$y_1y_6 - y_2y_5 + y_3y_4 = 0, \tag{16}$$

which is a useful check on the accuracy of the numerical integration. The closed differential system for $y_1 - y_6$ may be found by differentiating (15) and using (1) and it is the <u>linear</u> system

$$\begin{aligned} y_1' &= y_2, \\ y_2' &= y_3 + y_4, \\ y_3' &= a_1y_2 + y_5, \\ y_4' &= y_5, \\ y_5' &= a_2y_1 + a_1y_4 + y_6, \\ y_6' &= a_2y_2, \end{aligned} \tag{17}$$

where

$$\begin{aligned} a_1 &\equiv 2\alpha^2 + i\alpha R(f' - c), \\ a_2 &\equiv \alpha^4 + i\alpha^3R(f' - c) + i\alpha Rf'''. \end{aligned} \tag{18}$$

Since (17) is linear and homogeneous we may set $y_1 = 1$ when $\eta = \eta_1$ and it then follows from (8), (15) that the initial values for $y_1 - y_6$ at $\eta = \eta_1$ are

$$\begin{aligned} y_1 &= 1, \\ y_2 &= -(\alpha + p), \\ y_3 &= \alpha^2 + \alpha p + p^2, \\ y_4 &= \alpha p, \\ y_5 &= -\alpha p(\alpha + p), \\ y_6 &= \alpha^2p^2. \end{aligned} \tag{19}$$

We now integrate (17), (19) simultaneously with (3), (14) from $\eta = \eta_1$ to $\eta = 0$. At $\eta = 0$ the boundary condition (5) implies that y_1 should be zero there, so the eigenvalue c is iterated upon using a Newton-Raphson process until $y_1 = 0$ at $\eta = 0$ to within some specified tolerance on c in the usual manner. Thus the eigenvalue c is determined, its accuracy can be checked by repeating the process using instead of the Orr-Sommerfeld equation its Adjoint equation

$$\{D^2 - \alpha^2 - i\alpha R(f' - c)\}\{D^2 - \alpha^2\}\theta - 2i\alpha Rf''\theta' = 0, \qquad (20)$$

where θ is the Adjoint function. The difference between the two values of c so obtained is usually an excellent guide to their accuracy and if the discrepancy is not sufficiently small then more integration steps can be used and/or η_1 may be altered to a different value until the required accuracy is obtained.

Having thus found c and $y_1 - y_6$ we may now determine the eigenfunction φ by reasoning that there must exist constants λ, μ such that

$$\varphi = \lambda\varphi_1 + \mu\varphi_2,$$

so that also

$$\begin{aligned} \varphi' &= \lambda\varphi_1' + \mu\varphi_2', \\ \varphi'' &= \lambda\varphi_1'' + \mu\varphi_2'', \end{aligned} \qquad (21)$$

and

$$\varphi''' = \lambda\varphi_1''' + \mu\varphi_2'''.$$

The constants λ and μ may be eliminated from (21) in four different ways and if this is done then we obtain

$$\begin{aligned} y_1\varphi'' - y_2\varphi' + y_4\varphi &= 0, \\ y_1\varphi''' - y_3\varphi' + y_5\varphi &= 0, \\ y_2\varphi''' - y_3\varphi'' + y_6\varphi &= 0, \\ y_4\varphi''' - y_5\varphi'' + y_6\varphi' &= 0, \end{aligned} \qquad (22)$$

and so we may determine φ by integrating one of these equations from $\eta = 0$ to $\eta = \eta_1$. Since $y_1 = 0$ when $\eta = 0$ we cannot easily use either of the first two equations of (22) and so we must use either the third or fourth of (22). By trial and error we found that for the Blasius problem the third of (22) gave slightly more accurate results than the fourth equation.

Hence we determined φ by integrating

$$y_2\varphi''' - y_3\varphi'' + y_6\varphi = 0, \qquad (23)$$

from $\eta = 0$ to $\eta = \eta_1$ with the particular normalization $\varphi''(0) = (1 - i)R^{\frac{1}{2}}$. In order to maintain high accuracy we found it expedient to simultaneously integrate (17), occasionally resetting $y_1 - y_6$ to previously stored values, however, again to maintain accuracy - see the discussion on page 40 of [5]. The eigenfunction φ obtained by the compound matrix or 'normal' method described

FIGURE 1

EIGENFUNCTION of the ORR-SOMMERFELD EQUATION on [0,∞)

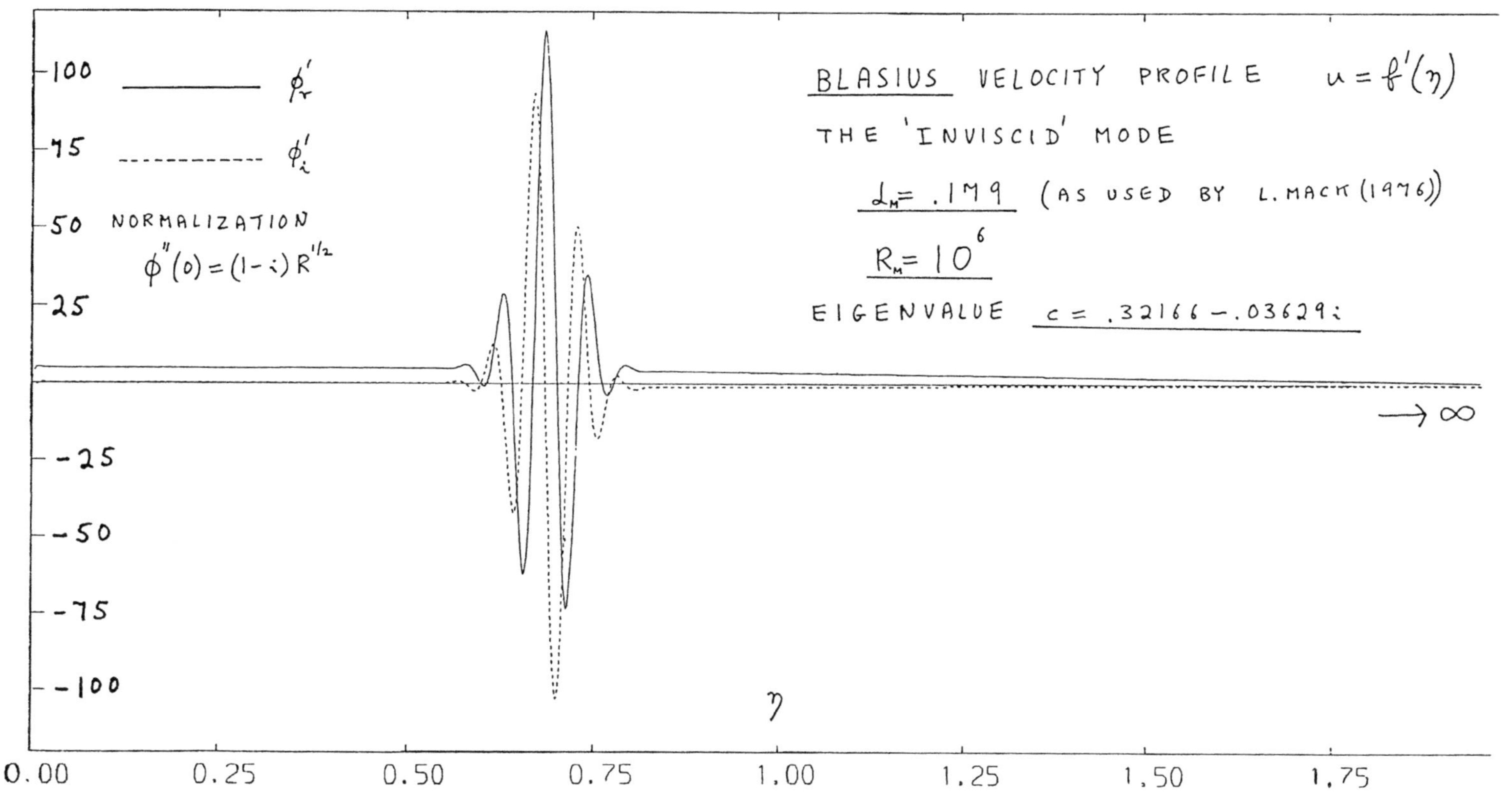

above is very accurate and this is essentially because the method monitors the normal to the subspace spanned by φ_1 and φ_2. Although φ_1, φ_2 span their subspace more and more badly as the integration proceeds note that the method never requires any detailed knowledge of φ_1, φ_2 - nor even which particular φ_1, φ_2 they are! For like $y_1 - y_6$ all of the coefficients in (23) are also components of the normal to the subspace which contains φ_1, φ_2. So the key thing in the method is the orientation of the subspace spanned by all those solutions which satisfy the known initial conditions at $\eta = \eta_1$ and not the way in which the subspace is characterized or labelled.

§4. Numerical Results

In order to search for a damped eigenfunction of the kind discussed by Tatsumi & Gotoh [2], (see Figure 3 of their paper), we concentrated our attention on the 'inviscid' eigenmode discovered by Mack [1] for the Blasius problem, (see Figure 3 of his paper). Mack used the orthonormalization method to calculate this particular eigenvalue for $\alpha_M = 0.179$ and for values of R_M up to about 50,000 although he did not obtain the corresponding eigenfunction. We have used the Compound Matrix Method to extend these calculations up to about $R_M = 5 \times 10^6$ and, in particular, when $R_M = 10^6$ then we find that $c = .32166 - .03629i$ and the eigenfunction is as given in Figure 1. For this eigenmode the limiting value of c as $R_M \to \infty$ is found to be $c_\infty = .32126 - .03671i$ by integrating Rayleigh's equation along a suitably chosen contour in the complex η-plane which passes below the singular point $\eta = \eta_s$ where $f'(\eta_s) = c_\infty$.

By comparing Figure 1 with Figure 3 of Tatsumi & Gotoh it may be seen that there is excellent qualitative agreement between the analysis and the numerical calculation. By varying R_M we were able to confirm that the thickness of the 'wavepacket' is of order $R_M^{-\frac{1}{4}}$ and that the number of oscillations is of order $R_M^{\frac{1}{4}}$ as predicted by the analysis. Note also the initial flip-up in the real part of φ' near $\eta = 0$ and the gradual decay towards infinity.

References

1. Mack, L.M. 1976 J, Fluid Mech. 73, 497.

2. Tatsumi, T. & Gotoh, K. 1971 Instability of Continuous Systems (Ed. H. Leipholz), 368.

3. Lin, C.C. 1955 The Theory of Hydrodynamic Stability. Cambridge University Press.

4. Ng, B.S. & Reid, W.H. 1980 J. Comput. Phys. 38, 275.

5. Davey, A. 1980 J. Comput. Phys. 35, 36.

On the Stability of Rotating Fluids

Knut S.Eckhoff

Department of Mathematics
University of Bergen
Bergen, Norway

Consider first two coaxial circular cylinders between which there is an inviscid and homogeneous fluid flowing in the azimuthal direction, i.e. the basic velocity field is given by

$$\underline{v} = v_0(r)\hat{\underline{\varphi}} \tag{1}$$

The stability problem for this basic flow (1) was first considered by Lord Rayleigh [1] who found that a necessary and sufficient condition for stability is that

$$\frac{d}{dr}(rv_0)^2 > 0 \tag{2}$$

for every r in the fluid.

The arguments used by Rayleigh [1] were not rigorous from a mathematical point of view. This was pointed out by Synge [2] who was able to treat the more general problem of a stratified, inviscid and incompressible fluid by the normal mode method restricted to axisymmetrical perturbations. In fact, Synge [2] found that the basic flow

$$\underline{v} = v_0(r)\hat{\underline{\varphi}} \quad , \quad \rho = \rho_0(r) \tag{3}$$

is stable if and only if for every r we have

$$\frac{d}{dr}(\rho_0 \ r^2 v_0^2) > 0 \tag{4}$$

Since (3) is analogous to a stratified fluid flowing horizontally in a gravitational field, we should expect that (3) cannot be stable unless

$$\frac{d}{dr}\rho_0 > 0 \tag{5}$$

That both (4) and (5) are necessary for stability will follow from the theory we shall present later. If (5) is satisfied it is easily seen that (4) is less restrictive with respect to v_0 than (2). In fact, for any given $v_0 \neq 0$ it is possible to satisfy (4) by choosing ρ_0 properly.

It is natural to ask whether the model which led to the conditions (4) and (5) is a proper model for a laboratory set-up. One obvious effect which has been neglected is the effect of gravity. The standard argument for neglecting gravity is that the inertial forces must be large compared to the force of gravity, which here only means that the fluid must be rotating sufficiently fast. If the axis of the rotating cylinders is perpendicular to the gravity field, this may be a valid argument. The argument becomes highly questionable if the axis is parallel to the gravity field, however, because then the inertial force is always perpendicular to the gravity. From the theory we shall present later it will follow that in the latter case it is never possible to neglect the effect of gravity as long as the fluid is stratified.

In order to make the scope of the theory as wide as possible, we shall consider a general axisymmetrical gravitational field, i.e. the field may be derived from a potential $V(r,z)$ in cylindrical coordinates. Furthermore we shall allow the fluid to be compressible; the model should therefore be appropriate also for studying the basic rotation behaviour of planetary and stellar atmospheres. The fundamental equations are

$$\begin{aligned} &\frac{\partial}{\partial t}\underline{v} + \underline{v}\cdot\nabla\underline{v} = -\rho^{-1}\nabla p + \nabla V \\ &\frac{\partial}{\partial t}\rho + \underline{v}\cdot\nabla\rho + \rho\nabla\cdot\underline{v} = 0 \\ &\frac{\partial}{\partial t}(p\rho^{-\gamma}) + \underline{v}\cdot\nabla(p\rho^{-\gamma}) = 0 \end{aligned} \tag{6}$$

We consider a basic flow of the form

$$\underline{v} = v_0(r,z)\hat{\underline{\varphi}} \quad , \quad \rho = \rho_0(r,z) \quad , \quad p = p_0(r,z) \tag{7}$$

Since (7) is assumed to satisfy (6), we have

$$-r^{-1}\, v_0^2 = -\rho_0^{-1}\,\frac{\partial p_0}{\partial r} + \frac{\partial V}{\partial r} \quad , \quad 0 = -\rho_0^{-1}\,\frac{\partial p_0}{\partial z} + \frac{\partial V}{\partial z} \tag{8}$$

Elimination of p_0 leads to

$$\left(r^{-1}\, v_0^2 + \frac{\partial V}{\partial r}\right)\frac{\partial \rho_0}{\partial z} - \frac{\partial V}{\partial z}\,\frac{\partial \rho_0}{\partial r} + 2r^{-1}\, v_0\,\frac{\partial v_0}{\partial z}\,\rho_0 = 0 \tag{9}$$

From this we see that in the considered basic flow (7) we may choose v_0 arbitrarily. Then ρ_0 is determined by (9) to within an arbitrary function of one variable. When both v_0 and ρ_0 have been chosen such that (9) is satisfied, p_0 is determined to within an additive constant by (8). This is the model for which the stability is studied in the paper by Eckhoff & Storesletten [3]; in the following the results of that paper will be outlined.

In order to study the stability properties of the basic flow (7), we perturb it by introducing the following expressions into (6)

$$\begin{aligned} \underline{v} &= \rho_0^{-\frac{1}{2}}\, u_r\, \underline{\hat{r}} + (\rho_0^{-\frac{1}{2}}\, u_\varphi + v_0)\underline{\hat{\varphi}} + \rho_0^{-\frac{1}{2}}\, u_z\, \underline{\hat{z}} \\ \rho &= \rho_0 + c_0^{-1}\,\rho_0^{\frac{1}{2}}\,(s_1 + s_2) \quad , \quad p = p_0 + c_0\,\rho_0^{\frac{1}{2}}\, s_2 \end{aligned} \tag{10}$$

Here $c_0 = (\gamma p_0/\rho_0)^{\frac{1}{2}}$ denotes the local sound speed. The vector $\underline{w} = \{u_r, u_\varphi, u_z, s_1, s_2\}$ then represents the perturbation superimposed on the basic flow (7), and the linearized equations for the perturbation are found to be

$$\frac{\partial \underline{w}}{\partial t} + A^1\,\frac{\partial \underline{w}}{\partial r} + A^2\,\frac{\partial \underline{w}}{\partial \varphi} + A^3\,\frac{\partial \underline{w}}{\partial z} + B\underline{w} = 0 \tag{11}$$

where

$$A^1 = \begin{Bmatrix} 0 & 0 & 0 & 0 & c_0 \\ 0 & 0 & 0 & 0 & 0 \\ 0 & 0 & 0 & 0 & 0 \\ 0 & 0 & 0 & 0 & 0 \\ c_0 & 0 & 0 & 0 & 0 \end{Bmatrix} , \quad A^2 = \begin{Bmatrix} r^{-1}v_0 & 0 & 0 & 0 & 0 \\ 0 & r^{-1}v_0 & 0 & 0 & r^{-1}c_0 \\ 0 & 0 & r^{-1}v_0 & 0 & 0 \\ 0 & 0 & 0 & r^{-1}v_0 & 0 \\ 0 & r^{-1}c_0 & 0 & 0 & r^{-1}v_0 \end{Bmatrix}$$

$$A^3 = \begin{Bmatrix} 0 & 0 & 0 & 0 & 0 \\ 0 & 0 & 0 & 0 & 0 \\ 0 & 0 & 0 & 0 & c_0 \\ 0 & 0 & 0 & 0 & 0 \\ 0 & 0 & c_0 & 0 & 0 \end{Bmatrix} , \quad B = \begin{Bmatrix} 0 & -2r^{-1}v_0 & 0 & \beta_1 & G_1 \\ 2r^{-1}v_0 - a & 0 & b & 0 & 0 \\ 0 & 0 & 0 & \beta_2 & G_2 \\ \alpha_1 & 0 & \alpha_2 & 0 & 0 \\ H_1 & 0 & H_2 & 0 & 0 \end{Bmatrix} \tag{12}$$

$$a = r^{-1}v_0 - \frac{\partial v_0}{\partial r} \quad , \quad b = \frac{\partial v_0}{\partial z}$$

$$\alpha_1 = c_0\rho_0^{-1}\frac{\partial\rho_0}{\partial r} - c_0^{-1}\left(r^{-1}v_0^2 + \frac{\partial V}{\partial r}\right) \quad , \quad \beta_1 = -c_0^{-1}\left(r^{-1}v_0^2 + \frac{\partial V}{\partial r}\right) ,$$

$$\alpha_2 = c_0\rho_0^{-1}\frac{\partial\rho_0}{\partial z} - c_0^{-1}\frac{\partial V}{\partial z} \quad , \quad \beta_2 = -c_0^{-1}\frac{\partial V}{\partial z} \quad , \tag{13}$$

$$G_1 = c_0^{-1}\left(\frac{\gamma}{2} - 1\right)\left(r^{-1}v_0^2 + \frac{\partial V}{\partial r}\right) \quad , \quad H_1 = r^{-1}c_0 + c_0^{-1}\left(r^{-1}v_0^2 + \frac{\partial V}{\partial r}\right) - \tfrac{1}{2}c_0\rho_0^{-1}\frac{\partial\rho_0}{\partial r} ,$$

$$G_2 = c_0^{-1}\left(\frac{\gamma}{2} - 1\right)\frac{\partial V}{\partial z} \quad , \quad H_2 = c_0^{-1}\frac{\partial V}{\partial z} - \tfrac{1}{2}c_0\rho_0^{-1}\frac{\partial\rho_0}{\partial z} .$$

Here $\underline{w}$ has been treated as a column vector. Since A^1, A^2, A^3 are symmetric matrices, we see that (11) is a symmetric hyperbolic system.

Since the coefficients in (11) depend both on r and z, the conventional normal mode method would not get us very far. However, there is another method available which is based on the so-called generalized progressing wave expansion method. This approach to the stability problem was first suggested by Eckhoff [4]; a rigorous mathematical justification has recently been given by Eckhoff [5]. A generalized progressing wave solution of (11) is a family of solutions of the form

$$\underline{w} = \underline{a}_0(r,\varphi,z,t)\exp\{i\omega\Phi(r,\varphi,z,t)\} + 0\left(\frac{1}{\omega}\right) \tag{14}$$

where ω is a frequency parameter and $i = \sqrt{-1}$. By substituting (14) into (11) and retaining only the leading order terms in ω^{-1}, we get

$$\{\Phi_t I + \Phi_r A^1 + \Phi_\varphi A^2 + \Phi_z A^3\}\underline{a}_0 = 0 \tag{15}$$

Since $\underline{a}_0 \neq 0$, (15) implies that

$$\det\{\Phi_t I + \Phi_r A^1 + \Phi_\varphi A^2 + \Phi_z A^3\} = 0$$

$$\Longleftrightarrow \{\Phi_t + r^{-1}v_0\,\Phi_\varphi\}^3[\{\Phi_t + r^{-1}v_0\,\Phi_\varphi\}^2 - c_0^2k^2] = 0 \tag{16}$$

where
$$k^2 = \Phi_r^2 + r^{-2}\,\Phi_\varphi^2 + \Phi_z^2 \tag{17}$$

Clearly, (16) is satisfied if and only if the phase function Φ

satisfies one of the following 3 equations

$$\Phi_t + r^{-1} v_0 \Phi_\varphi = 0 \tag{18}$$

$$\Phi_t + r^{-1} v_0 \Phi_\varphi + c_0 k = 0 \tag{19}$$

$$\Phi_t + r^{-1} v_0 \Phi_\varphi - c_0 k = 0 \tag{20}$$

Equation (18) corresponds to the internal gravity waves (inertial waves), while (19), (20) correspond to the acoustic waves. Since the latter are not expected to give rise to any instabilities, and since we only are going to establish necessary conditions for stability, we shall in the following confine our study to the case where Φ satisfies (18).

We introduce the notation

$$\xi^1 = \frac{\partial \Phi}{\partial r} \quad , \quad \xi^2 = \frac{\partial \Phi}{\partial \varphi} \quad , \quad \xi^3 = \frac{\partial \Phi}{\partial z} \ . \tag{21}$$

The ray equations associated with (18) are then

$$\begin{aligned} &\frac{dr}{dt} = 0 \quad , \quad \frac{d\varphi}{dt} = r^{-1} v_0 \quad , \quad \frac{dz}{dt} = 0 \ , \\ &\frac{d\xi^1}{dt} = r^{-1} \xi^2 a \quad , \quad \frac{d\xi^2}{dt} = 0 \quad , \quad \frac{d\xi^3}{dt} = -r^{-1} \xi^2 b \ . \end{aligned} \tag{22}$$

The solutions of (22) are readily found to be

$$\begin{aligned} &r = r_0 \quad , \quad \varphi = \varphi_0 + r_0^{-1} v_0 (r_0, z_0) t \quad , \quad z = z_0 \ , \\ &\xi^1 = \xi_0^1 + r_0^{-1} \xi_0^2 a(r_0, z_0) t \ , \ \xi^2 = \xi_0^2 \ , \ \xi^3 = \xi_0^3 - r_0^{-1} \xi_0^2 b(r_0, z_0) t \end{aligned} \tag{23}$$

where the subscript 0 refers to the initial values at $t = 0$. Since the equation (18) corresponds to a triple root in the characteristic equation (16), it follows from (15) that when the phase function Φ satisfies (18) the amplitude $\underline{a}_0$ must have the form

$$\underline{a}_0 = \sigma_1 \underline{r}_1 + \sigma_2 \underline{r}_2 + \sigma_3 \underline{r}_3 \tag{24}$$

Here $\sigma_1, \sigma_2, \sigma_3$ are scalar functions tc be determined, while $\underline{r}_1, \underline{r}_2, \underline{r}_3$ are eigenvectors corresponding to the triple root (18). It is not difficult to see that we may choose

$$\underline{r}_1 = k^{-1}\{r^{-1}\xi^2,\ -\xi^1,\ 0,\ \xi^3,\ 0\}$$
$$\underline{r}_2 = k^{-1}\{0,\ \xi^3,\ -r^{-1}\xi^2,\ \xi^1,\ 0\} \tag{25}$$
$$\underline{r}_3 = k^{-1}\{-\xi^3,\ 0,\ \xi^1,\ r^{-1}\xi^2,\ 0\}$$

When (14) satisfies (18), (24), (25), we see that (15) is fulfilled. By considering the terms next to the leading order in (11) when (14) is substituted, it is possible to show that $\sigma_1, \sigma_2, \sigma_3$ must satisfy the so-called transport equations which here become

$$\begin{aligned}
\frac{d\sigma_1}{dt} &= k^{-2}\{-r^{-1}\xi^1\xi^2 a - r^{-1}\xi^2\xi^3(\alpha_1+\beta_1)\}\sigma_1 \\
&+k^{-2}\{-r^{-1}\xi^1\xi^2(2b+\beta_1)+r^{-1}\xi^2\xi^3(2r^{-1}v_0-a+\alpha_2)\}\sigma_2 \\
&+k^{-2}\{(\xi^1)^2 b-\xi^1\xi^3(2r^{-1}v_0-a+\alpha_2)-(r^{-1}\xi^2)^2(b+\beta_1)+(\xi^3)^2\alpha_1\}\sigma_3, \\
\frac{d\sigma_2}{dt} &= k^{-2}\{r^{-1}\xi^1\xi^2(b-\alpha_1)-r^{-1}\xi^2\xi^3(2r^{-1}v_0-2a-\beta_2)\}\sigma_1 \\
&+k^{-2}\{r^{-1}\xi^1\xi^2(\alpha_2+\beta_2)+r^{-1}\xi^2\xi^3 b\}\sigma_2 \\
&+k^{-2}\{-(\xi^1)^2\alpha_2-\xi^1\xi^3(b-\alpha_1)+(r^{-1}\xi^2)^2(a+\beta_2)+(\xi^3)^2(2r^{-1}v_0-a)\}\sigma_3, \\
\frac{d\sigma_3}{dt} &= k^{-2}\{\xi^1\xi^3(2r^{-1}v_0-\beta_2)+(r^{-1}\xi^2)^2(b-\alpha_1)+(\xi^3)^2\beta_1\}\sigma_1 \\
&+k^{-2}\{-(\xi^1)^2\beta_2+\xi^1\xi^3\beta_1-(r^{-1}\xi^2)^2(a-\alpha_2)-(\xi^3)^2 2r^{-1}v_0\}\sigma_2 \\
&+k^{-2}\{-r^{-1}\xi^1\xi^2(\alpha_2+\beta_2)+r^{-1}\xi^2\xi^3(\alpha_1+\beta_1)\}\sigma_3.
\end{aligned} \tag{26}$$

along the rays (23). Thus the expressions (23) are supposed to be substituted into (26), which gives us a closed linear system of ordinary differential equations for the amplitude $\underline{\sigma} = \{\sigma_1, \sigma_2, \sigma_3\}$

$$\frac{d}{dt}\underline{\sigma} = A(t; r_0, z_0, \xi_0^1, \xi_0^2, \xi_0^3)\underline{\sigma} \tag{27}$$

According to the theory given by Eckhoff [4,5] we now have to study the stability properties of (27) for all possible values of the parameters $r_0, z_0, \xi_0^1, \xi_0^2, \xi_0^3$. This is done in Eckhoff & Storesletten [3] and leads to no less than 6 lemmas in order to cover all possible cases. When we want to draw the conclusions from these 6 lemmas to the stability problem for the basic flow (7) by the theory in Eckhoff [4,5], it turns out that we have to distinguish between the following two cases

a) V is independent of z, i.e. the external force field has no axial component.

b) $\partial V/\partial z \neq 0$ almost everywhere in the fluid, i.e. the external force field has an axial component.

For case a) the following result holds

Theorem 1. Let the potential for the external forces V be independent of z. In order that the basic flow (7) shall be stable it is necessary that the basic flow (7) is independent of z and that

$$\left(r^{-1} v_0^2 + \frac{\partial V}{\partial r}\right)\left\{\rho_0^{-1} \frac{\partial \rho_0}{\partial r} - c_0^{-2}\left(r^{-1} v_0^2 + \frac{\partial V}{\partial r}\right)\right\} \geq -2r^{-1} v_0\left(r^{-1} v_0 + \frac{\partial v_0}{\partial r}\right) \tag{28}$$

$$\left(r^{-1} v_0^2 + \frac{\partial V}{\partial r}\right)\left\{\rho_0^{-1} \frac{\partial \rho_0}{\partial r} - c_0^{-2}\left(r^{-1} v_0^2 + \frac{\partial V}{\partial r}\right)\right\} \geq 0 \tag{29}$$

hold everywhere in the fluid. If equality holds in (28) or (29) on some set of positive measure, it is furthermore necessary for stability of the flow (7) that

$$v_0 = \frac{\partial v_0}{\partial r} = \frac{\partial \rho_0}{\partial r} = \frac{\partial V}{\partial r} = 0 \tag{30}$$

holds almost everywhere on this set.

Theorem 1 is here formulated in a slightly different way than the corresponding result in Eckhoff & Storesletten [3], but it is not difficult to show that the two formulations are equivalent. For case b) the following result holds

Theorem 2. Let the potential for the external forces V be such that $\partial V/\partial z \neq 0$ almost everywhere. In order that the basic flow (7) shall be stable it is necessary that

$$\rho_0^{-1} \frac{\partial \rho_0}{\partial z} \frac{\partial V}{\partial z} - c_0^{-2}\left(\frac{\partial V}{\partial z}\right)^2 > 0, \tag{31}$$

$$r^{-1} v_0\left(r^{-1} v_0 + \frac{\partial v_0}{\partial r}\right) \geq \frac{\rho_0^{-1}\frac{\partial \rho_0}{\partial r} - c_0^{-2}\left(r^{-1} v_0^2 + \frac{\partial V}{\partial r}\right)}{\rho_0^{-1}\frac{\partial \rho_0}{\partial z} - c_0^{-2}\frac{\partial V}{\partial z}} r^{-1} v_0 \frac{\partial v_0}{\partial z}, \tag{32}$$

hold almost everywhere in the fluid. If equality holds in (32) on some set of positive measure, it is furthermore necessary for stability of the flow (7) that

$$v_0 = 0 \tag{33}$$

almost everywhere on this set.

The case where there are no external forces is covered by Theorem 1. If we set $V=0$ and take the incompressible limit $c_0 \to \infty$ in that theorem, the inequalities (28), (29) become when $v_0 \neq 0$

$$\begin{aligned} r^{-1} v_0^2 \rho_0^{-1} \frac{\partial \rho_0}{\partial r} + r^{-3} \frac{\partial}{\partial r}(r v_0)^2 &= r^{-3} \rho_0^{-1} \frac{\partial}{\partial r}(\rho_0 r^2 v_0^2) > 0 \\ r^{-1} v_0^2 \rho_0^{-1} \frac{\partial \rho_0}{\partial r} &> 0 \end{aligned} \tag{34}$$

Thus we have shown the necessity of the conditions (4), (5) to ensure the stability of the basic flow (7) for an incompressible fluid.

Professor C.-S.Yih has pointed out to the author that the results contained in Theorem 1 can be written in a way very similar to the above one also in the case where the fluid is compressible. In fact, by introducing the so-called potential density $\tilde{\rho}$ which may be defined by

$$\tilde{\rho}(r) = C \rho_0(r) p_0(r)^{-\frac{1}{\gamma}} \tag{35}$$

where $C > 0$ is a constant, and setting $V=0$, the inequalities (28), (29) may be written when $v_0 \neq 0$

$$\frac{d}{dr}(\tilde{\rho} r^2 v_0^2) > 0 \quad \& \quad \frac{d}{dr}\tilde{\rho} > 0 \tag{36}$$

When there is an external force field with an axial component, the relevant necessary conditions for stability are obtained from Theorem 2. The inequality (31) simply requires that the fluid must be stratified in a statically stable way in the axial direction. If we consider the case with no axial shear, i.e. the case where $\partial v_0/\partial z \equiv 0$, we see that the inequality (32) becomes when $v_0 \neq 0$

$$\tfrac{1}{2} r^{-3} \frac{\partial}{\partial r}(r v_0)^2 = r^{-1} v_0 \left(r^{-1} v_0 + \frac{\partial v_0}{\partial r} \right) > 0 \tag{37}$$

This is exactly Lord Rayleigh's criterion (2) for a homogeneous fluid! The inequality (37) concerns only the velocity profile; neither the stratification nor the external force field appear. This is radically different from (36), it is therefore never possible to neglect the effect of the external forces in this case.

Now, what can be said about the stability problem when $b = \partial v_0/\partial z \neq 0$ in some domain of the fluid? In case a) Theorem 1 shows that the basic flow then always is unstable. In case b) on the other hand, it seems possible that there may exist basic flows with axial shear which are stable. When $v_0 \neq 0$, the necessary condition (32) may be written

$$r^{-1} v_0\left(r^{-1} v_0 + \frac{\partial v_0}{\partial r}\right) > \frac{\rho_0^{-1}(\partial\rho_0/\partial r) - c_0^{-2}(r^{-1} v_0^2 + \partial V/\partial r)}{\rho_0^{-1}(\partial\rho_0/\partial z) - c_0^{-2}(\partial V/\partial z)}\, r^{-1} v_0 \frac{\partial v_0}{\partial z}$$

$$= \frac{\alpha_1}{\alpha_2} r^{-1} v_0 b = -\frac{2}{\alpha_2\beta_2}\left\{(r^{-1} v_0 b - \tfrac{1}{4}\alpha_2\beta_1)^2 - (\tfrac{1}{4}\alpha_2\beta_1)^2\right\} . \tag{38}$$

When the other necessary condition for stability (31) is satisfied, the right hand side in (38) is seen to be negative if and only if one of the following two conditions is satisfied

$$\text{(i)}\ \ 0 < r^{-1}\frac{\partial}{\partial z} v_0^2 < \alpha_2\beta_1 \ ,\quad \text{(ii)}\ \ \alpha_2\beta_1 < r^{-1}\frac{\partial}{\partial z} v_0^2 < 0 \ , \tag{39}$$

where

$$\alpha_2\beta_1 = -\left(\rho_0^{-1}\frac{\partial\rho_0}{\partial z} - c_0^{-2}\frac{\partial V}{\partial z}\right)\left(r^{-1} v_0^2 + \frac{\partial V}{\partial r}\right) \tag{40}$$

In these cases the necessary condition for stability (38) is less restrictive than Lord Rayleigh's criterion (37). Thus an axial shear may have a stabilizing influence in special cases.

When the axial shear does not satisfy one of the two conditions (39), the necessary condition for stability (38) becomes more restrictive with respect to the radial shear than Lord Rayleigh's criterion (37). If the axial shear is very strong compared to the stratification of the fluid, (38) becomes very restrictive indeed. In fact, if

$$\left|r^{-1}\frac{\partial}{\partial z} v_0^2\right| \gg |\alpha_2\beta_1| \ \ \& \ \ \left|r^{-1}\frac{\partial}{\partial z} v_0^2\right| \gg |\alpha_2\beta_2| \tag{41}$$

then it follows from (38) that the basic flow cannot be stable unless

$$r^{-3}\frac{\partial}{\partial r}(r v_0)^2 \gg \left|r^{-1}\frac{\partial}{\partial z} v_0^2\right| \tag{42}$$

In most cases therefore, an axial shear has a destabilizing influence on the basic flow. Furthermore we see that this destabilizing influence may be essential in many cases.

References

1. Lord Rayleigh: On the dynamics of revolving fluids. (1916) Scientific Papers 6, 447-453. Cambridge Univ. Press 1920.

2. Synge, J.L.: The stability of heterogeneous liquids. Trans. R. Soc. Can. 27 (1933) 1-18.

3. Eckhoff, K.S.; Storesletten, L.: On the stability of rotating compressible and inviscid fluids. J. Fluid Mech. 99 (1980) 433-448.

4. Eckhoff, K.S.: Stability problems for linear hyperbolic systems. Rep.No.54, Dept. Appl. Math., Univ. Bergen 1975.

5. Eckhoff, K.S.: On stability for symmetric hyperbolic systems,I. J. Diff. Eq. 40 (1981) 94-115.

Interactions of Axisymmetric and Non-Axisymmetric Disturbances in the Flow between Concentric Rotating Cylinders: Bifurcations near Multiple Eigenvalues+

R.C. Di Prima and J. Sijbrand*

Rensselaer Polytechnic Institute, Troy, N.Y., U.S.A.
University of Utrecht, Netherlands

We consider the instability of Couette flow between rotating concentric cylinders of infinite length. Let R_1, R_2 and Ω_1, Ω_2 denote the radii and angular velocities of the inner and outer cylinders, respectively; and let $\eta = R_1/R_2$ and $\mu = \Omega_2/\Omega_1$. We also introduce the Taylor number $T = -4A\Omega_1 d^4/\nu^2$ with $A = (\Omega_2 R_2^2 - \Omega_1 R_1^2)/(R_2^2 - R_1^2)$.

It is known from the linear stability analyis of Krueger, Gross, and DiPrima (1966) for η near one that (i) for $\mu = 0$ and increasing T, Couette flow first becomes unstable to an axisymmetric disturbance at $T = T_c(\mu = 0, \eta)$, (ii) in the absence of an axisymmetric disturbance Couette flow would become unstable to a non-axisymmetric disturbance at a value of T just slightly greater than T_c, and (iii) at μ sufficiently negative the critical value of T occurs for a non-axisymmetric disturbance. Thus, for fixed η near one there are parameter values (μ_1, T_1) for which axisymmetric and non-axisymmetric disturbances are simultaneously critical for Couette flow. We wish to study the bifurcation problem in the neighborhood of (μ_1, T_1), where we note that instability to an axisymmetric disturbance gives a steady bifurcation while instability to a non-axisymmetric disturbance gives a Hopf bifurcation.

In order to develop a mathematically tractable problem, we assume that the axial wavenumber of the axisymmetric and non-axisymmetric disturbances are the same, and we choose the wavenumber to be that corresponding to critical conditions for an axisymmetric disturbance. We let λ denote the dimensionless axial wavenumber (scaled with respect to R_2-R_1) and m denote the number of waves in the azimuthal direction. Dr. Peter Eagles has carried out the necessary calculations for $\eta = 0.95$ to determine the points (μ_1, T_1) for

+ This research was partially supported by the Army Research Office and the Fluid Mechanics Branch of the Office of Naval Research, and by the Netherlands Organization for the Advancement of Pure Research (Z.W.O.).

* Present address: Koninklijke/Shell Laboratorium, Amsterdam

simultaneous instability to an axisymmetric disturbance $(\lambda,0)$ and a non-axisymmetric disturbance (λ,m) with the results given in the Table.

Table

m	1	T_1	λ
1	-0.73976	11,973.87	3.482
2	-0.75123	12,319.01	3.507
4	-0.80284	14,015.01	3.628

Thus, for example, for $\eta = 0.95$ and $\mu = -0.73976$, Couette flow is simultaneously unstable to an axisymmetric disturbance with wavenumbers $(3.482,0)$ and a non-axisymmetric disturbance with wavenumbers $(3.482,1)$. We also note that at (μ_1, T_1) there are 6 critical modes with axial (Z) and azimuthal (θ) dependence as follows: $\cos \lambda Z$, $\sin \lambda Z$, $\exp(\pm im\theta)\cos\lambda Z$ and $\exp(\pm im\theta)\sin\lambda Z$. The first step in the bifurcation analysis is to represent the components of the disturbance by

$$\begin{aligned} &A_c u_c(r)\cos\lambda Z + A_s u_s(r)\sin\lambda Z + \\ &B_c v_c(r)\exp(im\theta)\cos\lambda Z + B_s v_s(r)\exp(im\theta)\sin\lambda Z + \\ &\bar{B}_c \bar{v}_c(r)\exp(-im\theta)\cos\lambda Z + \bar{B}_s \bar{v}_s(r)\exp(-im\theta)\sin\lambda Z + \Psi \end{aligned} \qquad (1)$$

Here the functions u_c, u_s, v_c, v_s denote the radial dependence of the respective modes. The quantities A_c, A_s, B_c, B_s are scalar functions of time with A_c, A_s varying in the real numbers and B_c, B_s in the complex domain. The residual component Ψ is orthogonal to the six critical modes.

We are now able to reduce the nonlinear partial differential equation to a finite- dimensional center manifold of dimension six; this reduction is performed by expressing Ψ in terms of the coefficients A_c, A_s, B_c, B_s. The nonlinear ordinary differential equations for these coefficients describe the 'flow' in the center manifold and are similar to those derived by Davey, DiPrima and Stuart (1968) for the case $\mu = 0$.

We give particular attention to the interaction of the three modes $\cos\lambda Z$ and $\exp(\pm im\theta)\sin\lambda Z$, which can give rise to Taylor-vortex and wavy-vortex flows which are often observed in experiments. So we put $A_s = B_c = 0$ and we obtain the following equations where terms of the order 7 have been omitted:

$$\frac{dAc}{dt} = a_0 A_c + a_1 A_c{}^3 + a_4 A_c |B_s|^2 + a_5 A_c{}^5 + a_6 A_c{}^3 |B_s|^2 + a_7 A_c |B_s|^4,$$
$$\frac{dBs}{dt} = b_0 B_s + b_1 B_s |B_s|^2 + b_4 B_s A_c{}^2 + b_5 B_s A_c{}^4 + b_6 B_s |B_s|^2 A_c{}^2 + b_7 B_s |B_s|^4, \qquad (2)$$

Complex conjugate equation for $\bar{B}_s$.

The scalar quantities a_j are real and the b_j are complex numbers. In the sequel we shall denote by b_{jr} and b_{ji} the real and imaginary parts of b_j. After performing the transformation $A_c = x$, $B_s = y \exp i(b_{0i}t+\phi)$ the equation for ϕ uncouples from those for x and y and we obtain

$$\frac{dx}{dt} = a_o x + a_1 x^3 + a_4 xy^2 + a_5 x^5 + a_6 x^3 y^2 + a_7 xy^4,$$
$$\frac{dy}{dt} = b_{0r} y + b_{1r} y^3 + b_{4r} x^2 y + b_{5r} x^4 y + b_{6r} x^2 y^3 + b_{7r} y^5. \qquad (3)$$

The coefficients a_j and b_{jr} depend on μ and T, while $a_0(\mu_1, T_1) = b_{0r}(\mu_1, T_1) = 0$.

It is now possible to study changes in the phase plane as a function of the two parameters $\mu-\mu_1$ and $T-T_1$. It is natural first to analyze system (3) with the fifth order terms left out. Keener (1976) and Langford and Iooss (1980) consider pairs of coupled ordinary differential equations with the same cubic nonlinearities as (3) by taking a_1, a_4, b_{1r}, b_{4r} fixed and varying a_0 and b_{0r}. The nature of the changes in the phase plane depends of course on the specific values of the 3rd order coefficients and the cubic systems of o.d.e.'s can be classified accordingly.

While the investigations by Keener and Langford-Iooss were carried out for arbitrary nonvanishing values of the 3rd order coefficients, it turns out that in the problem which is presently under discussion some of these coefficients are very small and change sign for values of μ, T very close to μ_1, T_1. For example, if we let μ take values close to μ_1 and compute $T = T_c(\mu,\eta = .95, \lambda = 3.482, m = 0)$ then $a_1(\mu,T) = 0$ for $\mu = -.68939$, which is indeed very close to $\mu_1 \doteq -.73976$. For this reason the analysis of (3) has to be taken to fifth order and the dependence of a_1, a_4, b_{1r}, b_{4r} on μ and T has to be taken into account up to first order in $\mu-\mu_1$ and $T-T_1$.

The numerical evaluation of the very complicated expressions for the fifth order coefficients have been carried out by Dr. Eagles and the analysis of the changes in the phase plane is in progress.

The full system of six amplitude equations is also studied, but only at cubic order in the amplitudes. For this case there does not appear to be a transformation that will reduce the order of the system.

For the case $\mu = 0$, we have noted that the axisymmetric and non-axisymmetric modes are not simultaneously unstable. However, it is possible to imbed this problem in a larger class of mathematical problems which we can analyze using the techniques of center manifold theory. In this way we present a rational basis (though admittedly complicated) for the analysis of the successive bifurcations from Couette flow to Taylor-vortex flow and then to wavy-vortex flow for increasing T with $\mu = 0$ as first studied by Davey, DiPrima, and Stuart (1968).

References

1. E.R. Krueger, A. Gross and R.C. DiPrima; 1966; On the relative importance of Taylor-vortex and non-axisymmetric modes in flow between rotating cylinders. J. Fluid Mech. 24, 521-538.

2. A. Davey, R.C. DiPrima, and J.T. Stuart; 1968; On the instability of Taylor vortices. J. Fluid Mech. 31, 17-52.

3. J.P. Keener; 1976; Secondary bifurcations in nonlinear diffusion reaction equations. Studies in Appl. Math. 55, 187-211.

4. W.F. Langford and G. Iooss; 1980; Interactions of Hopf and pitchfork bifurcations. To appear in Birkhäuser Lecture Notes, H.D. Mittelmann ed.

Stability of Supercritical Bénard Convection and Taylor Vortex Flow

E. L. KOSCHMIEDER

College of Engineering and
I. Prigogine Center for Statistical Mechanics
University of Texas, Austin

Summary

The stability of supercritical Bénard convection and supercritical Taylor vortices is discussed. There is still disagreement about the planform of the convective motions. The wave number of supercritical flow has been found to increase with the Rayleigh number. That has not yet been explained adequately. Periodic time-dependency has so far been observed only in very small containers. Supercritical axisymmetric Taylor vortices have a constant wave number but the wave number is nonunique. Wavy Taylor vortices are first strictly periodic, have two incommensurate frequencies at higher Taylor number, and a chaotic frequency spectrum if the Taylor number is increased a little further. Turbulent Taylor vortices at very high Taylor number are in the mean axisymmetric, have a clear spatial periodicity and are nonunique.

Introduction

Bénard convection is the consequence of the instability of a resting fluid layer heated uniformly from below, having therefore cold heavy fluid on top of warm light fluid. Taylor vortices are the consequence of the rotation of a fluid between two concentric cylinders. If the inner cylinder is rotated faster than the outer cylinder, the inner fluid rotates faster than the outer fluid and is therefore unstable to infinitesimal disturbances. Bénard convection and Taylor vortex flow are textbook examples of hydrodynamic stability. They have been studied intensively during the last decade and are often considered the key to the understanding of a much wider class of instabilities. Summaries of the theory of both instabilities can be found in Drazin and Reid [1]. Limitation of space does not make it possible to discuss details.

Bénard convection

The most informative result of linear theory of Bénard convection is the stability diagram Fig. 1. According to Fig. 1 the onset of convection occurs at the critical Rayleigh number Ra_c; below Ra_c all infinitesimal disturbances of the fluid decay.

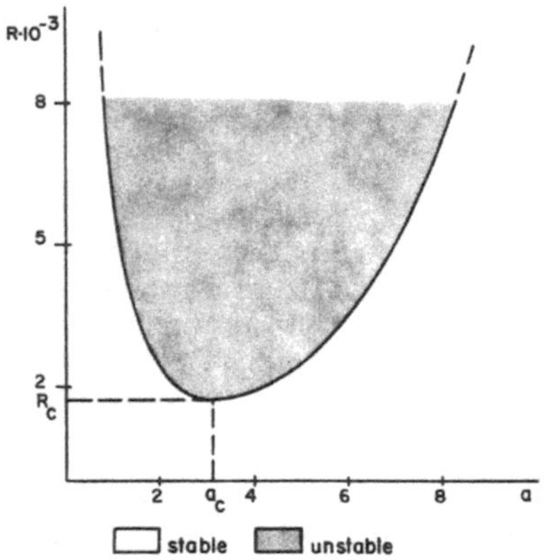

Fig. 1. Stability diagram of Bénard convection.

The Rayleigh number is a nondimensional measure of the applied (negative) vertical temperature gradient. It is

$$Ra = \frac{\alpha g \ \Delta T \ d^3}{\nu \kappa} \qquad (1)$$

As can be seen in Fig. 1, Ra_c is associated with the critical wave number a_c. The wave number is a nondimensional measure of the spatial structure of the instability. At Ra_c the horizontal section of one convective roll is practically as wide as the fluid layer is deep. The planform of the motions is determined by the membrane equation

$$\nabla_2^2 f(x,y) + a^2 f(x,y) = 0 \qquad (2)$$

The eigenvalue Ra_c is degenerate, there is an infinite number of patterns which obey eq. (2). All patterns which cover an infinite plane regularly are permitted, such as straight parallel rolls, square cells, rectangular cells, hexagonal cells, circular concentric cells etc. The multitude of possible patterns contradicts experience. Many attempts have been made to resolve the question of the preferred pattern. Earlier theories have searched for an intrinsic ability of the fluid to establish its own pattern via weakly nonlinear effects. Not much has come from these attempts; primarily because the hexagonal pattern, the prime candidate of the theories searching for an intrinsic selection mechanism, has lost much of its appeal after it was found that the famous hexagonal cells observed by Bénard are not caused by vertical instability but rather by surface tension gradients.

A different solution of the pattern selection problem emerged after experiments of Koschmieder [2] showed, that the pattern of convection, (of a Boussinesq fluid), reflects the shape of the lateral walls. He showed that in a circular container a pattern of circular concentric rolls forms, see Fig. 2. In a

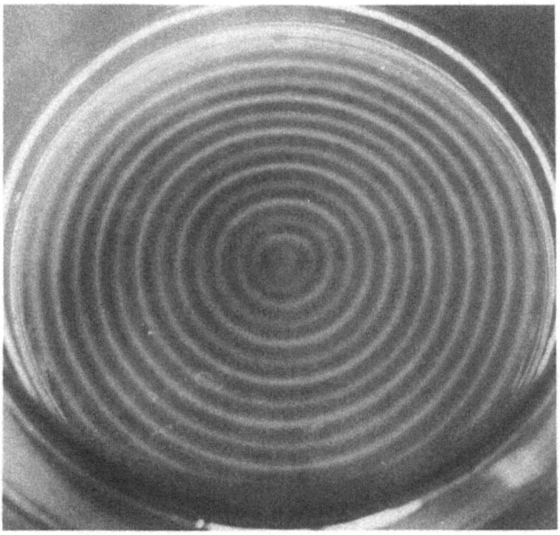

Fig 2. Circular concentric convective motion on a uniformly heated plate. After [2].

rectangular container he observed rolls parallel to the short side of the container, and in a square container square cells. Theories studying convection in bounded circular cylindrical containers by Charlson and Sani [3] and in rectangular boxes by Davis [4] are in agreement with the concept that the side walls determine the form of the pattern. It has, in particular, been confirmed by numerous experiments that in a rectangular box rolls form which are parallel to the short side. Other experiments have, however, resulted in patterns consisting of irregular rolls and cells, and it has been concluded from these experiments that the pattern, even in bounded containers, has a random structure. In order to prove that such irregular patterns are the result of the instability and not of nonuniformities of the temperatures in the horizontal boundaries it has to be shown that the orientation of the rolls in irregular patterns is random. This has not been done so far. To summarise, there is, after 20 years of debate, still disagreement about the form of the convective motion on a uniformly heated plate. All experiments are necessarily made in bounded containers, never on infinite planes. Small effects, such as weakly nonlinear effects or the consequences of the random distribution of infinitesimal disturbances compete with the necessarily finite effects of the walls. Agreement will only be reached if adequately performed experiments confirm theories which consider the basic experimental input.

Under supercritical conditions the wave number of the convective motions should be nonunique, at least according to linear theory, see Fig. 1. There is, for $Ra > Ra_c$, an entire range of unstable wave numbers around a_c. Experiments as reported in [2] and about a dozen other very varied experiments summarized in the review [5] have shown that under steady conditions the wave numbers are $< a_c$ and decrease with increasing $Ra > Ra_c$. Steady motions with wave numbers $> a_c$ do <u>not</u> occur. The results of the more specific modern measurements of the wave number are reproduced in Fig. 3. The ordinate in Fig. 3 is the wavelength λ instead of the wave number a, it is $\lambda = 2\pi/a$.

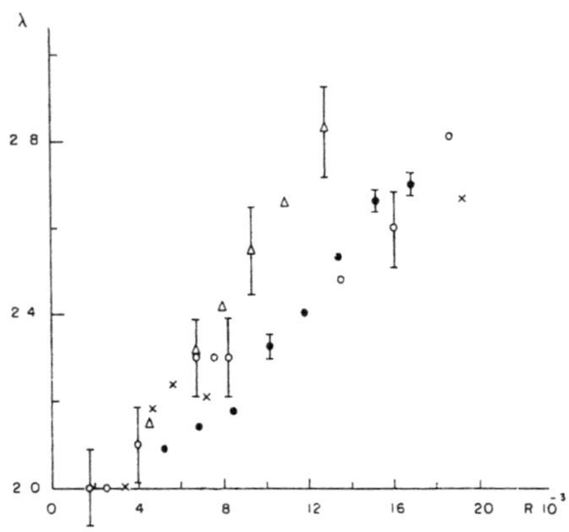

Fig. 3. Some observed supercritical wavelengths. After [5].

We mention finally that wave numbers $a > a_c$ can be produced either by rapid heating or through socalled controlled initial conditions, (Busse and Whitehead, [6]). The motions so initiated are, however, not stable, they change with time under otherwise constant conditions to motions with wave numbers $a < a_c$.

A theoretical explanation for the decrease of the supercritical wave numbers has so far not been found. Early theoretical work, summarized in [5], resulted invariably in the prediction that the wave number should increase with Ra. Two models proposed recently by Pomeau and Maneville [7] and Cross et al. [8] make the side walls responsible for the decrease of the supercritical wave number. Both models do not predict the slope of the change of the wave number, knowledge of which is crucial for the experimental verification of the results of these studies.

Laser-Doppler velocimeter measurements by Dubois and Bergé [9] have shown that the velocity of supercritical motions increases as the square root of $(Ra - Ra_c)/Ra_c$. Theoretical suggestions made a long time ago by Landau [10] and Hopf [11] have thereby been splendidly confirmed.

After all the work that has been done on the spatial structure of the convective instability, the temporal structure of supercritical convection has recently become a focal point of interest. An intense effort has been made in the last few years to find out whether the time-dependency begins with strictly periodic motions and how it proceeds from there on. The results obtained so far are somewhat puzzling. Gollub and Benson [12], for example, have observed sharp frequencies and their harmonics or subharmonics, but only in convection in containers with very small aspect ratios. So far only chaotic frequency spectra have been observed in shallow fluid layers. Actually Ahlers and Behringer [13] found that turbulence appears immediately after onset of convection in a shallow fluid layer of low Prandtl number.

The Taylor instability

In a classical paper G. I. Taylor presented a theory describing the instability of a viscous fluid layer between two concentric rotating cylinders and supported his theoretical results with convincing experimental evidence. The most informative result of linear theory is again the stability diagram, which is, if the gap between the cylinders is narrow, practically identical with the curve shown in Fig. 1, only that then the Rayleigh number is replaced by the Taylor number, which is a nondimensional measure of the rotation rate given by

$$Ta = \frac{4\Omega^2 d^4}{\nu^2} \tag{3}$$

There is again a critical Taylor number below which the fluid layer is stable to infinitesimal disturbances. At the critical Taylor number disturbances grow and the fluid forms axisymmetric toroidal rolls, examples of which can be seen in Fig. 4. The

Fig. 4. Supercritical nonunique Taylor vortices at $Ta = 9.1\ Ta_c$. After [19].

size of the toroidal vortices at Ta_c is given by the critical wave number a_c. In practice that means that the vortices are as high in vertical direction as they are deep in radial direction. In experiments, which work of course with finite columns, one observes that the instability does not occur spontaneously at Ta_c throughout the entire column, but forms first at the ends, just as in the case of convection. Very short columns, where the end effects become dominant, have been studied by Benjamin [14].

Under supercritical conditions the wave number should, according to the stability diagram of linear theory, be nonunique, just as in the Bénard case. The difficult nonlinear problem predicting the preferred wave number of moderately supercritical flow has not yet been solved. Stuart [15] has proposed the socalled shape assumption according to which the wave number should remain constant if Ta > Ta_c, for slightly supercritical Ta. Kirchgässner and Sorger [16] have shown rigorously that $a = a_c$ for Ta > Ta_c, likewise for slightly supercritical Ta. Progress has also been made predicting the range of possible axisymmetric supercritical wave numbers. Kogelman and DiPrima [17] have determined a new nonlinear stability diagram for slightly supercritical Ta which is shown, extrapolated over a large range of Taylor numbers in Fig. 5. Experiments of Burkhalter and Koschmieder [18] have shown that the wave number of axisymmetric Taylor vortices remains constant and equal to a_c over a wide range of Ta, if the Taylor number is increased very slowly from subcritical values on. Note that the constancy of the wave number of the Taylor instability differs markedly from the decrease of the wave number in the Bénard case, (see Fig. 3). Burkhalter and Koschmieder [19] have shown furthermore that stable axisymmetric vortices with an entire range of nonunique wave numbers can be realized through sudden or fast starts of the inner cylinder. A column of Taylor vortices obtained through a sudden start is shown at the left part of Fig. 4.

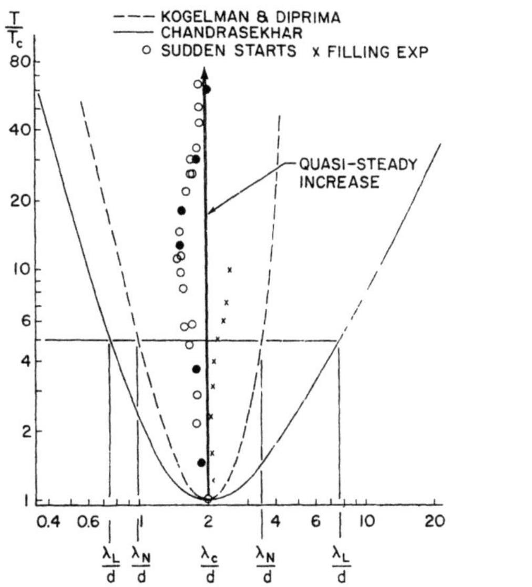

Fig. 5. Stability diagram of the Taylor instability, with linear and nonlinear stability diagram and experimental wavelengths. After [19].

We will now turn our attention to the temporal structure of Taylor vortices. The investigation of this topic began with the experiments of Coles [20], who discovered the wavy Taylor vortices. In wavy vortex flow an azimuthal periodicity of m waves is added to the original vertical periodicity of the toroidal vortices, see for example Fig. 6. Coles observed that in wavy vortex flow the axial wave number a, as well as

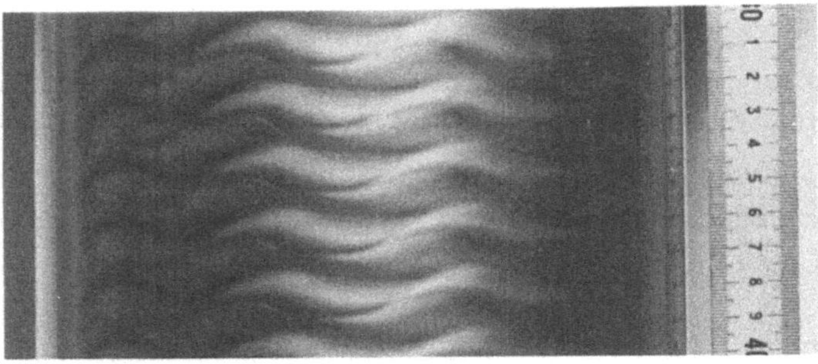

Fig. 6. Wavy Taylor vortices at $Ta > 8.5\ Ta_c$. After [24].

the azimuthal wave number m, are nonunique. The entire wavy pattern moves in azimuthal direction with a constant velocity of about 1/3 of the rotation rate of the inner cylinder. Davey, DiPrima and Stuart [21] developed the first theoretical description of wavy vortex flow.

The passing of the azimuthal waves over a fixed point in the gap is necessarily associated with a periodic variation of the velocity components. The frequency of this variation, which is the frequency of the azimuthal waves, has been measured with great accuracy by Fenstermacher, Swinney and Gollub [22], and independently with a different technique by Bouabdallah and Cognet [23]. The essential result of Swinney's measurements are shown in Fig. 7. We see first that the azimuthal waves appear with a sharp frequency ω_1. When the rotation rate of the inner cylinder is increased sufficiently a second, incommensurate frequency ω_2 appears. Increasing the rotation a little further leads to the formation of weak turbulence, indicated by the intensification of the noise at all frequencies, as is shown on the lowest part of Fig. 7.

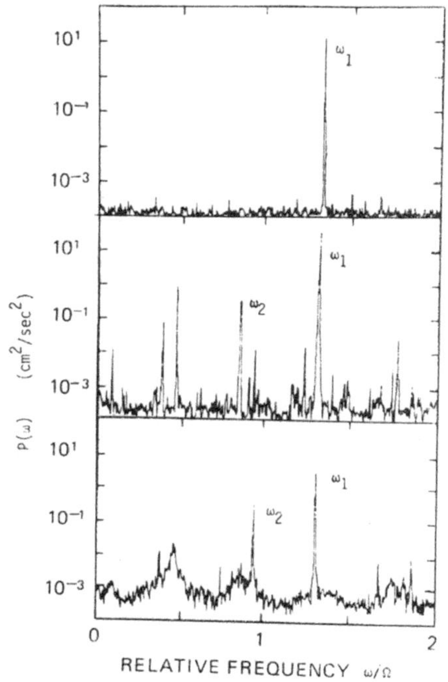

Fig. 7. Power Spectra of wavy Taylor vortices at three different Taylor numbers. After [22].

At Taylor numbers of order a thousand times Ta_c the instability is, of course, still present. Turbulent Taylor vortices are then, as shown by Koschmieder [24], in the mean axisymmetric and have a clear spatial periodicity. The wave number is, surprisingly, nonunique but always larger than the critical wave number.

References

1. Drazin, P. G. and Reid, W. H.: Hydrodynamic Stability. Cambridge University Press, 1981.

2. Koschmieder, E. L.: On convection on a uniformly heated plane. Beitr. Phys. Atmos. 39 (1966), 1-11.

3. Charlson, G. S. and Sani, R. L: Thermoconvective instability in a bounded cylindrical fluid layer. Int. J. Heat Mass Transfer 13 (1970), 1470-1496.

4. Davis, S. H.: Convection in a box: linear theory. J. Fluid Mech. 30 (1967), 465-478.

5. Koschmieder, E. L.: Bénard convection. Adv. Chem. Phys. 26 (1974), 177-212.

6. Busse, F. H. and Whitehead, J. A.: Instabilities of convection rolls in a high Prandtl number fluid. J. Fluid Mech. 60 (1971), 305-320.

7. Pomeau, Y. and Maneville, P.: Wavelength selection in Cellular Flows. Phys. Lett. 75A (1980), 296-298.

8. Cross, M. C., Daniels, P. G., Hohenberg, P. C. and Siggia, E. D.: Effect of sidewalls on wavenumber selection in Rayleigh-Bénard convection. Phys. Rev. Lett. 45 (1980), 898-901.

9. Dubois, M. and Bergé, P.: Experimental study of the velocity field in Rayleigh-Bénard convection. J. Fluid Mech. 85 (1978), 641-653.

10. Landau, L. D.: On the problem of turbulence. C. R. Acad. Sci. USSR, 44 (1944), 311.

11. Hopf, E.: Abzweigung einer Periodischen Lösung von einer Stationären Lösung eines Differentialsystems. Ber. Verh. Saechs. Akad. Wiss. Leipzig. Math-Phys. Kl. 94 (1942), 1-22.

12. Gollub, J. P. and Benson, S. V.: Many routes to turbulent convection. J. Fluid Mech. 100 (1980), 449-470.

13. Ahlers, G. and Behringer, R. P.: Evolution of Turbulence from the Rayleigh-Bénard Instability. Phys. Rev. Lett 40 (1978), 712-716.

14. Benjamin, R. B.: Bifurcation phenomena in steady flows of a viscous fluid. II. Experiments. Proc. R. Soc. London A 359 (1978), 27-43.

15. Stuart, J. T.: On the nonlinear mechanics of hydrodynamic stability. J. Fluid Mech. 4 (1958), 1-21.

16. Kirchgässner and Sorger, P.: Branching analysis for the Taylor problem. Quart. J. Mech. Appl. Math. 22 (1969), 183-190.

17. Kogelman, S. and DiPrima, R. C.: Stability of spatially periodic supercritical flows in hydrodynamics. Phys. Fluids 13 (1970), 1-11.

18. Burkhalter, J. E. and Koschmieder, E. L.: Steady supercritical Taylor vortex flow. J. Fluid Mech. 58 (1973), 547-560

19. Burkhalter, J. E. and Koschmieder, E. L.: Steady supercritical Taylor vortices after sudden starts. Phys. Fluids 17 (1974), 1929-1935.

20. Coles, D.: Transition in circular Couette flow. J. Fluid Mech. 21 (1965), 385-425.

21. Davey, A., DiPrima, R. C. and Stuart, J. T.: On the instability of Taylor vortices. J. Fluid Mech. 31 (1968), 17-52.

22. Fenstermacher, P. R., Swinney, H. L. and Gollub, J. P.: Dynamical instabilities and the transition to chaotic Taylor vortex flow. J. Fluid Mech. 94 (1979), 103-129.

23. Bouabdallah, A. and Cognet, G.: Laminar-turbulent transition in Taylor-Couette flow. In Laminar-Turbulent Transition, Springer 1980, 368-377.

24. Koschmieder, E. L.: Turbulent Taylor vortex flow. J. Fluid Mech. 93 (1970), 515-527.

Onset of Convection for the Bénard Problem in a Rotating Spherical Shell

P. CHOSSAT

Institut de Mathématiques et Sciences Physiques

Université de Nice

NICE, FRANCE

Summary

The conditions of appearance and stability of stationnary and periodic convective flows in a self-gravitating sphericall shell $D = \{X \,/\, 0 \leq R_1 \leq |X| \leq R_2\}$ are presented in the case when D rotates slowly about an axis, and $R_1/R_2 \approx 0.3$. Using rotationnal symmetry properties of the model, we show that the solutions have the form of rotating waves and interact to give rise to secondary bifurcations.

The mathematical model.

The initial equilibrium state is purely conductive and without motion in the rotating frame. For this initial state we assume the temperature gradient and the gravitationnal field to have the form grad $T = \beta(r)\underline{r}$ and $g = \gamma(r)\underline{r}$, where $\beta(r) = \beta_1/r^3 + \beta_2$ and $\gamma(r) = \gamma_1/r^3 + \gamma_2$. This corresponds to a uniform distribution of heat sources in D and to a self-gravitating shell. These fields are invariant by the full group of rotations in $\mathbb{R}^3$, which is noted SO(3).

Now let $\underline{u}$ be the velocity field and Θ the perturbation of temperature. The equations for $(\underline{u}, \Theta)$ are taken in the classical Boussinesq approximation [1] and with the boundary conditions $u \equiv 0$ and $\Theta \equiv 0$ on ∂D (free-type boundary conditions may also be considered as in [2]). After an anequate choice of scale, the domain is defined by

$$D = \{X \in \mathbb{R}^3 \,/\, 0 < \eta \leq |X| \leq 1\}, \text{ where } \eta = R_1/R_2, \text{ and}$$

the problem takes the form of an operator equation in a suitable Hilbert space H :

$$(1)\qquad \begin{cases} \dfrac{dZ}{dt} = \mathscr{L}_{\lambda,\varepsilon}\, \underline{Z} + M(\underline{Z},\underline{Z}) \ , \quad \underline{Z} = (\underline{u},0) \\ \underline{Z}(t) \in \mathscr{D} \hookrightarrow H. \end{cases}$$

The operator $\mathscr{L}_{\lambda,\varepsilon}$ is linear : $\mathscr{L}_{\lambda,\varepsilon} = A + \lambda B + \varepsilon C$, where A has domain $\mathscr{D}$ and B , C are bounded in H. The parameter λ is the square root of the Rayleigh number and ε is the Taylor number (proportionnal to the speed of rotation of D). The operator -A is self-adjoint and positive and we assume B to be self-adjoint This is the case if $\beta_1 = \gamma_1 = 0$ or $\beta_2 = \gamma_2 = 0$ (for instance) The operator C is skew-symmetric. The operator M(.,.) is bilinear symmetric. It was shown in [2] that the problem (1) enters in a classical frame [6] for the stationnary and Hopf's bifurcation analysis (and the analysis of stability for the bifurcated solutions).

The bifurcation problem.

It is well-known, as a consequence of the positiviness of -A, that the state of pure conduction (trivial solution of the problem (1)) is stable for $|\lambda|$ and $|\varepsilon|$ small enough : all the eigenvalues of $\mathscr{L}_{\lambda,0}$ will pass through 0($\mathscr{L}_{\lambda,0}$ is self-adjoint, so that its eigenvalues are real). At this critical value the trivial solution looses its stability while new stationnary solutions may occur ("stationnary bifurcation").

The fundamental property of equation (1) is its invariance by rotationnal symmetries : Let T_g be the representation of SO(3) in H, defined by $T_g\underline{Z}(X) = (g\underline{u}(g^{-1}X), \Theta\,(g^{-1}X))$. Then

$$(2)\qquad \begin{cases} \mathscr{L}_{\lambda,0}\, T_g\, \underline{Z} = T_g\, \mathscr{L}_{\lambda,0}\, \underline{Z} \\ M(T_g\underline{Z},T_g\underline{Z}) = T_g M(\underline{Z},\underline{Z}) \text{ whatewer} \\ g \in SO(3)\,,\ \underline{Z} \in \mathscr{D}\ ; \end{cases}$$

$$(3)\qquad C\, T_\varphi \underline{Z} = T_\varphi C\, \underline{Z} \text{ whatewer } \varphi \in C_\infty,\ \underline{Z} \in H,$$

where C_∞ denotes the group of rotations about the axis Oz (C is the operator corresponding to the Cotiolis force in the equation of motion).

A first consequence of (2) and (3) for the linear part is the following theorem, proved in [3] .

Theorem 1. (i) The kernel V_o of $\mathscr{L}_{\lambda_o,0}$ is an invariant space associated to the irreducible representation $T_g^{(\ell)}$ of SO(3), where ℓ depends on the thickness η of the shell ($\ell \longrightarrow \infty$ as $\eta \longrightarrow 1$). Hence dim $V_o = 2\ell+1$.

(ii) There exist $\ell+1$ values $\lambda_j(\varepsilon) = \lambda_c + k_j\varepsilon^2+O(|\varepsilon|^4)$, $k_j > 0$ such that :

1) $\lambda_o(\varepsilon)$ is a stationnary bifurcation point of (1),

2) $\lambda_j(\varepsilon)$ $(j\neq 0)$ are Hopf's bifurcation points of (1) when $\varepsilon\neq 0$.

In the following we shall suppose $\eta \approx 0.3$, while corresponds to $\ell=2$, as shown by numerical computations. This is the case for the convection in the Earth's core (for example). The case $\varepsilon=0$ (non-rotating shell) was solved in [3]. When $\varepsilon\neq 0$, the part (ii) of the theorem means that at each critical $\lambda_j(\varepsilon)$, there are only two simple conjugated eigenvalues of $\mathscr{L}_{\lambda,\varepsilon}$ on the imaginary axis. It could therefore be possible to use the classical techniques to compute the bifurcated solutions [6] . However this wouldn't permit to solve the singularity at $\varepsilon=0$, i.e. to get the full diagram of bifurcation close to the point $(\lambda,\varepsilon) = (\lambda_c,0)$.
The idea is to transform the Hopf's bifurcation problem to a stationnary bifurcation problem, and then to apply the Lyapunov-Schmidt procedure in order to reduce the equation (1) to an equation in the kernel V_o. The resolution of this bifurcation equation will completely solve the problem.
Let T_φ denote the representation of C_∞ (group of rotations about Oz) in H, and suppose that we look for solutions of (1) of the form

(4) $\quad \underline{z}(t) = T_{\omega t}\, \underline{\zeta}(t), \quad \underline{\zeta}(t)\in\mathscr{D}$.

From the invariance properties (2) and (3), equation (1) is formally equivalent to

$$(5) \qquad \frac{d\zeta}{dt} + \omega L\underline{\zeta} = \mathscr{L}_{\lambda,\varepsilon}\, \underline{\zeta} + M(\underline{\zeta},\underline{\zeta}),$$

where L is the infinitesimal generator of T_φ in H. The stationnary solutions of (5) correspond to $T = \frac{2\pi}{\omega}$ - periodic solutions of (1) - the so-called rotating waves -, and T_1-periodic solutions of (5) correspond to quasi-periodic solutions of (1) with periods T and T_1. Bifurcation of such solutions was studied in an abstract context by Renardy [5] . Therefore, if we look for solutions (4), the bifurcation problem defined by the theorem (1) reduces to a stationnary bifurcation problem for the equation (5). It can be shown that the Lyapunov-Schmidt method applies to (5) at the critical values $\lambda = \lambda_c$ and $\varepsilon=0$, so that the problem reduces to an equation in V_o :

(6) $$0 = G(\lambda,\varepsilon,\omega\underline{X}),$$

where $G : \mathbb{R}^3 x V_o \longrightarrow V_o$ is analytical in a neighborhood of $(\lambda_c,0,0,0)$, and $\underline{X}$ is the orthogonal projection of $\underline{\zeta}$ on V_o in H. Moreover it was proved in [4] , using the center manifold theorem, that the stability of the bifurcated solutions was given by the (negative) sign of the real part of the eigenvalues of $\mathscr{A}_{\lambda,\varepsilon} = D_X G(\lambda,\varepsilon,\omega_j,\underline{X}_j)$, where $(\omega_j,\underline{X}_j)$ are the bifurcated solutions.

Because the eigenvalues of $\mathscr{L}_{\lambda_j,\varepsilon}$ on the imaginary axis are simple ($\varepsilon\neq 0$), we will be sure that the only solutions branching at $\lambda_j(\varepsilon)$ are rotating waves (there is only are branch passing through each bifurcation point). Note that the purely stationnary solutions ($\omega=0$) can be considered as "special" rotating waves.

The problem is now to compute the mapping G. Because of the invariance properties (2) and (3), it has a quite simple structure, which can be determined by using, on one hand, the Sattinger's rule [6] for computing covariant mappings by a symmetry group, and on the other hand the particular properties of the operators for the Bénard problem [2] .

The computations where performed in the case $\ell=2$, as we said, and by taking a particular basis $\{\xi_k\}_{-2 \leq k \leq 2}$ of the complexified space Vo+iVo, called the "canonical basis" [6] , which has the property $\underline{\xi}_{-k} = (-1)^k \ \bar{\underline{\xi}}_k$.

We set $\underline{X} = \sum_{k=-2}^{2} x_k \xi_k$ and $G = \sum_{k=-2}^{2} g_k \xi_k$. We have

$x_{-k} = (-1)^k \bar{x}_k$ and $g_{-k} = (-1)^k \bar{g}_k$ (we look for real solutions).

Then, setting $X = (x_o, x_{\pm 1}, x_{\pm 2})$, we get

(7) $g_k(\lambda\ \varepsilon\,\omega\, X) = \tilde{g}_k(\lambda, X) + \varepsilon h'_k(\lambda, \varepsilon, X) + \omega h''_k(\lambda, \varepsilon, \omega, X)$, $-2 \leq k \leq 2$,

where $\tilde{g}_k(\lambda, X) = a(\lambda, |X|^2, p(X))\, x_k + b(\lambda, |X|^2, p(X))\, B_k(K)$

a and b being real analytical functions of λ , $|X|^2$ ($|X|^2 = x_o^2 + 2|x_1|^2 + 2|x_2|^2$) and $p(X)$ (homogeneous polynomial of degree 3). $B_k(X)$ is an homogeneous polynomial of degree 2. The polynomials $p(X)$ and $B_k(x)$ doesn't depend on any coefficient.

We give the first terms of the series expansions (those which are needed for the results) :

1) $a(\lambda, |X|^2, p(x)) = a_1(\lambda - \lambda_c) + a_2|X|^2 + \ldots,$
   $a_1 > 0, \quad a_2 < 0,$

2) $b(\lambda, |X|^2, p(X)) = b_2|X|^2 + \ldots$

3) $h'_k(\lambda, \varepsilon, X) = (\gamma_k \varepsilon + k i \beta)\, x_k + \ldots,\ \gamma_k < 0,\ \beta > 0.$

Remark 1 : the ... part doesn't necessary mean "higher order terms".

Remark 2 : the coefficients can be computed numerically.

Remark 3 : In the theorem 1, $k_j = -a_1^{-1} \gamma_j$.

Diagrams of bifurcation and stability [4] .

The time-dependance of the rotating waves is given by

$T_{\omega t}\, \underline{X}' = x_o \underline{\xi}_o + x_1 e^{i\omega t}\, \underline{\xi}_1 + x_1\, e^{-i\omega t} \bar{\underline{\xi}}_1 + \bar{x}_2 e^{i\omega t} \bar{\underline{\xi}}_2 + x_t\, e^{-2i\omega t} \bar{\underline{\xi}}_2.$

Theorem 2. (i) $\lambda_o(\varepsilon)$ is a bifurcation point of purely stationnary solutions $\underline{X}_o = x_o\, \underline{\xi}_o$. These solutions are axisymmetric about the axis Oz (the convective flows are confined in the meridian planes). The bifurcation is slightly transcritical with a turning point at $\tilde{\lambda}_o(\varepsilon) = O(\varepsilon^4)$.

ii) $\lambda_1(\varepsilon)$ is a bifurcation point of rotating waves $(\omega_1, \underline{X}_1)$, with $\omega_1 = \varepsilon\beta + O(\varepsilon)$. The bifurcation is either subcritical or supercritical, depending on the sign of a coefficient.

iii) $\lambda_2(\varepsilon)$ is a supercritical bifurcation point of rotating waves $(\omega_2, \underline{X}_2)$, $\omega_2 = 2\varepsilon\beta + O(\varepsilon)$, $\underline{X}_2$ having no componants on $\{\xi_1, \xi_{-1}\}$.

iv) A secondary (supercritical) bifurcation of solutions of type $(\omega_2, \underline{X}_2)$ appears on the branch of stationnary solutions $\underline{X}_o$, at a point

$$\hat{\lambda}_o(\varepsilon) = \lambda_c - a_1^{-1} a_2 \left(\frac{2(\gamma_2-\gamma_o)}{3\, b_2}\right)^{2/3} \varepsilon^{4/3} + O(\varepsilon^2).$$

v) If $\lambda_1(\varepsilon)$ is either the smallest or the greatest critical value, and under a supplementary condition on the sign of a coefficient, quasi-periodic solutions may bifurcate from the branches $(\omega_2, \underline{X}_2)$. The primary period would be $T_2 = \frac{2\pi}{\omega_2}$ and the secondary period $\hat{T} = \frac{2\pi}{\hat{\omega}}$, with $\hat{\omega} = O(\varepsilon^{5/3})$.

The stability of these solutions depends on the sign of b_2 and $\gamma_j - \gamma_k$ $(j \neq k)$. The following diagrams are drawn in the plane (λ, x_o). The full lines correspond to stable solutions and the dotted lines to unstable solutions. We have supposed $b_2 < 0$ (the case $b_2 > 0$ is obtained by changing x_o in $-x_o$). The solutions (i), (ii) and (iii) are noted (0), (1) and (2). We suppose here that the turning point of (0) is negative and that (1) is supercritical (if (1) were subcritical and $\lambda_1(\varepsilon)$ the smallest critical value, all the bifurcated solutions would be unstable).

Remark. Numerical computations of the coefficients γ_j (which are not yet published) seem to show that the critical values are $\lambda_2(\varepsilon) < \lambda_1(\varepsilon) < \lambda_o^{\cdot}(\varepsilon)$ $(\varepsilon \neq 0)$, at least for a larger scale of values of the Prandtl number (which enters in the operators).

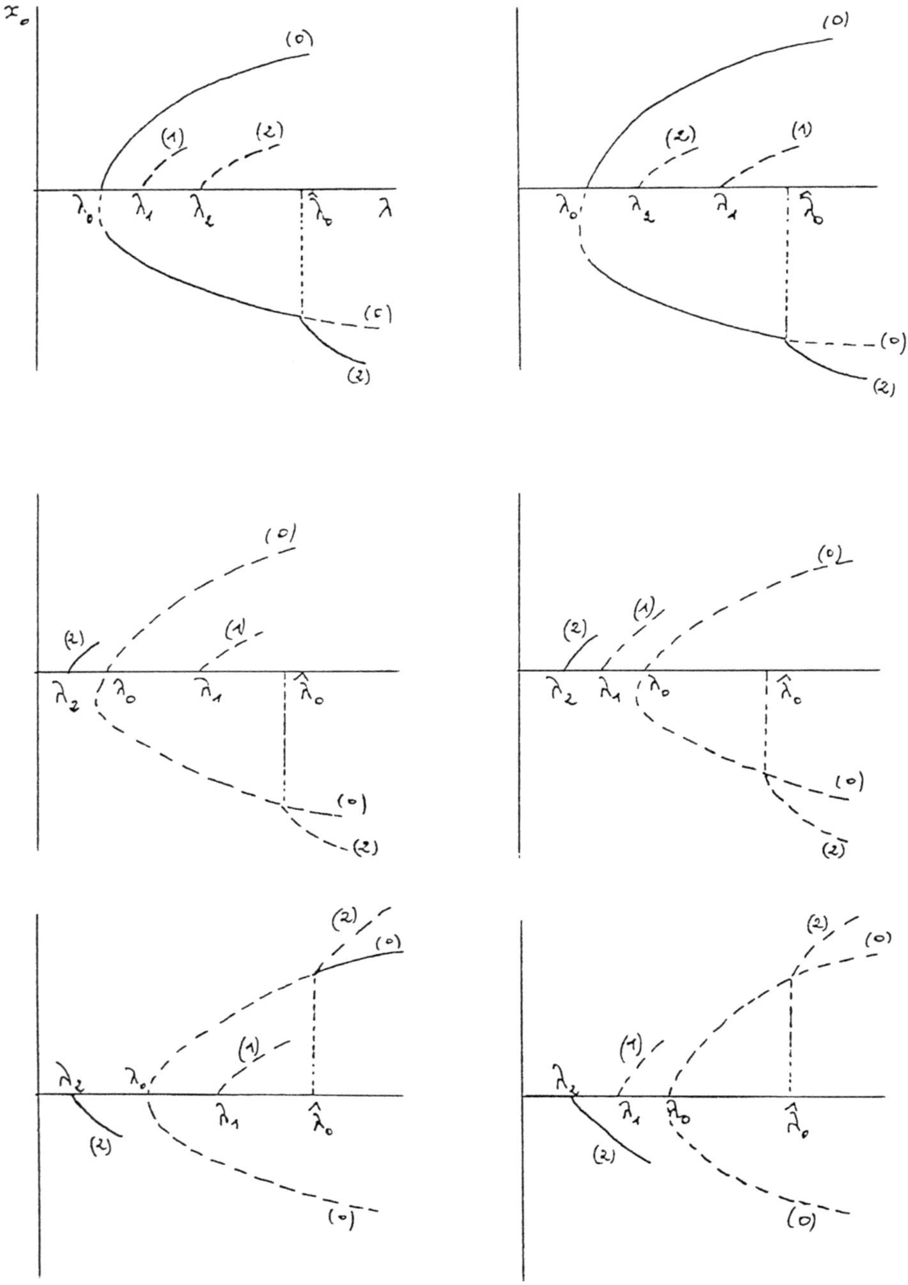

References.

[1] CHANDRASEKHAR S. Hydrodynamic and hydromagnetic stability Oxford U. Press, London (1961).

[2] CHOSSAT P. Thèse de 3ème cycle, U. de Nice (1977).

[3] CHOSSAT P. SIAM J. of Appl. Math., 37, 3 (1979) pp. 624-647.

[4] CHOSSAT P. Preprint.

[5] RENARDY M. Preprint.

[6] SATTINGER D. Lectures Notes in math, n° 762 (1979) Springer Verlag.

Bifurcation in the Bénard Problem with a Free and Deformable Surface

J. W. THOMAS AND SCOTT W. TENTO

Department of Mathematics
Colorado State University
Fort Collins, Colorado 80523 USA

1. Introduction

In 1900, Henri Bénard [1] noticed that when a horizontal layer of fluid was heated from below, at first the fluid remained at rest, with heat being transferred upward through the fluid by conduction. When a critical temperature difference between the top and bottom surfaces was reached, cellular convection began.

Later experimentalists confirmed and extended Bénard's work (see Block [2] and Koschmieder [3]). Typically, horizontal cross sections of the observed convection cells were hexagons, rectangles, or infinite strips.

Rayleigh [4] gave the first mathematical description of the Bénard problem. In [5] Schluter, Lortz and Busse gave a comprehensive linearized stability analysis explaining the onset of convection. Rabinowitz [6] and Iudovich [7] have obtained rigorous bifurcation results for the Bénard problem, and Fife and Joseph [8], [9] have rigorously studied the problem in a setting more general than the other works.

All the above references to mathematical papers have formulated the Bénard problem with buoyancy as the force driving convection. It has been shown [2] that in Bénard's particular experiments it was not buoyancy but rather the variation of the coefficient of surface tension with temperature that drove convection. However, for a wide range of experimental conditions buoyancy is the primary cause for the onset of convection.

Also, all of the mathematical works mentioned above have assumed that the surfaces of the fluid layer were nondeformable. In this paper, we obtain a bifurcation result for the Bénard problem where the upper fluid surface is allowed to deform. Izakson and Iudovich [10] studied the free surface problem without a Banach space setting. The proof of the important Proposition 3.1 is essentially due to Izakson and Iudovich.

A survey of the experimental studies of the Bénard problem is given in [3]. Chandrasekhar [11] is the classical reference for linearized stability studies of the Bénard problem. Kirchgässner [12] and Sattinger [13] give surveys for rigorous treatments of bifurcation problems in fluid mechanics in general.

2. Formulation of the Problem

We will consider a layer of fluid which extends infinitely in the horizontal (x and y) directions. The fluid is bounded below by a rigid perfectly conducting surface, and the upper surface is free and deformable. The bottom surface is held at a constant temperature T_0, and the top surface is held at the constant temperature T_1, $T_0 > T_1$. We consider only time-independent solutions, and apply the Boussinesq approximation. If the z-axis points downward and the conduction solution is subtracted away, the equations describing the fluid layer take the form

$$(\underline{u}\cdot\underline{\nabla})\underline{u} = -\underline{\nabla}p - R\theta\underline{e} + \nabla^2\underline{u} \tag{2.1}$$

$$Pr\,(\underline{u}\cdot\underline{\nabla})\theta = \nabla^2\theta - w \tag{2.2}$$

$$\underline{\nabla}\cdot\underline{u} = 0 \tag{2.3}$$

$$\underline{u} = \underline{0},\ \theta = 0 \text{ on } z = 1 \tag{2.4}$$

$$Pr\theta + h = 0 \text{ on } z = h(x, y) \tag{2.5}$$

$$\underline{u}\cdot\underline{n} = 0 \text{ on } z = h(x, y) \tag{2.6}$$

$$(-p - Fh - R/Pr\ h + \tfrac{1}{2}R/Pr\ h^2)\underline{n} + 2D\cdot\underline{n} = \Sigma J\underline{n} \text{ on } z=h(x,y), \tag{2.7}$$

where $\underline{u} = (u, v, w)^T$ is the scaled velocity vector, p is the scaled pressure, θ is the scaled temperature, h(x, y) is the scaled free surface, $\underline{n}$ is the normal to the free surface, D is the rate of deformation tensor, $\underline{e} = (0, 0, 1)^T$, R is the Rayleigh number and is proportional to the temperature difference $T_0 - T_1$, Pr is the Prandtl number, F is the Froude number, Σ is a constant proportional to the coefficient of surface tension, and J is the mean curvature of the free surface. The solution of (2.1) to (2.7) corresponding to the conduction solution is given by $\underline{u}(x, y, z) = \underline{0}$, $p(x, y, z) = \theta(x, y, z) = 0$, and $h(x, y) = 0$. It should be noted that problem (2.1) to (2.7) differs from the usual Bénard problem in that the boundary conditions (2.6) and (2.7) are both nonlinear. Also observe that the Rayleigh number appears in the boundary condition (2.7) as well as in the equations.

For any nontrivial solution the physical domain in which the fluid is confined deforms. Hence the domain of definition of the members of the underlying function space changes, and we do not have a Banach space setting. To obtain a Banach space setting we map the physical space onto the space $\Omega = \{(\bar{x}, \bar{y}, \bar{z}): \bar{x}, \bar{y} \in \mathbb{R}, 0 \leq \bar{z} \leq 1\}$ using the mapping $\bar{x} = x$, $\bar{y} = y$, $\bar{z} = z - (1 - z)h/(1 - h)$. This technique is equivalent to the method of domain perturbation developed by Joseph and Fosdick in [14].

The result of transforming the problem (2.1) to (2.7) is quite lengthy and can be found in [15]. If we supress the bar notation, the linearization of the fully mapped problem is given by

$$\nabla^2 \underline{u} - \underline{\nabla} p - R\theta \underline{e} = \underline{0} \tag{2.8}$$

$$\nabla^2 \theta - w = 0 \tag{2.9}$$

$$\underline{\nabla} \cdot \underline{u} = 0 \tag{2.10}$$

$$\underline{u} = \underline{0},\ \theta = 0 \text{ on } z = 1 \tag{2.11}$$

$$u_z = v_z = w = 0 \text{ on } z = 0 \tag{2.12}$$

$$-p + (R + F\ Pr)\theta + Pr\Sigma(\theta_{xx} + \theta_{yy}) + 2w_z = 0 \text{ on } z = 0. \tag{2.13}$$

3. Bifurcation

To show that a nontrivial solution of the fully mapped problem bifurcates from the conduction solution, we seek to use a theorem of Crandall. See Theorem 1 of [16]. To obtain bifurcation from Crandall's theorem, we must show that there is a value, R_c, of R for which the linearization (2.8) to (2.13) has a one-dimensional null space.

To reduce the null space of (2.8) to (2.13) to a one-dimensional space, it is usual to seek solutions exhibiting some type of symmetry in the x,y-plane. Let us look for solutions of the form:

$$u(x, y, z) = \sum_{j=0}^{\infty} \sum_{k=0}^{\infty} A_{jk}(z) \sin jax \cos kby \tag{3.1}$$

$$v(x, y, z) = \sum\sum B_{jk}(z) \cos jax \sin kby \tag{3.2}$$

$$w(x, y, z) = \sum\sum C_{jk}(z) \cos jax \cos kby \tag{3.3}$$

$$\theta(x, y, z) = \sum\sum D_{jk}(z) \cos jax \cos kby \tag{3.4}$$

$$p(x, y, z) = \sum\sum E_{jk}(z) \cos jax \cos kby. \tag{3.5}$$

When the forms (3.1) to (3.5) are substituted into equations (2.8) to (2.13), we obtain a family of systems of ordinary differential equations. These systems reduce to the following family of problems:

$$L_{jk}^3 D_{jk}(z) = -c_{jk}^2 R D_{jk}(z) \tag{3.6}$$

$$D_{jk}(1) = L_{jk} D_{jk}(1) = L_{jk} D'_{jk}(1) = 0 \tag{3.7}$$

$$L_{jk} D_{jk}(0) = L_{jk} D''_{jk}(0) = 0 \tag{3.8}$$

$$c_{jk}^2 D_{jk}(0)-(R+F\ Pr-c_{jk}^2 Pr\Sigma)^{-1}(L_{jk}D_{jk}'''(0)-c_{jk}^2 L_{jk}D_{jk}'(0)) = 0, \qquad (3.9)$$

where $c_{jk}^2 = j^2a^2 + k^2b^2$ and $L_{jk} = (\frac{d^2}{dz^2} - c_{jk}^2)$. Here after we shall suppress the subscripts j and k.

We now establish a proposition.

Proposition 3.1. The problem (3.6) to (3.9) possesses a smallest, positive algebraically simple eigenvalue with respect to the space $L^2[0, 1]$.

Proof: We construct the Green's function for the problem (3.6) to (3.9). It is convenient to write the problem in the equivalent form

$$L^2C(z) = c^2RD(z) \qquad (3.10)$$

$$LD(z) = -C(z) \qquad (3.11)$$

$$C(1) = C'(1) = D(1) = C(0) = C''(0) = 0 \qquad (3.12)$$

$$c^2D(0)-(R+F\ Pr-c^2Pr\Sigma)^{-1}(3c^2C'(0)-C'''(0))=0. \qquad (3.13)$$

A straightforward calculation gives the symmetric Green's function for the problem

$$L^2C = c^2RD$$

$$C(0) = C''(0) = C(1) = C'(1) = 0,$$

which we denote by $G_1(s, z)$. This function is given in [10] and [15]. We let $G_2(s, z)$ denote the symmetric Green's function for

$$LD = -C,\ D(0) = D(1) = 0.$$

Now the original sixth order problem (3.6) to (3.9) can be written as the following integral equation

$$D(z) = R\int_0^1\Bigg\{\int_0^1 c^2G_1(s, t)G_2(z, s)ds + (R+F\ Pr-c^2Pr\Sigma)^{-1}(\sinh c)^{-1}\sinh c(1-z) \qquad (3.14)$$
$$[3c^2G_{1,z}(0,t)-G_{1,zzz}(0,t)]\Bigg\}D(t)dt.$$

Izakson and Iudovich [10] using Krein's theory of oscillating kernels showed that the kernel $\int_0^1 G_1(s, t)G_2(z, s)ds$ is positive. They also showed that the term $3c^2G_{1,z}(0, t) - G_{1,zzz}(0, t)$ is positive. So if we require $R+F\ Pr - c^2Pr\Sigma$ to be a fixed positive number (this may be done while still allowing R to vary freely by adjusting the parameters F, Pr and Σ), we see that the whole kernel in (3.14) is positive for all t and z in [0, 1]. Krein's theory of positive operators [17] implies that there is a smallest, positive, algebraically simple (in $L^2[0, 1]$) eigenvalue of (3.14). Hence there is a smallest, positive, algebraically simple eigenvalue of the original problem (3.6) to (3.9).

Q.E.D.

Note that for each value of $c_{jk}^2 = j^2a^2 + k^2b^2$ we get an eigenvalue $R(c_{jk})$. If we allow a and b to vary continuously, we can consider c as a continuously varying parameter. Numerical evidence indicates that R(c) is a convex function. We assume that this is the case, and we denote the minimum value of R(c) by R_c. Without loss of generality we may assume that R_c is attained in the problem (3.6) to (3.9) when $j = k = 1$.

In [15] the authors apply the Crandall theorem to obtain the following results.

THEOREM 3.1. The full nonlinear system obtained by mapping the problem (2.1) to (2.7) admits $(R_c, \underline{0})$ as a bifurcation point at which a branch of nontrivial solutions possessing rectangular symmetry appears. (To save space, we omit mention of the Banach spaces used.)

If we seek solutions of the full nonlinear problem exhibiting hexagonal symmetry, we again obtain the sixth-order equation (3.6) to (3.9). Then the above discussion may be repeated.

4. Perturbation Series

Next we will show how to calculate the terms of the perturbation series for the nontrivial solution. Let us denote the full nonlinear differential equations by F and the boundary conditions by B. The linear part of F is given by equations (2.8) to (2.10), and the linear part of B is given by (2.11) to (2.13). The mappings F(U) and B(U) are analytic and may be written in series form:

$$F(U) = (L_0 + rL_1)U + N_2(U, U) + r\bar{N}_2(U, U) + \dots = 0 \tag{4.1}$$

$$B(U) = (B_0 + rB_1)U + B_2(U, U) + r\bar{B}_2(U, U) + \dots = 0. \tag{4.2}$$

The real number r is given by $R - R_c$, and N_m, $\bar{N}_m$, B_m and $\bar{B}_m$ are m-linear forms.

Analyticity of F and B and Crandall's theorem imply that the nontrivial solution branch is analytic and we may write

$$U = \varepsilon U_1 + \varepsilon^2 U_2 + \dots, \quad \text{and} \tag{4.3}$$

$$r = \varepsilon r_1 + \varepsilon^2 r_2 + \dots, \tag{4.4}$$

where $U = (u, v, w, \theta, p)^T$, and the free surface h has been eliminated from the problem using the boundary condition (2.5).

If we substitute (4.3) and (4.4) into the differential equation (4.1) and (4.2) and collect powers of ε, we obtain the following problem for the coefficient of ε,

$$L_0U_1 = 0$$

$$B_0U_1 = 0.$$

This is the problem (2.8) to (2.13). For the rectangular case this problem was reduced to the sixth-order differential equation (3.6) to (3.9) with $j = k = 1$. We may solve the equation by elementary means. By reversing the reduction which leads to equation (3.6), we can analytically calculate all five components of the vector U_1.

Next, the coefficient of the ε^2 term gives the problem,

$$L_0U_2 + r_1L_1U_1 + N_2(U_1, U_1) = 0 \tag{4.5}$$

$$B_0U_2 + r_1B_1U_1 + B_2(U_1, U_1) = 0. \tag{4.6}$$

Since U_1 is now known, (4.5) and (4.6) is a nonhomogeneous partial differential equation for U_2 with nonhomogeneous boundary conditions. A change of variables gives an equivalent problem of the form

$$L_0U_2 = -r_1L_1U_1 + N_2(U_1, U_1) + r_1V_1 + V_2 \tag{4.7}$$

$$B_0U_2 = 0, \tag{4.8}$$

where V_1 and V_2 result from eliminating $r_1B_1U_1 + B_2(U_1, U_1)$ from (4.6).

The problem (4.7) and (4.8) may be reduced to a nonhomogeneous sixth-order ordinary differential equation with constant coefficients. The nonhomogeneity will be combinations of elementary functions and so U_2 can be found analytically. The number r_1 is found by applying to equation (4.7) the eigenfunction of the operator L_0^*.

Continuing in this way, it is possible to analytically calculate any finite number of terms of the series (4.3) and (4.4).

References

1. Bénard, H: Les Tourbillons Cellulaires dans une Nappe Liquide, parts 1 and 2, Rev. Generale Sci. Pures Appl., 11 (1900), 1261,1309.
2. Block, M.J.: Surface Tension as the Cause of Bénard Cells and Surface Deformation in a Liquid Film, Nature 178 (1956), 650.
3. Koschmieder, E.L.: Bénard Convection, Advances in Chemical Physics (Ed. I.Prigogine and S.A.Rice) 26, Wiley (1974).

4. Rayleigh, Lord: On Convection Currents in a Horizontal Layer of Fluid, When the Higher Temperature is on the Under Side. Phil.Mag. 32 (1916), 529.

5. Schluter,A.; Lortz,D.; Busse,F.H.: On the Stability of Steady Finite Amplitude Cellular Convection. J.Fluid Mech. 23 (1965), 129.

6. Rabinowitz,P.H.: Existence and Nonuniqueness of Rectangular Solutions of the Bénard Problem. Arch.Rat.Mech.Anal. 29 (1968), 32.

7. Iudovich,V.I.: On the Origin of Convection. P.M.M. 30, No. 6 (1966), 1193.

8. Fife,P.: The Bénard Problem for General Fluid Dynamical Equations and Remarks on the Boussinesq Equations. Indiana Univ. Math.J. 20 (1970), 303.

9. Fife,P; Joseph,D.D.: Existence of Convective Solutions of the Generalized Bénard Problem Which Are Analytic in Their Norm. Arch.Rat.Mech. Anal. 33 (1969), 116.

10. Izakson,V.K.; Iudovich,V.I.: Convection Development in a Liquid Layer with a Free Boundary. Mekhanika Zhidkosti i Gaza 3, no. 4 (1968), 23.

11. Chandrasekhar,S.: Hydrodynamic and Hydromagnetic Stability. Oxford Univ. Press (1961).

12. Kirchgässner,K.: Bifurcation in Nonlinear Hydrodynamic Stability. SIAM Review 17 (1975), 652.

13. Sattinger,D.H.: Topics in Stability and Bifurcation Theory. Univ. of Minnesota Press (1972).

14. Joseph,D.D.; Fosdick,R.L.: The Free Surface on a Liquid Between Cylinders Rotating at Different Speeds. Part 1, Arch.Rat.Mech.Anal. 49 (1973), 321.

15. Tento,S.W.: The Bénard Problem with a Free and Deformable Surface. Dissertation, Colorado State University (1981).

16. Crandall,M.G.: An Introduction to Constructive Aspects of Bifurcation and the Implicit Function Theorem. Applications of Bifurcation Theory, Academic Press (1977).

17. Krasnosel'skii,M.A.: Positive Solutions of Operator Equations. P. Noordhoff, Ltd., Groningen (1964).

MIX
Papier aus verantwortungsvollen Quellen
Paper from responsible sources
FSC® C105338

If you have any concerns about our products,
you can contact us on
ProductSafety@springernature.com

In case Publisher is established outside the EU,
the EU authorized representative is:
Springer Nature Customer Service Center GmbH
Europaplatz 3, 69115 Heidelberg, Germany

Printed by Libri Plureos GmbH
in Hamburg, Germany